PHYSICAL CAPITAL DEVELOPMENT AND ENERGY TRANSITION IN LATIN AMERICA AND THE CARIBBEAN

PHYSICAL CAPITAL DEVELOPMENT AND ENERGY TRANSITION IN LATIN AMERICA AND THE CARIBBEAN

José Alberto Fuinhas
Matheus Koengkan
Renato Santiago

Elsevier
Radarweg 29, PO Box 211, 1000 AE Amsterdam, Netherlands
The Boulevard, Langford Lane, Kidlington, Oxford OX5 1GB, United Kingdom
50 Hampshire Street, 5th Floor, Cambridge, MA 02139, United States

Library of Congress Cataloging-in-Publication Data
A catalog record for this book is available from the Library of Congress

British Library Cataloguing-in-Publication Data
A catalogue record for this book is available from the British Library

ISBN: 978-0-12-824429-6

For information on all Elsevier publications
visit our website at https://www.elsevier.com/books-and-journals

Publisher: Joe Hayton
Acquisitions Editor: Graham Nisbet
Editorial Project Manager: Leticia M. Lima
Production Project Manager: Omer Mukthar
Cover Designer: Victoria Pearson

Typeset by SPi Global, India

Contents

Introduction

This book is about two current and prominent issues for the Latin America and the Caribbean (LAC) region: physical capital development and energy transition. The choice to analyse these two aspects in the region is easy to explain. First, given that there is the idea that the LAC suffers from a lack of physical capital investment, it becomes increasingly interesting to investigate the accuracy of such hypothesis and analyse the impact that LAC capital stock (public and private) has had on the region's development. This analysis should be done not only through the evaluation of the effects from LAC physical capital on the regional growth, but also through the evaluation of its effects on income inequality (given that it is one of the region's most worrying problems and can be an obstacle for a sustainable growth path) and on energy intensity (given the increased worries with the region's future energy demand and energy security). Second, due to the tremendous renewable energy potential that is usually associated with LAC and due to the considerable political and investment efforts that the region has been made in order to change the paradigm from fossil sources to renewable energy (the LAC region has one of the largest percentages of renewables in the energy mix compared with the other world regions), it becomes especially interesting to evaluate this region's energy transition process in order to see what has been successfully achieved so far and what needs to be improved. Additionally, making the connection with the previous subject, it is also interesting to explore the contribution of the various types of capital stock for the promotion of renewable energy in this region. Moreover, given the weight that fossil fuels can have on these economies, it is also advisable to uncover the relationship that both types of energy (fossil and renewable) have with economic growth and, finally, enquire about the effects that this transition has had on factors such as environmental degradation and health.

In sum, this book aims to present results and conclusions regarding these two vital themes to be able to develop several policy implications that could help the regional policymakers develop strategies and policies for the sustainable development of the LAC region. Thus, this book is divided into two major sections. In the first we investigate how the LAC countries capital stock is affecting the development of the region, whereas in the second we primarily address the LAC countries' energy transition process.

Is the state of Latin America and the Caribbean capital stock affecting the development of the region?

As previously expressed, the first four chapters of this book addresses the impact that the LAC capital stock (public and private) have had on the region's development. More precisely, in the first chapter, we analysed the evolution of the physical capital (public and private) in the LAC region from the beginning of 1970s to the end of the second decade of the new millennium recurring to data from the IMF 'Investment and Capital Stock Dataset'. From our analysis, we were able to conclude that the economic conjuncture always influenced the evolution of LAC public and private capitals. Particularly, we observed that during the 1970–2017 period, the LAC public investment was always relatively low. In contrast, the private investment revealed to be much more volatile, with picks and breaks of substantial magnitude, depending on whether the region was going through a boom or a bust. Regarding the LAC public and private capital stocks as a percentage of GDP, we saw that, although increased from 1970 to 2017, the evolution was slow and, in some decades, it was nearly constant. In sum, we verify that there really seems to exist a lack of physical capital investment in the LAC, which can be very worrisome. It can prevent the region from being more competitive and from achieving the desired sustainable growth path. In this sense, we conclude that it is truly necessary that the LAC countries increase their physical capital investments in order to pursue their growth and development goals.

In the second chapter, we analysed the relationship between public capital stock, private capital stock, and economic growth for a group of 30 LAC countries from 1970 to 2014. The panel vector autoregression (PVAR) methodology and the panel dynamic ordinary least squares (PDOLS) and panel fully modified ordinary least squares

(PFMOLS) estimators were used to achieve the goals of our analysis. The results from the estimations point to a positive effect from both public and private capital on the long-run economic growth of this group of countries. However, in the short run, we verified that in addition to the adverse effect it seems to have on growth, public capital also appears to crowd out private capital. In general, these findings suggest that although LAC governments should continue to support public and private investment projects, some changes should be made in the planning and execution of such investments to avoid undesirable effects, especially the ones that were found in the short run.

After the analysis of the relationship between capital stock and economic growth, in the third chapter, we focused our analysis on the effects that the LAC capital stock (public and private) had on the income inequality levels of 18 countries from the LAC region, over a period ranging from 1995 to 2017. The panel autoregressive distributed lag (PARDL) model was used to conduct our analysis, which allowed us to decompose the variables' effects into their short-run and long-run components. The results from all the estimated models were unanimous, indicating that, in the short run, capital stock (public and private) had an enhancing effect on the income inequality levels of these countries. This probably suggests that the investments are being made in the already prosperous/wealthiest areas. Regarding the long run, it seems that as we move forward in time, the effects from capital stock (public and private) on income inequality appear to vanish, which could be an indication of the need for more significant (long-term planned) investment. We should state that these outcomes were also verified when we included dummy variables in the models to correct the detected outliers. Overall, these results seem to indicate that for these countries to reduce their income gap, they should increase their physical capital investment simultaneously as they improve/change the management and the selection criteria of these same investments (especially thinking in the neediest areas).

In order to conclude the analysis of the impact of capital stock on the LAC region development, after the analysis of its relationship with economic growth and income inequality, in the fourth chapter we turned our attention to the analysis of the impact that the LAC capital stock (public and private) has had on the region's energy intensity trend over 45 years (1970–2014). This analysis was based on a panel of 21 countries from the LAC region. Again, as in the previous chapter, we used a PARDL model in order to distinguish the short- and long-run impacts of the variables on the LAC energy intensity. In addition to the PARDL estimation, the convergence of this region in terms of energy intensity was also analysed through the convergence analysis and the club clustering algorithm. Finally, an ordered logit regression model was also estimated in order to test if private and public capital stocks were able to influence the formation of converge clubs. According to the results from our estimations, there is an enhancing effect from both types of physical capital on the LAC long-run energy intensity. The results also point that neither public nor private capital affects the club convergence, which means that the conclusion that LAC governments should increase the investment and acquisition of new physical capital with higher energy efficiency (as well as improve the exiting capital) could be extended to all LAC countries. This suggestion will help the LAC diminish (even more) its energy intensity and, subsequently, calm down the regional energy demand and energy security worries.

Essays on the Latin America and the Caribbean countries' energy transition process

The LAC countries' energy transition process is analysed in the last five chapters of this book. The Latin America region's energy transition process and its initiatives and challenges were approached in the fifth chapter. In the Latin America region, the energy transition process began in the 1970s, or more precisely in 1973, in Brazil and Paraguay with the Itaipu treaty that resulted in the construction of very large hydropower, the Itaipu dam (from 1974 to 1984). Other energy transition initiatives arose in the region in the following decades, such as the Proalcool programme in 1975 in Brazil after the first oil shock in 1973. As is known, several initiatives have encouraged the energy transition process in several countries from the Latin America region. However, these initiatives suffered from structural problems in some Latin America regions, such as Argentina, Brazil, and Mexico. Nevertheless, there is a success of energy transition in the region. Paraguay was able to handle the energy transition initiatives and challenges.

The effect of public, private, and public–private partnership's capital stock on renewable energy's installed capacity (a proxy of renewable energy investment) is investigated in the sixth chapter. Data for 19 countries from the LAC region, from 1990 to 2015, and the Quantiles via Moments methodological approach are used. The preliminary tests results indicate (i) low multicollinearity in the variables, (ii) cross-sectional dependence, (iii) stationarity in some variables, (iv) fixed effect in all models, and (v) presence of serial correlation up to the second order. Indeed, these results proved to be promising and adequate to advance with the Quantile via Moments regression model. The estimated model indicates that the public and public–private partnership capital stock positively affects renewable energy's installed capacity. In contrast, the private capital stock does not cause any effect on renewable energy's installed capacity.

The energy transition effect on economic growth and consumption of nonrenewable energy is investigated in the seventh chapter. Data for five Mercosur countries for the period from 1981 to 2014 and the PVAR methodology are used. The estimated model indicates that renewable energy consumption (a proxy of energy transition) increases economic growth and decreases nonrenewable energy consumption. Moreover, the Granger-causality Wald test results indicate a bidirectional relationship between energy consumption (from both renewable and fossil sources) and economic growth. It suggests that countries' economic growth depends on fossil fuels. There is also evidence of substitutability in energy consumption from renewable and fossil sources in periods of drought. The evolution of globalisation has had a positive indirect effect on the Mercosur countries' consumption of renewable energy. The results and analysis of this research can be useful for local governments, not only as a basis for further examinations of the nexus between economic growth and energy consumption but also for designing new policies to increase energy consumption from renewable sources and promote economic development.

The energy transition impact on the LAC region's environmental degradation was investigated for the period from 1990 to 2014, in the eighth chapter. A panel nonlinear autoregressive distributed lag (PNARDL) approach was used in the form of an unrestricted error correction model. The preliminary tests indicated the presence of low-multicollinearity, cross-sectional dependence in all variables in natural logarithms, the presence of the stationarity in some variables, and the presence of fixed effects. Indeed, these results proved to be promising and adequate to advance with the PNARDL model regression. The empirical results indicate an asymmetric effect of the ratio of renewable energy on fossil energy, both in the short run (impacts) and the long run (elasticities). Indeed, it negatively impacts, −0.0601 on positive variations and 0.0792 on negative variations, in the short run. In the long run, the impact was revealed to be −0.0281 on positive variations and 0.0339 on negative variations.

The ninth chapter ends the analysis of the energy transition process of the LAC region. The effect of the energy transition on deaths from air pollution is investigated. Data for 19 countries from the LAC region, for the period from 1995 to 2016, and a PARLD model are used. The estimated model indicates that the energy transition and economic growth decrease deaths from air pollution, while the urbanisation and international tourism increase them. Therefore, the energy transition's capacity to decrease these deaths is related to the rapid investment growth in renewables energy technologies from 1990 to 2016. The energy consumption from these technologies decreases air pollution in the region and their components, such as carbon monoxide and other gases. This evidence demonstrates that the LAC region is in the right way in the energy transition process.

CHAPTER

1

A brief history of physical capital in Latin America and the Caribbean since the 1970s

JEL codes E22, F21, O54

1.1 Introduction

Capital stock is one of the primary inputs of the classic production function, representing the available physical capital at any given moment in a certain economy. It is, without a doubt, an extremely important asset for countries' expectations of growth and development. It represents not only infrastructure assets (e.g. roads, bridges, railroads, airports, tunnels, etc.) but also other types of physical capital such as vehicles, machines, and tools, which are used in the production of the goods and services that populations need. Unlike other production resources (such as materials), physical capital is usually not destroyed in the production process. However, it suffers from the gradual declining of its value over time (depreciation). Given this last explanation, capital stock at any given moment can be obtained by the value of the existing capital (minus the value of its depreciation) plus the value of new investments.

As stated, physical capital is essential for growth, given its great influence on the various stages of production, and so it is usually assumed that countries need to grant a certain level of capital investment in order to attain a stable growth path. Let us imagine a country that lacks roads for example and that does not invest in the maintenance of the existing ones (i.e. does not account for the effect of depreciation). This factor may have a negative effect on its production process as it will probably increase the difficulty (and the costs) of obtaining the necessary resources for production; it will also hamper the connection between production and the final consumer, thus harming the country's overall economic output.

In this example, we see the major role played by public investment in physical capital supply, mainly in the provision of public infrastructure and utilities which can be used by all the productive fabric for exercising their activity. It is important to note that the importance of the public provision of physical capital does not end with the construction of roads, highways, or airports, because the provision of social infrastructure, such as schools or hospitals, can also have an important effect on the country's development and growth, mainly given its positive effects on human capital.

However, as is sometimes affirmed, there are numerous countries whose governments do not have the degree of economic slack and monetary accommodation needed to make such investments. Usually, this is more likely to happen in the case of developing countries which face large fiscal consolidation problems and are hostages of their huge public debts. This induces the private sector to participate hugely in the provision of infrastructure in several countries, with governments often opting to privatise certain infrastructure sectors.

Regarding the sample of this book, if we follow some earlier reports from the United Nations, the IMF, and the World Bank (Faruqee, 2016; Lardé and Sánchez, 2014; Perrotti, 2011; Fay and Morrison, 2007), it can be seen that there is an increasing suspicion that Latin America and the Caribbean (LAC) is suffering from an "infrastructure gap." This could be (and could have been) harmful for the economic sustainability of this region and its development and competitivity. Following the advice of these international organisations, the region will probably need to raise their investment in physical capital in the near future, or there is a risk that this gap may gradually hamper the region's growth.

This fact raises some doubts about the capacity of LAC governments to invest in new physical capital as well as maintaining the existing capital stock. There are several possible reasons for this to happen [e.g. low degree of

https://doi.org/10.1016/B978-0-12-824429-6.00003-6

economic slack and monetary accommodation, low levels of quality and efficiency in investment management, low total factor productivity (TFP), among others]. However, the truth is that capital stock is essential to a country's economic activity, and so it becomes crucial for these countries to start raising their investment levels to escape the present situation and promote development and growth.

Before we proceed with the analysis of the way in which the LAC capital stock has affected the development of the region, we should firstly focus on the description of the historical evolution of physical capital in the LAC, together with the regional economic situation, starting with the 1970s.

1.2 Latin America and the Caribbean Physical Capital in the 1970s

As is known, LAC economies have always been affected by periods of booms and busts, as is natural in the economic world. However, the propensity of this region to be affected by external shocks, mainly due to the weight of exploitation and exportation of natural resources in these economies, together with the persistent political and social problems that the region faces (e.g. political instability, corruption, high inequality levels), makes the effects of such booms and busts far more exacerbated and more frequent than in other regions of the world. This makes it more difficult for these countries to follow a stable and sustainable growth path.

In accordance with the data from the World Bank in Fig. 1.1, the 1970s was a decade in which this region was able to achieve a period of fairly high growth. The growth rates of the region were always above 3%. Following the IMF Working Paper of Zettelmeyer (2006), in the 1970s, Latin America had a superior per capita growth than other developing regions [namely South Asia (SA), the Middle East/North Africa (MENA), and Sub-Saharan Africa (SSAfr)], just behind the East Asia and Pacific (EAP) region. However, it should also be stated that not all countries from the LAC experienced such positive effects. According to the book by Loayza et al. (2005, p. 9), published by the World Bank, there was a number of Caribbean countries whose growth rates had shown a decreasing trend since the 1960s: "*The Bahamas, Barbados, Belize, and the small island countries share in common a decreasing trend in growth rates since the 1960s and 1970s.*" Despite this observation, according to the same book, the 1970s was undoubtedly a decade of solid growth for the LAC region in general. Knowing this, the question now is whether the countries of this region took advantage of this period to invest in their physical capital.

Looking at Fig. 1.2, with the most recent data from the "Investment and Capital Stock Dataset" of IMF (2017) for the LAC countries, we see that the levels of both types of capital stocks (public and private) grew over time. This is to be expected, given that it is difficult to see a situation where the value of depreciation surpasses the value of new investments. However, if we look at Fig. 1.3, it can be seen that levels of public and private capital stocks as a percentage of GDP stayed almost constant during this decade.

As can be observed, public capital stock was always around 70% of GDP, whereas private capital stock was around 140% of GDP. As De Jong et al. (2018, p. 5543) stress: "*As a percentage of GDP, public capital stocks are generally either flat or*

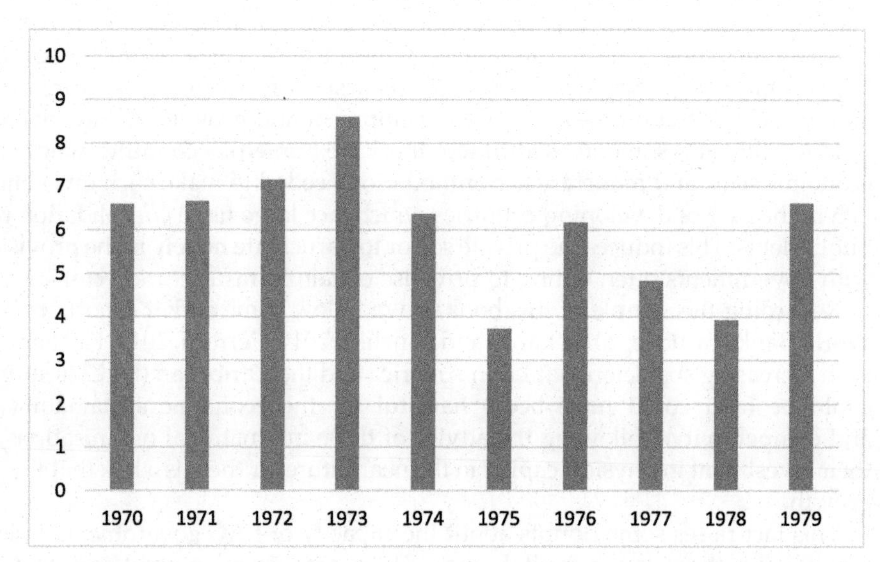

FIG. 1.1 GDP growth (annual %) in the LAC (1970–79). This graph was created by the authors and was based on GDP growth (annual %) data from the World Bank: https://data.worldbank.org/indicator/NY.GDP.MKTP.KD.ZG.

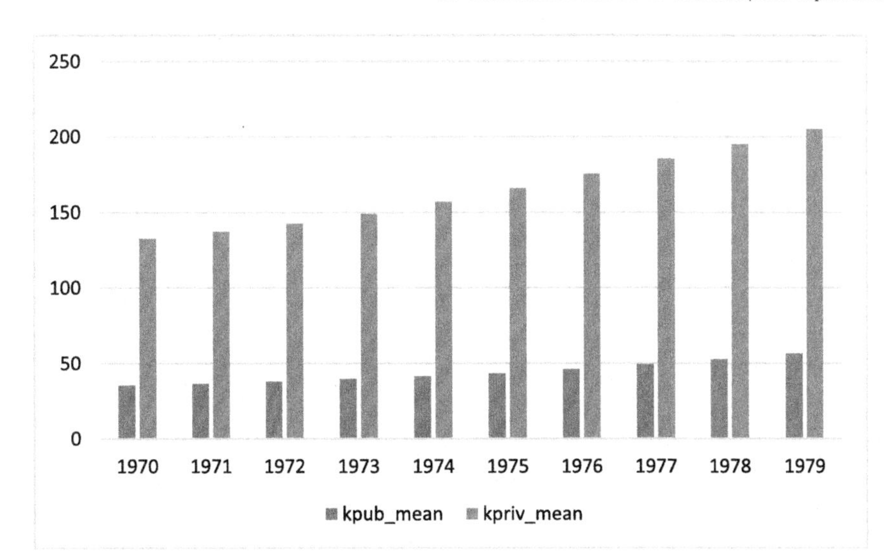

FIG. 1.2 Public capital stock (kpub_mean) and private capital stock (kpriv_mean) (in billions of constant 2011 international dollars) in the LAC (1970–79). This graph was created by the authors and was based on the data from the "Investment and Capital Stock Dataset" of IMF (2017). The blue bars (gray colour bars in print version) represent the mean of public capital stock in billions of constant 2011 international dollars for the LAC, whereas the orange bars (dark gray colour bars in print version) are the mean of private capital stock in billions of constant 2011 international dollars for the same group of countries.

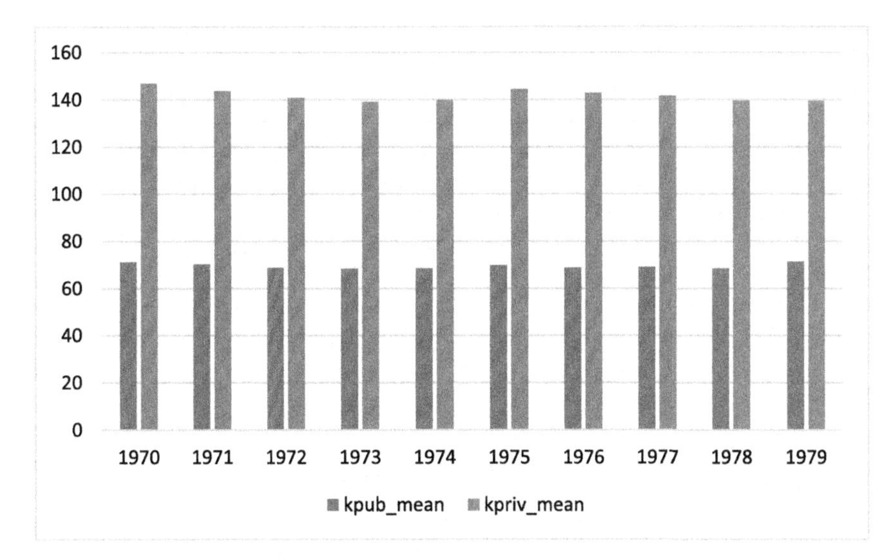

FIG. 1.3 Public capital stock (kpub_mean) and private capital stock (kpriv_mean) (% of GDP) in the LAC (1970–79). This graph was created by the authors and was based on the data from the "Investment and Capital Stock Dataset" of IMF (2017). The blue bars (gray colour bars in print version) represent the mean of public capital stock as a percentage of GDP for the LAC, and the orange bars (dark gray colour bars in print version) represent the mean of private capital stock as a percentage of GDP for the same group of countries.

BOX 1.1

Investment and capital stock dataset.

The "Investment and Capital Stock Dataset" of IMF (2017) is composed of data on public capital stock and investment, private capital stock and investment, and on public-private partnerships (PPPs) capital stock and investment, for a sample of around 170 countries from 1960 to 2017. The capital stock data was constructed based on the methodology applied by Kamps (2006) and Gupta et al. (2014), according to the perpetual inventory method (PIM). To conduct the analysis in this chapter we collected data on public capital stock and investment and private capital stock and investment for the following LAC countries: Antigua and Barbuda, Argentina, The Bahamas, Barbados, Belize, Bolivia, Brazil, Chile, Colombia, Costa Rica, Dominica, Dominican Republic, Ecuador, El Salvador, Grenada, Guatemala, Haiti, Honduras, Mexico, Nicaragua, Panama, Paraguay, Peru, St. Kitts and Nevis, St. Lucia, St. Vincent and the Grenadines, Uruguay, and Venezuela.

BOX 1.2

LAC public-private partnerships (PPPs).

Today it is usual to see cooperation between the public and private sectors in several economic sectors (primarily in infrastructure projects). This can be especially useful in situations where the governments have a low degree of economic slack. In this scheme, public assets and/or services are provided through an arrangement (long-term contract) between a government entity and a private party, with the latter bearing a substantial risk and management responsibility in exchange for a certain remuneration. The graph in Fig. 1.14 was created by the authors recurring to IMF (2017) "Investment and Capital Stock Dataset", and it shows the mean of the PPPs capital stock (kppp_mean) and the mean of the PPPs investment (ippp_-mean), both as a percentage of GDP, for the LAC region. The data on PPSs is more scarce compared to the data on public capital stock and investment and private capital stock and investment, although we were able to collect PPPs data from 1985 to 2017 for Argentina, Belize, Bolivia, Brazil, Colombia, Costa Rica, Dominica, Dominican Republic, Ecuador, El Salvador, Grenada, Guatemala, Haiti, Honduras, Mexico, Nicaragua, Paraguay, Peru, St. Lucia, and Venezuela. The data for Antigua and Barbuda, the Bahamas, Barbados, Chile, Panama, St. Kitts and Nevis, St. Vincent and the Grenadines, and Uruguay were not available. As can be seen, the PPPs in the LAC escalated during the 1990s and in the first half of the 2000s, primarily due to the macroeconomic stabilisation and fiscal consolidation programmes that were adopted by the majority of the countries from this region. Given their debt problems, the LAC governments had to decrease their public expenses, which led them to see the PPP arrangements as a possible solution to soften the effects of the decrease in public investment. Greater liberalisation and the change in the political strategy of the region, which encouraged privatisation, are also factors that led to the increased collaboration between the public and private sectors. More recently, in accordance with IDB (2017, p. viii), during the period 2006–15, the LAC region invested *"US$361 billion in around 1000 PPP infrastructure projects, mostly in energy and transport"*, and that *"The PPP market in the region is highly concentrated in Brazil, followed at a significant distance by Mexico and Colombia, while Honduras leads in PPP investment relative to GDP."*

falling." This could mean that there is a lack of investment from governments in order to bear the existing capital stock, which could also be extended to the case of private capital, given that it also presents a similar flat evolution (Box 1.1).

To construct their "Investment and Capital Stock Dataset," the IMF uses data on gross fixed capital formation to construct their investment flow series, which is combined with data on the initial capital stock, and on the depreciation rate, to construct their capital stock data. If we decide to look only at "investment", instead of looking for the levels of public and private capital stocks, then as shown in Fig. 1.4 we can see the evolution of the public investment (gross

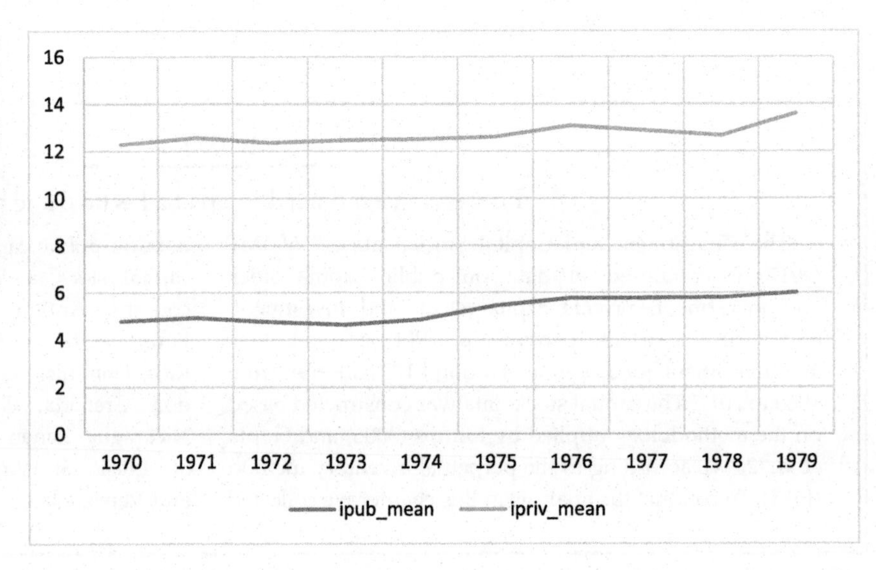

FIG. 1.4 Public investment (ipub_mean) and private investment (ipriv_mean) (% of GDP) in the LAC (1970–79). This graph was created by the authors and was based on the data from the "Investment and Capital Stock Dataset" of IMF (2017). The blue line (gray colour line in print version) represents the mean of public investment as a percentage of GDP for the LAC and the orange line (dark gray colour line in print version) represents the mean of private investment as a percentage of GDP for the same group of countries.

fixed capital formation) and private investment (gross fixed capital formation), both as a percentage of GDP, in the LAC during the 1970s.

During the 1970s, both public and private investment had periods of increase and decrease. However, it can be seen that, by the end of the decade (1979), both types of investment achieved a maximum of 13.6% of GDP in the case of private investment, and 6% of GDP in the case of public investment, which shows a moderate effort by the LAC to increase their investment in fixed assets. During this decade, LAC governments focused on industrialising their economies and took out large-value loans from diverse commercial banks and international lending institutions to support their growth [particularly for infrastructure development (Henderson et al., 2000)]. In fact, following Grosse and Goldberg (1996), between 1970 and 1981, the Latin American countries' external debt grew at an impressive rate of 27% per year. Private enterprise followed a similar borrowing strategy.

The level of borrowing increased with the oil price shocks that occurred in this decade, especially after the 1973 oil crisis and with the 1973–75 world recession (which lowered the commodity prices), in order to soften the negative effects that these situations were causing in LAC economies.

If we look at Fig. 1.4, it can be seen that public investment started to increase again in 1974, probably based on the borrowed money. However, as the World Bank (1995) stresses, there were a lot of physical capital investments which were poorly planned, primarily in the public sector, with low social rates of return, and which, in the end, were mainly inefficient. It is also worth noting that in this decade, most LAC governments were still of a considerable size and had tremendous weight in various economic sectors. This characteristic could have led to situations where private investment was crowded out as in the case of infrastructure provision and development, where the private sector had a lower level of participation (World Bank, 1995).

As could be expected, this strategy was far from being sustainable. With the slowdown of the world economy, the increase in the global interest rates, the increase in the debt services, and another energy crisis in 1979, the conditions were created for the terrible decade of the 1980s.

1.3 Latin America and the Caribbean physical capital in the 1980s

When entering the 1980s, LAC countries started to feel the effects of their high debt levels and that their macroeconomic situation was becoming more and more affected by the countries' debt-service problems. This period was mainly characterised by large balance of payments deficits, high inflation, and unemployment, a phenomenon which is now known as stagflation. However, in the final years of the 1970s and at the beginning of the 1980s, LAC governments saw this situation as temporary, and continued to borrow.

The poorly accomplished economic policy of the region, with only modest progress in structural policies (in which we include those focused on public infrastructure) and with failed stabilisation policies (Loayza et al., 2005), combined with the effects from external shocks, gave rise to several harmful effects [e.g. a drop in investment, domestic savings,

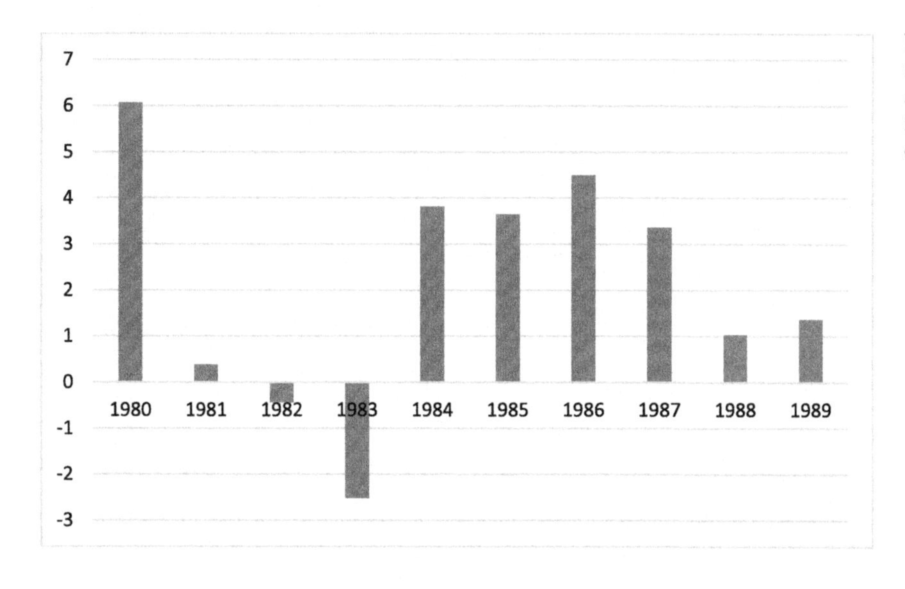

FIG. 1.5 GDP growth (annual %) in the LAC (1980–89). This graph was created by the authors and was based on GDP growth (annual %) data from the World Bank: https://data.worldbank.org/indicator/NY.GDP.MKTP.KD.ZG.

and capital flight (Atkins, 1999)], increasing the pressure associated with their debt. One alarming aspect was the massive drop in regional growth rates (Fig. 1.5).

As can be seen, the growth rate of the LAC region dropped from 6% in 1980 to less than 0.4% in 1981. Moreover, in the following years (1982, 1983), the economic contraction of the region worsened with negative growth rates. The LAC region had already been undergoing a period of loss of economic spark since 1973. However, it had never seen such a significant drop in its growth rate in the 1970s and 1960s. The "Lost Decade" had finally arrived.

At the turn of the decade, the United States increased their interest rates in order to fight inflation. This action placed the LAC countries in a difficult situation. Given their already high debt, the increase in debt service payments meant that they became incapable of repaying the debt. When loans started to be denied, the LAC governments began to feel powerless as to how to overcome their debt servicing problem. On 12 August 1982, Mexico declared that it could no longer meet its external debt-servicing obligations (in Fig. 1.6 we present Mexico's external debt stocks as a percentage of the gross national income). After a brief period, several LAC countries announced that they were in the same situation. The debt problem had turned into a debt crisis.

In response to this situation, the LAC region started to change its economic policy paradigm, passing from favouring heavy state interventionism and being inwardly oriented, to an increased trust in markets and openness. During the 1980s and 1990s, the region adopted a series of reforms focused on economic stabilisation, privatisation, and liberalisation. These reforms were later coined as the "Washington Consensus" (Williamson, 1990), and were primarily promoted by the IMF, the World Bank, and the US Treasury. Bearing all these in mind now lets us see how the

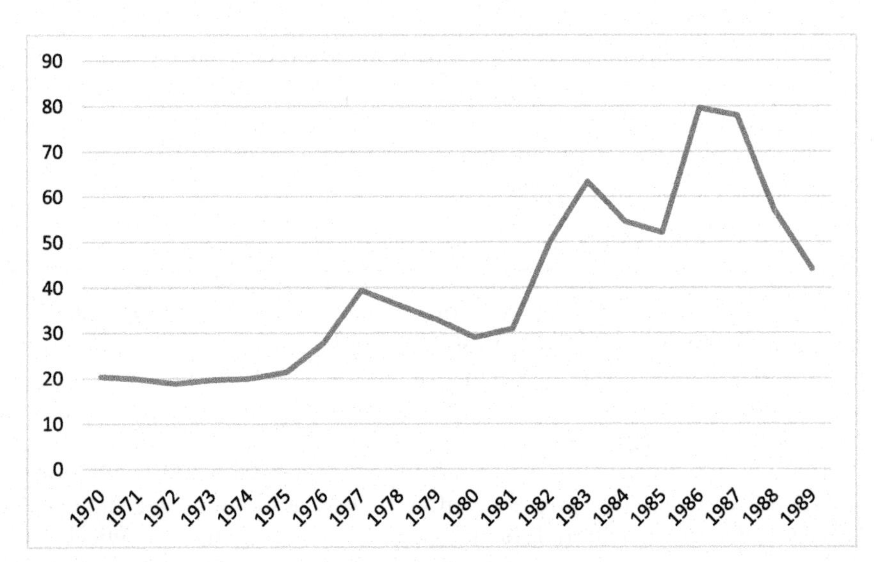

FIG. 1.6 External debt stocks (% of GNI) in Mexico (1970–89). This graph was created by the authors and was based on the external debt stocks (% of GNI) data from the World Bank: https://data.worldbank.org/indicator/DT.DOD.DECT.GN.ZS.

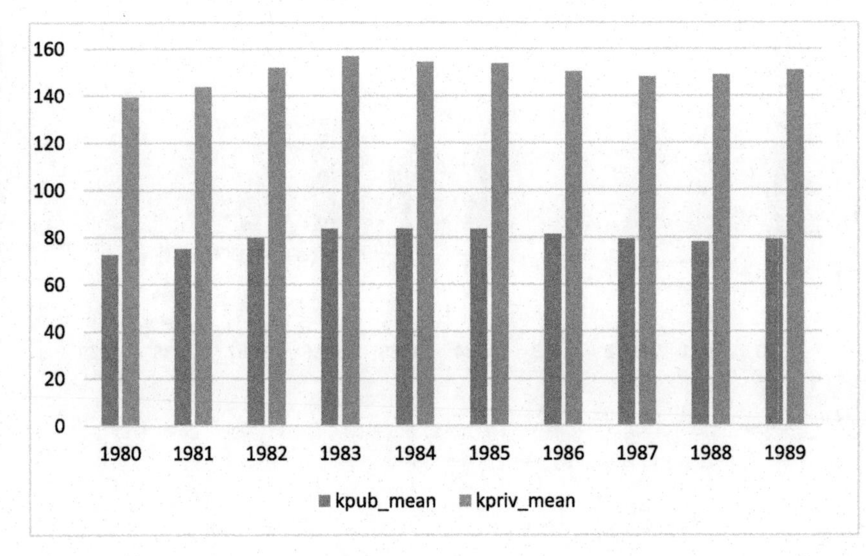

FIG. 1.7 Public capital stock (kpub_mean) and private capital stock (kpriv_mean) (% of GDP) in the LAC (1980–89). This graph was created by the authors and was based on the data from the "Investment and Capital Stock Dataset" of IMF (2017). The blue bars (gray colour bars in print version)represent the mean of public capital stock as a percentage of GDP for the LAC, and the orange bars represent the mean of private capital stock as a percentage of GDP for the same group of countries.

public and private capital evolved during the 1980s, starting with the graph of the public and private capital stocks as a percentage of GDP (Fig. 1.7).

At first appearance, we see a positive evolution from the public and private capital stocks as a percentage of GDP, with both types of stocks, public and private, achieving a maximum of 83.6% and 156.8% of GDP in 1983, respectively. Despite the decreasing trend that followed, the shares of public and private capital were, in general, higher than that in the previous decade. Looking at Fig. 1.8, we see that this information could be misleading and that these higher shares may be related to some negative features.

The fact is that, as we have already stated, both public and private capital (as a stock) rarely decrease. As can be seen from the blue and orange lines, the levels from both types of the stock increased during this decade. However, due to the aforementioned issues, regional GDP in this decade registered a poor performance compared to that in the 1970s, with negative growth rates in 1982 and 1983. This means that the increase in public and private capital stocks, as a percentage of GDP, was mainly due to the decrease in the regional output rather than the fact that there was a greater investment. In Fig. 1.9, we can see public investment and private investment, both as a percentage of GDP, in LAC between 1980 and 1989.

As can be seen, there was a break in both types of investment in the region during the 1980s. On the one hand, to reduce the LAC countries' deficits and debt levels (fiscal consolidation), governments made large expenditure cuts, namely on social programmes and infrastructure projects (Birdsall et al., 2010). On the other hand, as we have already

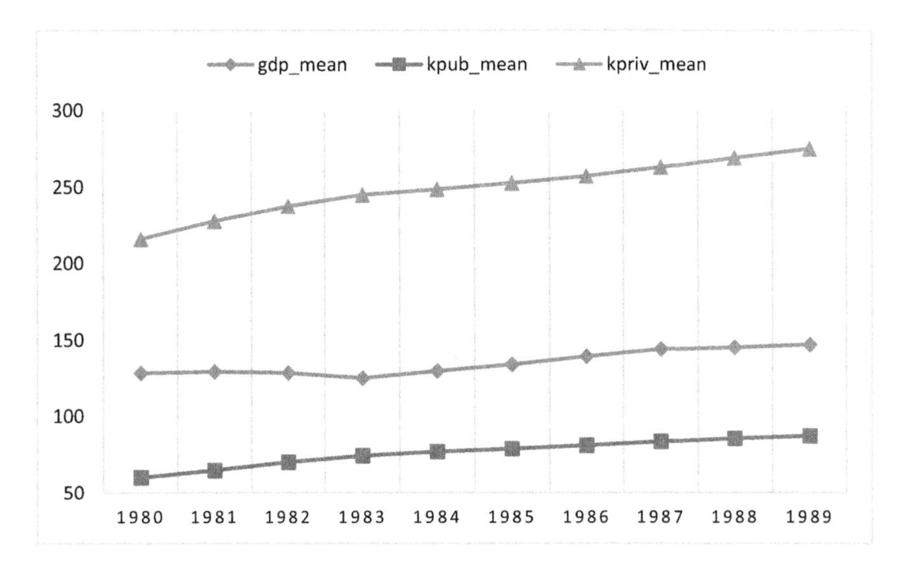

FIG. 1.8 Public capital stock (kpub_mean), private capital stock (kpriv_mean), and gross domestic product (gdp_mean) (in billions of constant 2011 international dollars)–LAC (1980–89). This graph was created by the authors and was based on the data from the "Investment and Capital Stock Dataset" of IMF (2017). The blue line (square line in print version) represents the mean of public capital stock in billions of constant 2011 international dollars, the orange line (triangle line in the print version) represents the mean of private capital stock in billions of constant 2011 international dollars, and the green line (diamond line in print version) represents the mean of gross domestic product in billions of constant 2011 international dollars, all for the LAC.

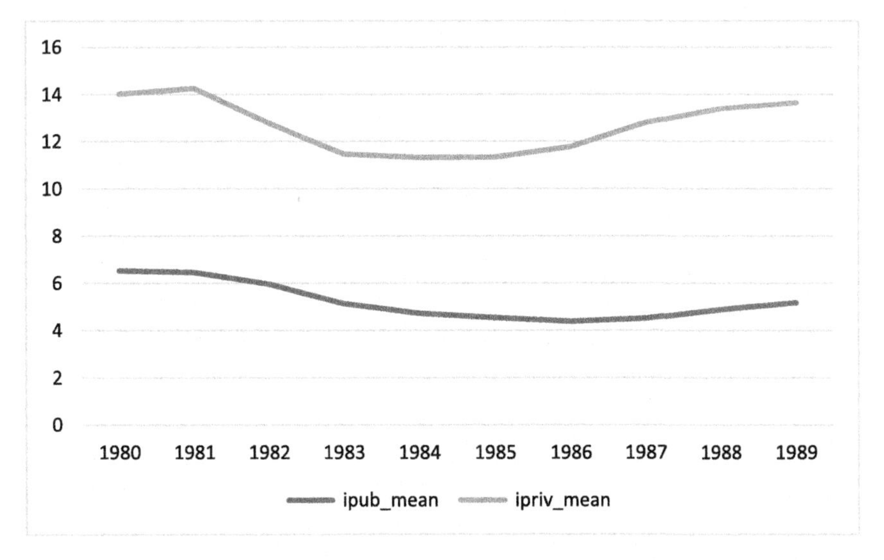

FIG. 1.9 Public investment (ipub_mean) and private investment (ipriv_mean) (% of GDP) in the LAC (1980–89). This graph was created by the authors and was based on the data from the "Investment and Capital Stock Dataset" of IMF (2017). The blue line represents the mean of public investment as a percentage of GDP for the LAC and the orange line represents the mean of private investment as a percentage of GDP for the same group of countries.

stressed, due to the debt crisis, the LAC region suffered from considerable capital flight, with foreign private investors looking for regions with reduced risk and a more stable economic panorama (compared to the LAC) to accommodate their investments (Harris and Nef, 2008). As was expected, this undesirable situation produced adverse effects on the development and economic growth of the LAC region (Ramirez and Nazmi, 2003; Pastor, 1990), contributing to an increase in the divergence of this region towards similar developing countries such as the East Asian (EA) countries (UN, 2017). For example, during the 1980s, with the slowdown on the LAC investment, the infrastructure gap of the region was exacerbated, and the difference between the infrastructure levels of the LAC and the EA increased markedly (Easterly and Servén, 2003).

Given the deterioration in the LAC region economic during the 1980s, regional governments knew that an effort should be made in order to reverse this adverse situation in the following decade. The strategy was to continue with a strong macroeconomic restructuration, with LAC governments abandoning the old import substitution industrialisation scheme and continuing to make reforms in favour of more open and market-orientated economies. In fact, before the turn of the decade, the Brady plan introduced in the region (in 1989) as a debt restructuration plan focused on debt relief, without excluding these countries from international capital markets. Basically, if the LAC governments continued with their market liberalisation reforms, they would be able to get their commercial bank debts reduced, as well as have new money from banks and international institutions (Frenkel, 1989). In the next section, we will see what happened to the LAC public and private capital during the 1990s, with all the reforms implemented in the region.

1.4 Latin America and the Caribbean physical capital in the 1990s

In spite of the reforms made in the 1980s, it was in the 1990s that the region totally embraced the "Washington Consensus" guidelines. Following Birdsall et al. (2010, pp. 7–8), policies were created in order to foment "*macroeconomic stabilisation (e.g. fiscal discipline to avoid high inflation, tax reform to broaden the tax base and positive real interest rates to overcome financial repression) and outward orientation (e.g. the elimination of import quotas and low and uniform import tariffs and a competitive exchange rate to induce nontraditional export growth and the removal of barriers to FDI)*," as well as to foment the market orientation of these economies by "*removing the entrepreneurial function of the state (e.g. privatisation of state enterprises); freeing and enabling markets (via deregulation, the strengthening of property rights, moderate marginal tax rates, low and uniform import tariffs a level playing field for foreign and domestic firms); and complementing markets (via the reorientation of public expenditures to primary education, health and infrastructure, both for growth and to improve the distribution of income).*"

With all the structural reforms which were applied in the first half of the 1990s, the LAC region managed to control their high inflation problem (Fig. 1.10), registering a good performance concerning the macroeconomic stabilisation of the region. Again, following Birdsall et al. (2010), these reforms also enabled the re-entry of private capital inflows into the region, thus alleviating the serious problem of capital flight encountered in the previous decade.

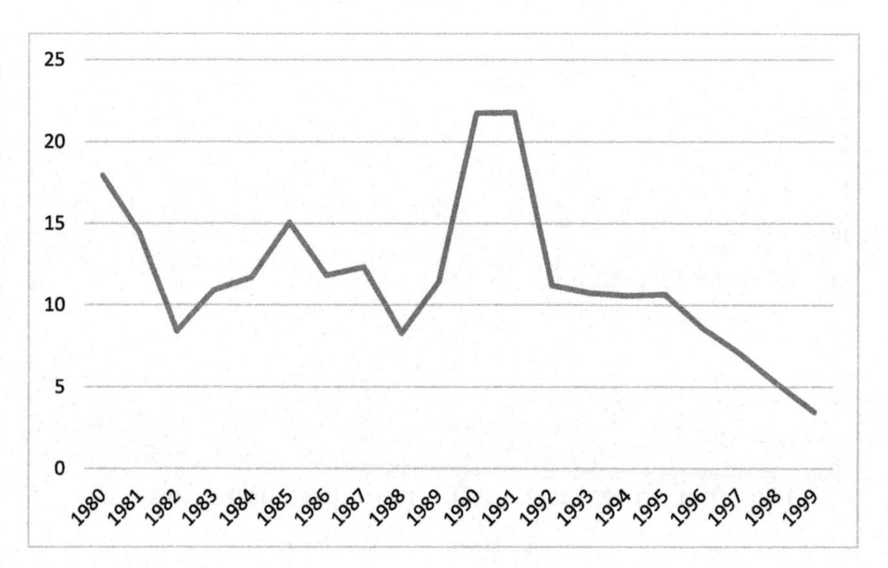

FIG. 1.10 Inflation and consumer prices (annual %) in the LAC (1980–99). This graph was created by the authors and was based on inflation and consumer price (annual %) data from the World Bank: https://data.worldbank.org/indicator/FP.CPI.TOTL.ZG.

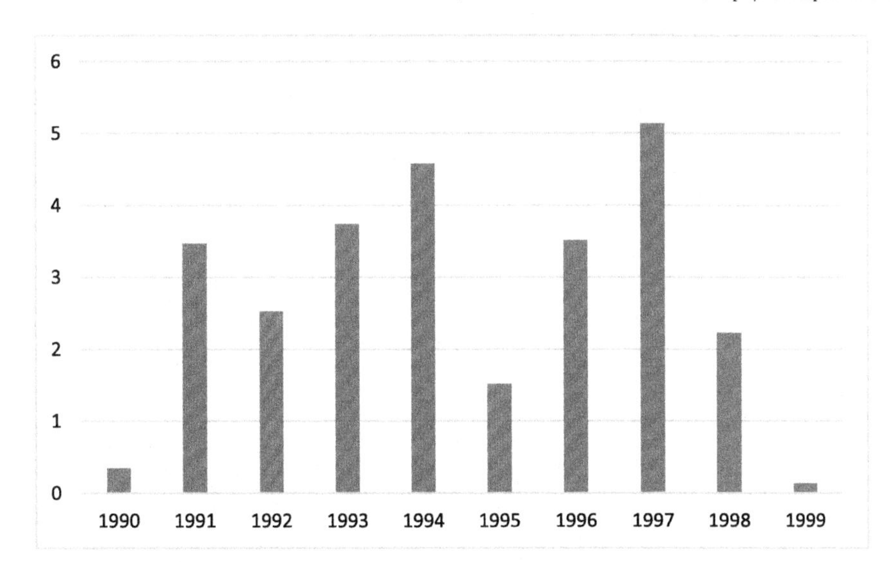

FIG. 1.11 GDP growth (annual %) in the LAC (1990–99). This graph was created by the authors and was based on GDP growth (annual %) data from the World Bank: https://data.worldbank.org/indicator/NY.GDP.MKTP.KD.ZG.

In fact, as can be seen in Fig. 1.11, the 1990s were a decade of recovery for the LAC region, which registered an improvement in growth rates compared with the previous decade. This improvement was primarily felt in the countries that conducted economic stabilisation reforms (Loayza et al., 2005). In accordance with Loayza et al. (2005), this growth recovery was mostly generated through improvements in the region's TFP rather than the capital accumulation (human and physical capital).

As can be seen in Fig. 1.12, public and private capital stocks as a percentage of GDP were once again relatively constant during this decade, with public capital stock at around 79% of GDP while private capital stock was around 156% of GDP. Compared to the previous decade, it can be seen that public capital stock as a percentage of the GPD remained at the levels of the end of the 1980s, while private capital stock as a percentage of GDP was relatively higher than that in the previous decade. This time, given the positive growth rates of the region, it is possible that the increased percentage (mainly concerning private capital stock) was registered through the rise in investment.

As is noticeable in the graph in Fig. 1.13, after 1992, there was an increase in LAC private investment, which maintained this positive trend until 1998 (in 1999 it fell, accompanying the regional growth rate). This positive trend can be mainly attributed to the reforms of the first half of the 1990s which allowed the re-entrance of private capital inflows into the LAC region, primarily due to the consensus that privatisation and liberalisation should be encouraged. For

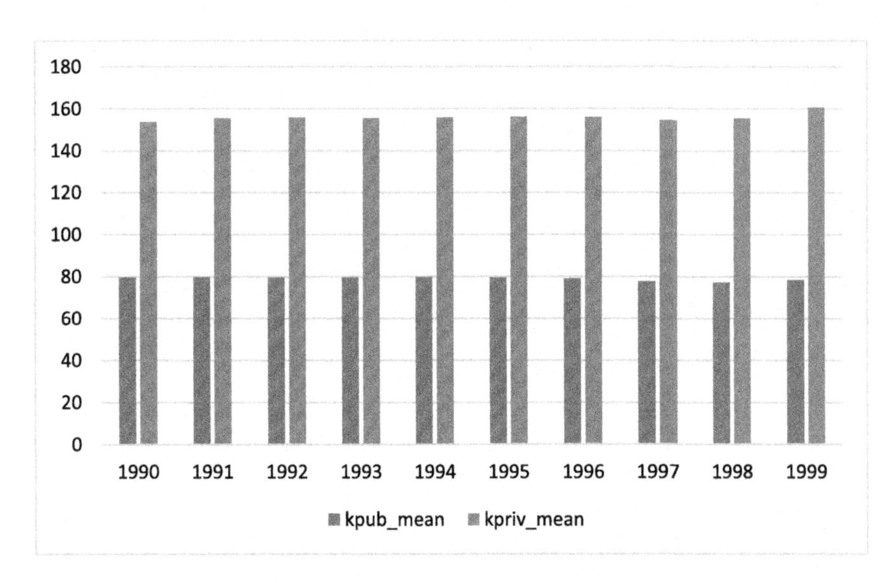

FIG. 1.12 Public capital stock (kpub_mean) and private capital stock (kpriv_mean) (% of GDP) in the LAC (1990–99). This graph was created by the authors and was based on the data from the "Investment and Capital Stock Dataset" of IMF (2017). The blue bars (gray colour bars in print version) represent the mean of public capital stock as a percentage of GDP for the LAC, and the orange bars (dark gray colour bars in print version) represent the mean of private capital stock as a percentage of GDP for the same group of countries.

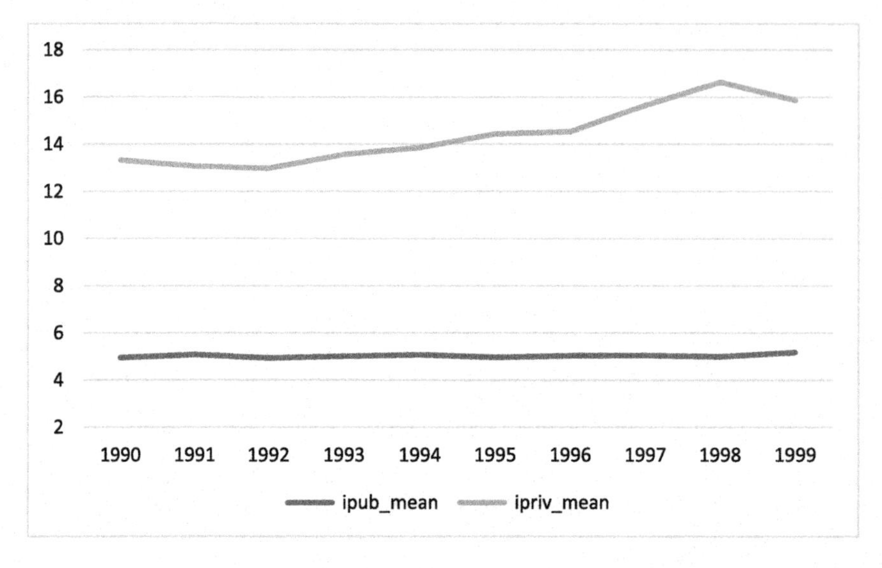

FIG. 1.13 Public investment (ipub_mean) and private investment (ipriv_mean) (% of GDP) in the LAC (1990–99). This graph was created by the authors and was based on the data from the "Investment and Capital Stock Dataset" of IMF (2017). The blue line (gray colour line in print version) represents the mean of public investment as a percentage of GDP for the LAC and the orange line (dark gray colour line in print version) represents the mean of private investment as a percentage of GDP for the same group of countries.

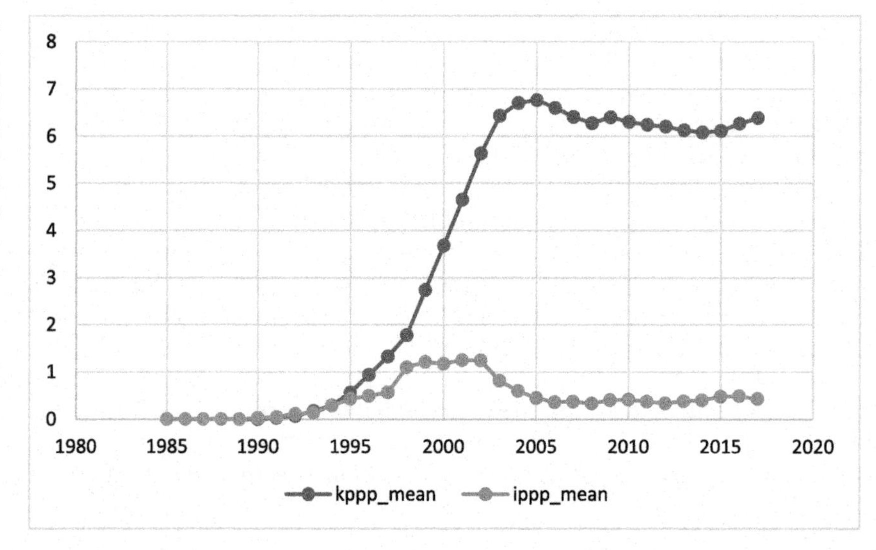

FIG. 1.14 Public-private partnership (PPP) capital stock (kppp_mean) and investment (ippp_-mean) (% of GDP) in the LAC (1985–2017). This graph was created by the authors and was based on the data from the "Investment and Capital Stock Dataset" of IMF (2017). The blue line (gray colour line in print version) represents capital stock, with the orange line (dark gray colour line in print version) representing investment (Box 1.2).

example, Calderón and Servén (2010, p. 29) stated that "*Following the opening up of infrastructure activities, private investment commitments rose from $10 billion in 1990 to over $70 billion in 1998.*"

Concerning public investment, it remained constant during the 1990s (around 5% of GDP), primarily due to the fiscal consolidation and economic stabilisation strategies that were followed which led to the subsequent cuts in government expenditure. Despite the positive trend in private investment during this decade and the benefits that it could produce (e.g. new ideas and technologies), following Rozas (2010), it seems that it did not contribute greatly to the gross fixed capital formation in the infrastructure sector. According to the same author, although private investment had a considerable impact on the areas of telecommunications and energy, it did not manage to compensate public investment cuts in other areas. As ECLAC (2002, p. 24) underlines: "*private-sector suppliers generally concentrated on higher-income or lower-risk sectors.*" Moreover, following Faruqee (2016, p. 86), "*Although privatisations were particularly important in LAC in the late 1990s, and concessions remain important today, Asia has experienced a much larger proportion of greenfield investment,*" which is a type of foreign investment directly linked to new physical capital. In fact, according to Fay and Morrison (2007, p. 16), the LAC infrastructure gap with the EA countries "*grew by a huge margin between 1980 and 1997.*"

With the arrival of the new millennium, the LAC region was once again hit by financial problems. The Asian financial crisis (1997) did not take long to spread its effects to other parts of the globe, and the LAC region, which had seen its integration into the world globalisation process increase in the 1990s, was soon affected by it. At the turn of the millennium, several LAC countries entered a recession and, once again, the region went through a turbulent period; the growth rate of the LAC stayed at 0.14% in 1999 (Fig. 1.11). This situation, together with the growing disappointment with the results from the reforms (e.g. failed in the promotion of sustainable growth and improved living standards), placed the LAC region in difficult circumstances again.

1.5 Latin America and the Caribbean physical capital in the new millennium

Despite the positive growth trend of the first half of the 1990s, the start of the new millennium showed that some problems, such as the volatility of the LAC region to external shocks, continued to persist in the region. Following Birdsall et al. (2010, p. 27): "*Volatility is estimated to have reduced the region's historical growth rate by one percentage point, with particularly strong negative impacts on investment in infrastructure and human capital, and especially detrimental impacts on poverty and inequality.*" As we stressed above, in 1997 the Asian financial crisis happened and shortly afterwards, its adverse effects were seen to spread across the globe. By the late 1990s, the LAC region started on a path of economic recession, followed by the upsurge of a banking and currency crisis in several of the region's countries (e.g. the Brazilian and Argentinian crises of 1998–99 and 2001–02, respectively). This situation led some scholars to call the period between 1998 and 2002 the "lost half-decade" (e.g. Ocampo, 2009).

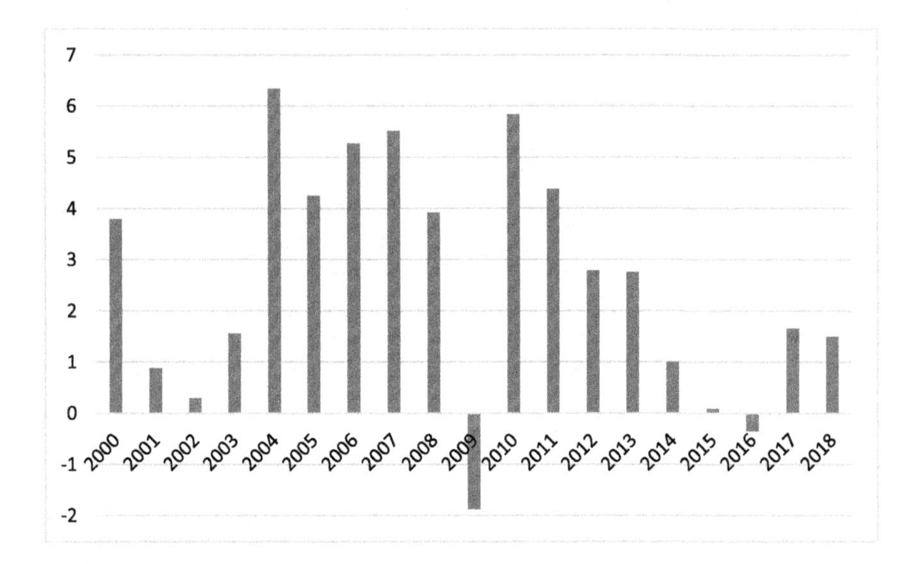

FIG. 1.15 GDP growth (annual %) in the LAC (1990–99). This graph was created by the authors and was based on GDP growth (annual %) data from the World Bank: https://data.worldbank.org/indicator/NY.GDP.MKTP.KD.ZG.

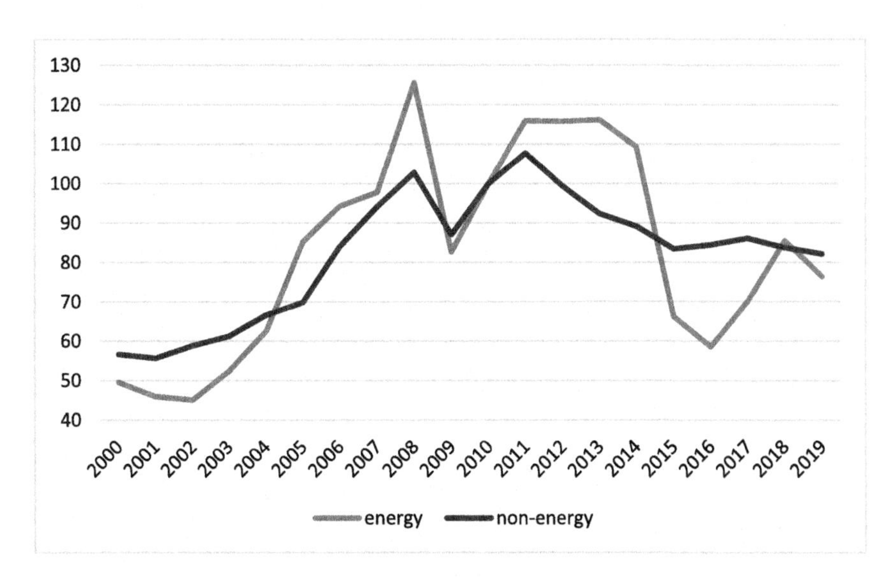

FIG. 1.16 Energy and nonenergy commodity price index (annual indices, 2010 = 100, real 2010 US dollars) during the 2000–19 period. This graph was created by the authors and was based on the "World Bank Commodity Price Data (The Pink Sheet)" from the World Bank: https://www.worldbank.org/en/research/commodity-markets. The blue line (gray colour line in print version) represents the index values for energy commodity prices, with the red line (dark gray colour line in print version) representing the index values for nonenergy commodity prices.

However, after this period of low growth, the LAC region was able, once more, to start on a path to recovery in 2003 (Fig. 1.15). From 2003 to 2007, the LAC region underwent a period of economic boom that had not been seen since the 1970s. According to Ocampo (2009, p.704), this boom was built on a combination of *"high commodity prices, booming international trade, exceptional financing conditions, and high levels of remittances."* However, the world economy has once again shaken in 2008 by a global financial crisis which, among other negative effects, pushed the prices of commodities down (Fig. 1.16).

Due to the effects of the financial crisis, and mainly due to the drop in commodity prices, the LAC region's growth rate dropped to negative levels once again in 2009, around −1.9% (Fig. 1.15). Nevertheless, contrary to expectations and projections, in the following years, the region registered a quick recovery—in 2010, the regional growth rate reached 5.8% (Fig. 1.15). According to the World Bank (2011, p. 12), this recovery was the product of the changes that were made in the economic policy of the region and which *"converted the region's traditional factors of external shock amplification (weak currencies, weak fiscal processes, and weak banking systems) into shock absorbers (flexible and credible currencies, stronger public finances, and well-capitalised and liquid banking systems)."* In this document, World Bank (2011) points out another important factor in the LAC's quick recovery: the Chinese connection. The increased link between China and the LAC countries (which grew considerably during the 2000s) helped the region to overcome the effects of the 2008 crisis, not only because of augmented China-LAC trade and direct Chinese foreign investment but also because of the impact of China on international commodity prices. The stimulus packages that China launched in response to the 2008 crisis had a positive effect on its import demand, allowing international commodity prices to rise again (Li et al., 2012).

Given the end of the commodity prices boom in 2014, and the slowdown of Chinese and United States growth in the most recent years, the LAC region growth rates started to decline: 1% in 2014, 0.09% in 2015, and −0.3 in 2016 (Fig. 1.15). Following the "Preliminary Overview of the Economies of Latin America and the Caribbean" (ECLAC, 2019), in the 2014–20 period, the LAC region was estimated to have registered the slowest growth in the past four decades, mostly due to falls in the pace of their economic activity, in GDP per capita, and in investment, consumption, and exports.

After this review of the region's economic performance in the new millennium, let us now turn our focus to the evolution of public and private capital in the LAC between 2000 and 2017 (which is the year up to which we have available data), starting with the analyses of the progression of public and private capital stock shares as a percentage of GDP during this period (Fig. 1.17).

In the graph in Fig. 1.17, it can be seen that, overall, the shares of private capital stock as a percentage of GDP were higher than that in the previous decade, registering an increase until 2002, a decrease in the following years until 2007, and increasing again in the period 2008–09. After 2009, until 2017, the share of the private capital stock as a percentage of GDP was relatively constant, with the mean being around 173% of GDP. Regarding the shares of public capital stock as a percentage of GDP, they increased from 2000 until 2002, decreased until 2008, and since then, they have been increasing again until 2017. However, the shares were very similar to those from the 1990s, except for the periods of 2002–03 and 2005–17, where they surpassed 80% of GDP.

Again, these data raise doubts regarding the actual investment in physical capital, given that the decrease in both shares was found in a period of economic boom for the LAC region, whereas the shares for periods of economic

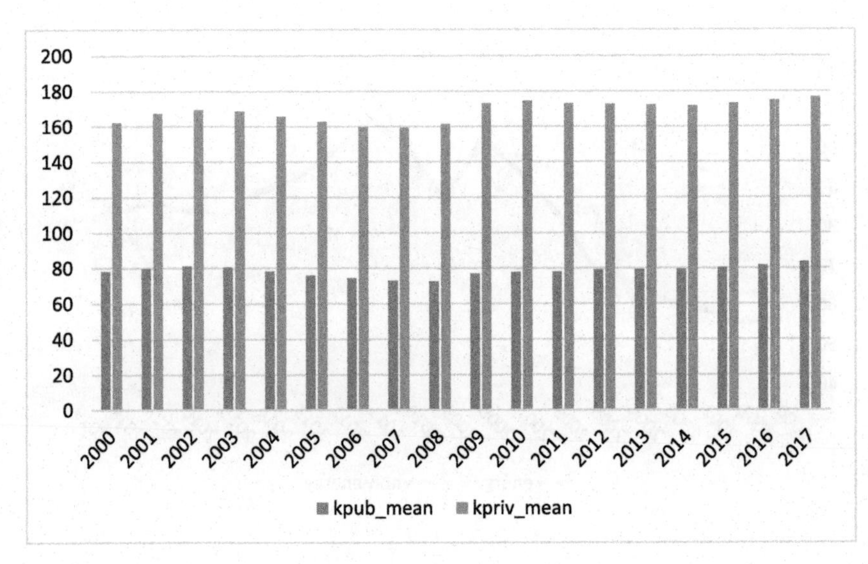

FIG. 1.17 Public capital stock (kpub_mean) and private capital stock (kpriv_mean) (% of GDP) of the LAC (2000–17). This graph was created by the authors and was based on the data from the "Investment and Capital Stock Dataset" of IMF (2017). The blue bars (dark gray colour bars in print version) represent the mean of public capital stock as a percentage of GDP for the LAC, and the orange bars (dark gray colour bars in print version) represent the mean of private capital stock as a percentage of GDP for the same group of countries.

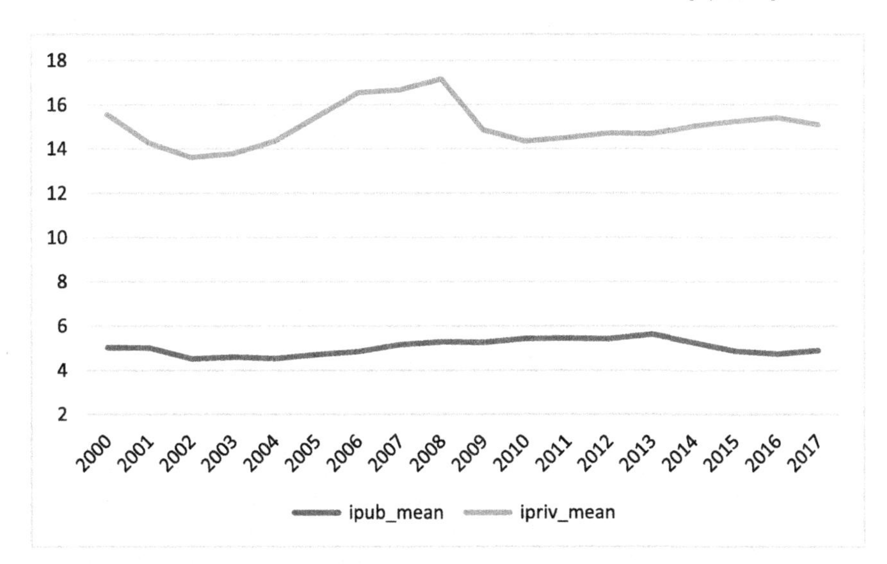

FIG. 1.18 Public investment (ipub_mean) and private investment (ipriv_mean) (% of GDP) in the LAC (2000–17). This graph was created by the authors and was based on the data from the "Investment and Capital Stock Dataset" of IMF (2017). The blue line (gray colour line in print version) represents the mean of public investment as a percentage of GDP for the LAC and the orange line (dark gray colour line in print version) represents the mean of private investment as a percentage of GDP for the same group of countries.

turbulence (e.g. 2002, 2009) are relatively higher. Therefore, in Fig. 1.18, we can see the public and private investment data, as a percentage of GDP, during the same time span.

As was to be expected, during the end of the "lost half-decade" (1998–2002), private investment (% of GDP) dropped, due to the negative economic performance of the region at that time which, once again, led to a wave of capital flight. Public investment (% of GDP) also diminished during the same period. However, the magnitude of its decline was much smaller than that in the case of private investment, not only because of these countries' financial conditions, which were better than that in the 1970s but also because public investment in the previous years was far from being ambitious, as were the general government expenses.

Throughout the commodity prices boom, which started in 2003, private investment grew again, at least until the beginning of the global financial crisis (2008–09)—in 2008, it achieved a maximum of 17% of GDP—when it contracted again until 2010. After 2010, private investment grew again, even after the end of the commodity prices boom (2014), only to decrease again in 2017. Concerning public investment, it can be seen that, after the decrease at the turn of the millennium and the subsequent stagnation (until 2004), it followed a moderate growth trend until 2013, when it returned to a decreasing trend. We should note that, since 2000, public investment has always been around the 5% of GDP, and even more impressive is that, even when the global financial crisis struck the LAC region (2009), public investment did not fall markedly. In fact, it rapidly recovered (2008: 5.29%; 2009: 5.25%; 2010: 5.43%). This situation can be linked with the economic success during the commodity prices boom, which allowed LAC governments to improve their public finances and design countercyclical responses to the global financial crises with public investment programmes, namely for public infrastructure (Lardé, 2016).

In Figs. 1.17 and 1.18, it can be seen that public and private capital stock shares increased in periods of decreased economic growth and investment and decreased in periods of enhanced growth and investment. This probably means that both LAC public and private investment are far from the value that they should be to guarantee a substantial reduction in the regional infrastructure gap.

For example, private capital stock shares were around 78% of GDP in 2000, 79% in 2001, and 81% in 2002, whereas public investment shares were around 5% of GDP in 2000 and 2001 and 4.5% in 2002. At the same time, the LAC growth rates were 3.8%, 0.8%, and 0.3% for the same respective years. In spite of the decreasing trend in both growth and investment, LAC public capital stock as a percentage of GDP increased from 2000 to 2002 and, conversely, decreased from 2003 to 2007 (from 80% of GDP to 73%) during the commodity boom, when LAC growth rates recovered and public investment increased. This could mean that, although investment increased during the period between 2003 and 2007, it was not proportional to the increase in GDP, and so the region could not maintain the public capital stock share from the previous period (2000–02), which was mainly influenced by the contraction of GDP rather than due to increased investment. This could also be applied to the case of private capital stock share, and also to other periods which were analysed (as we have already seen). The overall conclusion is that the LAC region must increase its investment levels, especially concerning physical capital. As Cavallo and Powell (2019, p. 97) say: "*Latin America*

BOX 1.3

Global infrastructure outlook.

The Global Infrastructure Outlook, produced by the Global Infrastructure Hub (a G20 organisation) together with the Oxford Economics, forecasts the infrastructure investment needs and gaps up to the year 2040 for 56 countries from five regions (Africa, Americas, Asia, Europe, and Oceania), covering seven sectors (energy, telecommunications, transport: airports, transport: ports, transport: rail, transport: road, and water). From the analysis of the countries' current investment trends and needs, Oxford Economics can produce an estimate of these countries' investment gap. This tool can be accessed through https://outlook.gihub.org. Among the 56 countries included in the outlook, there are eight countries from the LAC (Argentina, Brazil, Chile, Colombia, Ecuador, Mexico, Paraguay, and Uruguay). In Table 1.1, we present the estimates of these countries' investment gap.

From the data present in the previous table, we see that Brazil is the country with the biggest investment gap (1.2 trillion $US) among these eight LAC countries, whereas Uruguay has the lowest investment gap (10 billion $US). Again, this gap represents the difference between the estimate of what would be invested under the current trends until 2040 and the countries' investment needs, which means that Brazil for example will need to invest 1.2 trillion $US more up to 2040, in light of the current investment trend, in order to close the gap. Moreover, among the seven sectors, we see that the "transport: road" sector is the one with the biggest investment gap in Brazil, with a value of 852 billion $US. Finally, in a more positive view, we also note that there are some sectors in which the investment gap is 0 for some of these countries ("telecommunications" and "transport: ports" in Argentina, "transport: ports" and "transport: road" in Paraguay, and "water" in Uruguay).

TABLE 1.1 Infrastructure investment gap.

	All sectors	Energy	Tele-communications	Transport: Airport	Transport: Ports	Transport: Rail	Transport: Road	Water
Argentina	$358 B	$39 B	0	$833 M	0	$15 B	$302 B	$1 B
Brazil	$1.2 T	$109 B	$17 B	$31 B	$71 B	$102 B	$852 B	$7 B
Chile	$53 B	$6 B	$8 B	$886 M	$3 B	$35 B	$65 M	$20 M
Colombia	$100 B	$10 B	$5 B	$2 B	$5 B	$497 M	$75 B	$1 B
Ecuador	$55 B	$6.1 B	$1.9 B	$1.2 B	$1.5 B	$4.6 B	$40 B	$228 M
Mexico	$544 B	$39 B	$19 B	$3.4 B	$2.7 B	$14 B	$464 B	$1.6 B
Paraguay	$17 B	$12 B	$1 B	$2.7 B	0	$1.4 B	0	$31 M
Uruguay	$10 B	$2.7 B	$2.1 B	$289 M	$687 M	$3.7 B	$27 M	0

Notes: T, B, and M represent trillion $US, billion $US, and million $US, respectively.
Source: Oxford Economics.

and the Caribbean invests too little in infrastructure. Investment is low not only in comparison to other regions, it is also low in relation to the region's need to close large infrastructure gaps" (Box 1.3).

1.6 Conclusion

As is known, the LAC region has always been affected by booms and busts which influence its economic performance and its economic strategies, especially due to the propensity of this region to be affected by external shocks. That said, recent decades have seen the LAC countries experiencing periods of economic growth, like those from the 1970s, the beginning of the 1990s, and the 2003–14 commodity boom, and periods of economic depression, such as those from the 1980s and the period of 1998–2002.

From our analysis, we noted that the LAC public and private capital have always been influenced by the economic conjuncture in each of these periods. For example, we saw that in the 1970s, the LAC countries took advantage of the positive economic conjuncture to increase the public and private capital of the region, whereas, in the 1980s, public and private capital investment dropped sharply, mainly due to the debt crisis that struck the region; this led to private capital flight and the accentuated reduction in public investment. With the macroeconomic stabilisation and liberalisation plans, which were implemented in the region after the 1980s crisis, private capital returned to the region in the 1990s, mainly due to several economic sectors, such as the infrastructure sector, opening up to private participation. However, due to the macroeconomic stabilisation and fiscal consolidation strategies, the LAC public capital investment continued to be very cautious in the 1990s, with the region showing relatively low public investment levels in this decade. During the period 1998–2002, the LAC passed through another crisis which, once again, negatively affected its capital investment. However, after this tough period, the region went back to economic recovery, primarily due to the commodity price boom (2003–14). Public and private investment grew again until the global financial crisis of 2008. Despite the drop in private investment after the crisis hit the region, LAC governments did not drop their investments, following a countercyclical strategy. Due to the high commodity prices and the increase in the link between the LAC and China during the 2000s, the region was able to overcome the global financial crisis. In 2010 the region already seemed to have recovered from the impacts of the crisis, at least, looking at its growth rate. LAC public investment only decreased again with the end of commodity prices boom in 2014, whereas private investment, which returned to a growing trend after 2010, never returned to the levels of the years prior to the financial crisis. Overall, it can be seen that during the period 1970–2017, LAC public investment was always relatively low, only passing 6% of GDP a few times (1979, 1980, 1981). In contrast, the evolution of private investment was much more volatile, with peaks (in good economic times) and breaks (in times of crisis) of large magnitude.

If we look at LAC public and private capital stocks as a percentage of GDP, it can be seen that, despite the increase from 1970 to 2017, the evolution was slow and, in some decades, it was nearly constant. In addition, the fact that some of the highest shares were registered in years of economic deceleration also raises suspicions regarding the lack of capital investment in the LAC region. This lack of investment is worrisome, given that it can prevent the region from being more competitive and from achieving a sustainable growth path. This is especially serious in the case of LAC infrastructure.

As already mentioned, there is an infrastructure gap in LAC, especially when compared with similar developing (emerging) countries such as the EA (or EAP) countries—see for example the difference between the LAC and EAP public investment shares since the 1970s shown in Fig. 1.19. This fact can harm the competitiveness of this region, not only in terms of production but also in terms of investment absorption, given that enterprises will prefer to invest in countries where there is a satisfactory supply of infrastructure, allowing them to carry out their activities without further limitations. Infrastructure is also important to connect regions and populations to areas where there are greater economic opportunities, but the reverse is also true that is allowing an enhanced economic dynamic to be created in regions which were previously more isolated. Moreover, there is also the case of social infrastructure (e.g. schools, hospitals), which positively impacts human capital and allows greater equality in access to education and health. Finally, regarding sustainability, if the region wants to enter on a greener economy path, as energy security

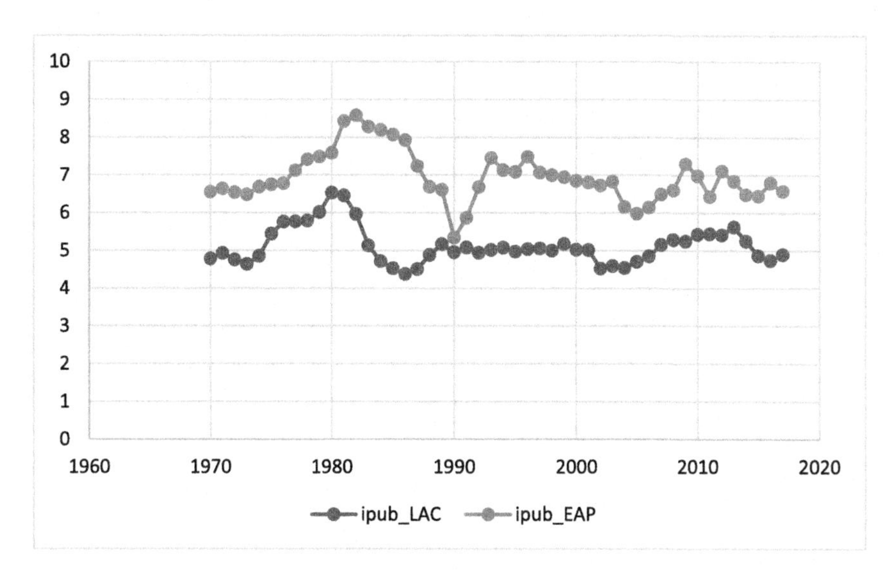

FIG. 1.19 Public investment (% of GDP) mean for LAC (ipub_LAC) and EAP (ipub_EAP) during 1970–2017 period. This graph was created by the authors and was based on the data from the "Investment and Capital Stock Dataset" of IMF (2017). The blue line (gray colour line in print version) represents the mean of public investment as a percentage of GDP for the LAC, and the orange line (dark gray colour line in print version) represents the mean of public investment as a percentage of GDP for the EAP (Cambodia, China, Indonesia, Korea, Lao P.D.R., Malaysia, Mongolia, Myanmar, Philippines, Singapore, Thailand, and Vietnam).

and environmental degradation problems demand, investment should probably be made in the maintenance of existing capital and the substitution of the outdated capital for more energy-efficient capital.

Given all these previous assumptions, we believe that it would be important to the LAC to increase their public investment levels in physical capital namely in infrastructure development, in order to promote the growth, equity, and sustainability of the region. However, private investment should not be forgotten, given that it can help governments which have financing problems or are still under strong macroeconomic stabilisation programmes.

References

Atkins, G., 1999. Latin America and the Caribbean in the International System. Routledge, New York, ISBN: 978-0813333830.

Birdsall, N., de la Torre, A., Caicedo, F.V., 2010. The Washington Consensus: Assessing a Damaged Brand. Policy Research Working Paper No. 5316, World Bank, Washington, DC, https://doi.org/10.1596/1813-9450-5316.

Calderón, C., Servén, L., 2010. Infrastructure in Latin America. Policy Research Working Paper No. 5317, World Bank, Washington, DC, https://doi.org/10.1596/1813-9450-5317.

Cavallo, E., Powell, A., 2019. Building Opportunities for Growth in a Challenging World. 2019 Latin American and Caribbean Macroeconomic Report, Inter-American Development Bank, Washington, DC, https://doi.org/10.18235/0001633.

De Jong, J.F.M., Ferdinandusse, M., Funda, J., 2018. Public capital in the 21st century: as productive as ever? Appl. Econ. 50 (51), 5543–5560. https://doi.org/10.1080/00036846.2018.1487002.

Easterly, W., Servén, L., 2003. The Limits of Stabilization: Infrastructure, Public Deficits, and Growth in Latin America. Latin American Development Forum, World Bank, Washington, DC, https://doi.org/10.1596/978-0-8213-5489-6.

ECLAC, 2002. A Decade of Light and Shadow. Latin America and the Caribbean in the 1990s. Economic Commission for Latin America and the Caribbean (ECLAC), Santiago. Available at: http://hdl.handle.net/11362/13048.

ECLAC, 2019. Preliminary Overview of the Economies of Latin America and the Caribbean 2019. Economic Commission for Latin America and the Caribbean (ECLAC), Santiago. Available at: http://hdl.handle.net/11362/45001.

Faruqee, H., 2016. Regional Economic Outlook, April 2016, Western Hemisphere Department: Managing Transitions and Risks. International Monetary Fund, Washington, DC, https://doi.org/10.5089/9781498329996.086.

Fay, M., Morrison, M., 2007. Infrastructure in Latin America and the Caribbean: Recent Developments and Key Challenges. Directions in Development-Infrastructure. World Bank, Washington, DC, https://doi.org/10.1596/978-0-8213-6676-9.

Frenkel, M., 1989. The international debt problem: an analysis of the Brady plan. Intereconomics 24, 110–116. https://doi.org/10.1007/BF02928561.

Grosse, R., Goldberg, L.G., 1996. The boom and bust of Latin American lending, 1970–1992. J. Econ. Bus. 48 (3), 285–298. https://doi.org/10.1016/0148-6195(96)00015-X.

Gupta, S., Kangur, A., Papageorgiou, C., Wane, A., 2014. Efficiency-Adjusted public capital and growth. World Development 57, 164–178. https://doi.org/10.1016/j.worlddev.2013.11.012.

Harris, R., Nef, J., 2008. Capital, Power, and Inequality in Latin America and the Caribbean. Rowman & Littlefield, Lanham, MD, ISBN: 978-0742555242.

Henderson, J.D., Delpar, H., Brungardt, M.P., Weldon, R.N., 2000. A Reference Guide to Latin American History. M.E. Sharpe, New York, ISBN: 978-1563247446.

IDB, 2017. Evaluation of Public-Private Partnerships in Infrastructure. Inter-American Development Bank, Washington, DC. Available at: https://publications.iadb.org/en/evaluation-public-private-partnerships-infrastructure.

IMF, 2017. Estimating the Stock of Public Capital in 170 Countries. Fiscal Affairs Department, International Monetary Fund, Washington, DC. Available at: https://www.imf.org/external/np/fad/publicinvestment/pdf/csupdate_aug19.pdf.

Kamps, C., 2006. New estimates of government net capital stocks for 22 OECD countries, 1960–2001. IMF Econ. Rev. 53, 120–150. https://doi.org/10.2307/30035911.

Lardé, J., 2016. Latin America's Infrastructure Investment Situation and Challenges. FAL Bulletin No. 347, Economic Commission for Latin America and the Caribbean (ECLAC), Santiago. Available at: http://hdl.handle.net/11362/40849.

Lardé, J., Sánchez, R., 2014. The Economic Infrastructure Gap and Investment in Latin America. FAL Bulletin No. 332, Economic Commission for Latin America and the Caribbean (ECLAC), Santiago. Available at: http://hdl.handle.net/11362/37381.

Li, L., Willett, T.D., Zhang, N., 2012. The effects of the global financial crisis on China's financial market and macroeconomy. Econ. Res. Int. https://doi.org/10.1155/2012/961694.

Loayza, N., Fajnzylber, P., Calderon, C., 2005. Economic Growth in Latin America and the Caribbean: Stylized Facts, Explanations, and Forecasts. World Bank, Washington, DC, https://doi.org/10.1596/0-8213-6091-4.

Ocampo, J.A., 2009. Latin America and the global financial crisis. Cambr. J. Econ. 33 (4), 703–724. https://doi.org/10.1093/cje/bep030.

Pastor, M., 1990. Capital flight from Latin America. World Dev. 18 (1), 1–18. https://doi.org/10.1016/0305-750X(90)90099-J.

Perrotti, D., 2011. The Economic Infrastructure Gap in Latin America and the Caribbean. FAL Bulletin No. 293, Economic Commission for Latin America and the Caribbean (ECLAC), Santiago. Available at: http://hdl.handle.net/11362/36339.

Ramirez, M.D., Nazmi, N., 2003. Public investment and economic growth in Latin America: an empirical test. Rev. Dev. Econ. 7 (1), 115–126. https://doi.org/10.1111/1467-9361.00179.

Rozas, P., 2010. América Latina: Problemas y Desafíos del Financiamiento de la Infraestructura. Revista CEPAL No. 101. Available at:, pp. 59–83. https://www.cepal.org/es/hojasinformativas/america-latina-problemas-desafios-financiamiento-la-infraestructura.

UN, 2017. Chapter III—The end of the Golden Age, the debt crisis and development setback. In: World Economic and Social Survey 2017: Reflecting on Seventy Years of Development Policy Analysis. UN, New York. Available at: https://wess.un.org/chapter3/.

Williamson, J., 1990. Latin American Adjustment: How Much Has Happened? Institute for International Economics, Conference Volume, Washington, DC, ISBN: 978-0881321258.

World Bank, 1995. Meeting the Infrastructure Challenge in Latin America and the Caribbean. Directions in Development. World Bank, Washington, DC, https://doi.org/10.1596/0-8213-3028-4.

World Bank, 2011. Latin America and the Caribbean's Long-Term Growth: Made in China? LAC Semiannual Report. World Bank, Washington, DC. Available at: https://doi.org/10.1596/26674.

Zettelmeyer, J., 2006. Growth and Reforms in Latin America; A Survey of Facts and Arguments. IMF Working Papers No. 06/210, International Monetary Fund, Washington, DC, https://doi.org/10.5089/9781451864700.001.

CHAPTER

2

The relationship between public capital stock, private capital stock, and economic growth in the Latin American and Caribbean countries: A matter of complementarity

JEL codes E22, F21, O54

2.1 Introduction

As was previously seen (Chapter 1), LAC public and private investment has had a far from stable path, varying according to the macroeconomic situation of the region, with the economic booms and busts that we are already accustomed to in the region. Overall, we can say that LAC physical capital investment has always been below expectations. In some sectors, such as the infrastructure sector, the lack of appropriate investment is more obvious, with the harmful effects from this shortage being increasingly noticed by the economies of the LAC region. Following Cavallo and Powell (2019), the lack of new capital can cost the LAC up to 1 percentage point of the GDP growth, which may rise to 15 percentage points if this gap continues as it is in the next 10 years. Moreover, according to the report "Regional Economic Outlook, April 2016, Western Hemisphere Department: Managing Transitions and Risks" (Faruqee, 2016), the LAC's competitiveness is compromised by the state of the regional infrastructure, meaning that the region needs increased progress or else there is the risk that infrastructure shortfalls gradually hamper the region's growth.

Despite the role that private capital can play in closing this gap, it is known that the *"majority of current infrastructure investment is public"* (Cavallo and Powell, 2019; p. 68). As the IMF (2017; p. 1) stresses: *"public investment is a key input in the creation of a network of physical assets over time, including economic infrastructure (roads, airports, electric utilities, etc.) and social infrastructure (public schools, hospitals, prisons, etc.)"*. This accordingly puts pressure on LAC governments to increase their investment levels. This pressure increases with the need of these countries to reach the so-called sustainable development goals (SDGs) (Castellani et al., 2019).

Given the above facts, it is natural that LAC governments need to develop growth-promoting strategies in this field. With the purpose of helping the policymakers from this region, in this chapter, we discuss the positive and negative effects of public investment on economic growth and analyse the relationship between public capital stock and economic growth in the LAC, in order to see if the rise in public investment will benefit the region's long- and -short-term growth. Additionally, for public capital we also include private capital stock in our estimation to explore the relationship that it has with growth as well as with public capital.

2.2 Public capital stock and economic growth

There is little doubt around the idea that public capital stock is an essential input to a country's economic activity (see Fig. 2.1). Among the various positive effects, we can say that it enables an increase in aggregate output, enhances the economy's physical and financial resources, reduces private sector costs (Erden and Holcombe, 2006), contributes

https://doi.org/10.1016/B978-0-12-824429-6.00008-5

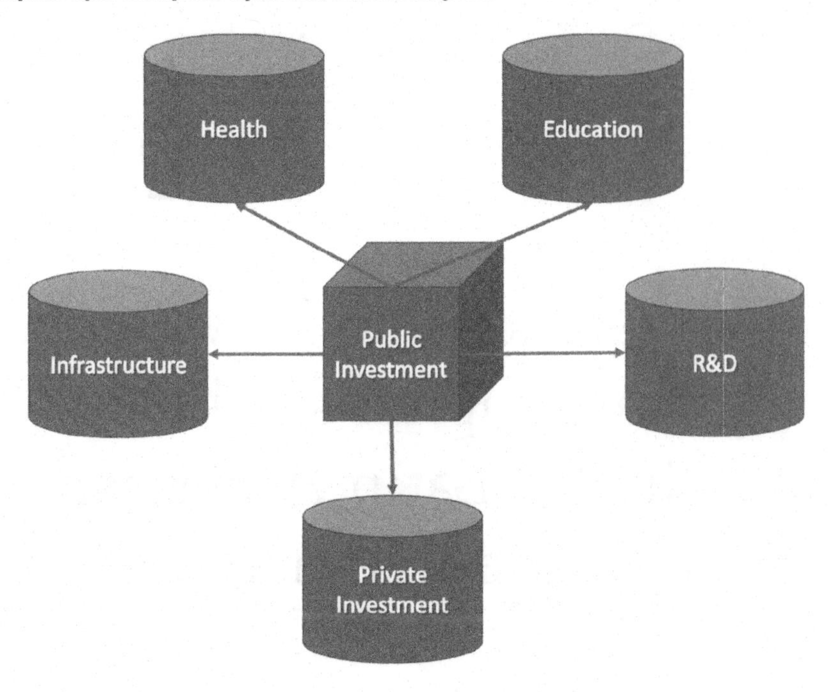

FIG. 2.1 Public investment and main acting areas (created by the authors).

to the advancement and maintenance of human capital (Ramirez and Nazmi, 2003) and contributes to higher long-run growth and an increase in the aggregate demand in the short run (Barbiero and Darvas, 2014). As the LAC region has seen a decrease in its public investment since the 1980s (see Chapter 1), it is pertinent to understand if these positive effects were lost. More precisely, it will be interesting to analyse how the region's public capital levels have affected the region's growth.

The drop in public investment, however, seems to have been a worldwide tendency. Following the report "Is It Time for an Infrastructure Push? The Macroeconomic Effects of Public Investment" (IMF, 2014), public capital stock as a share of output has declined significantly in the last 30 years, worldwide. This generates problems, especially for the least developed countries which, given the lack of investment, see their infrastructure levels still far from the levels of the countries considered as advanced. However, as these countries develop, their necessity for public investment increases (as conversely, they need enhanced public investment to be able to develop). Despite the possible positive effects that public investment would have on their economic growth, the high public debt and lack of investment efficiency that many of these countries face (e.g. the LAC countries) make it advisable to weigh several factors before advancing with public investment programmes. Additionally, the pressure surrounding public investment strategies in developing countries has also increased with a substantial number of failed projects in the past (Gupta et al., 2014).

Even though countries need roads, airports, schools, and hospitals (among others) to grant a sound and sustainable economic performance, there are several factors that may lead public investment to have a biased effect on the economy. For public investment to have beneficial effects, factors such as the degree of economic slack and monetary accommodation of the countries, their investment efficiency (public investment management), and how they will finance the public investment (IMF, 2014) should be taken into account. The distorting effects of higher taxes (associated with public investment financing), the small macroeconomic multiplier of public investment, and the crowding-out effect of public debt on the capital formation (Liaqat, 2019; Perotti, 2004; Fisher and Turnovsky, 1998) are some of the reasons for governments to be more cautious with their expenditures. In the case of the LAC countries, the borrowing strategy that their governments followed led the region into a debt crisis in the 1980s (see Fig. 2.2), which negatively affected their public investment (which was consistently reduced), as was all public expenditure. This was done in accordance with the economic adjustment and stabilisation programmes that were implemented in the region in order to overcome the crisis effects. Following Ramirez and Nazmi (2003), the excessive reduction in public investment seems to have had a detrimental impact on the economic growth of Latin American countries. Given the past mistakes, we believe that it is especially important today that governments of these countries develop sustainable financing methods and strategies before proceeding with an indiscriminate increase in public investment.

Additionally, in accordance with the law of diminishing returns, the effects of increased public investment on economic growth can also depend on the existing capital stock. Following Romp and De Haan (2007; p. 9), "an increment to the public capital stock would have a small (large) output effect if the capital stock in the previous period was large

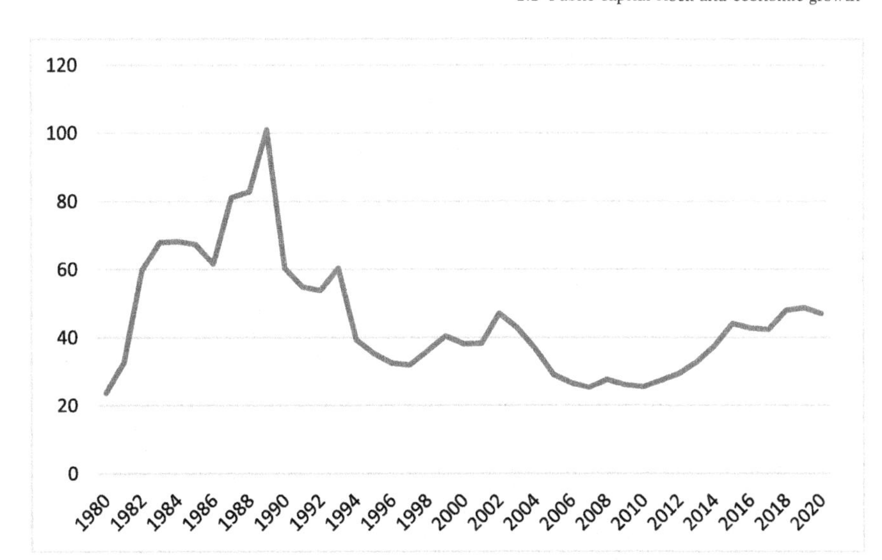

FIG. 2.2 External debt (% of GDP) of LAC (1980–2020). This graph was created by the authors and was based on the data from the "World Economic Outlook 2019" database by the IMF: https://www.imf.org/external/pubs/ft/weo/2019/02/weodata/index.aspx.

(small)". By this law, one can imagine that an increase of the LAC public capital, especially infrastructure, could bring a large output effect to the region, given that there is a shortage of this input nowadays. Finally, the effect of public capital on growth can also depend on the relationship between public and private capital, given that if they act as substitutes instead of complements, it can generate a "crowding-out" effect on investment of the private sector (Erden and Holcombe, 2006). This alerts us to the fact that LAC public investment must be articulated with the private initiative so that they do not compete with each other. This will allow the private sector to reduce costs and to increase the marginal productivity of private capital and, subsequently, help the region to achieve higher growth rates (Box 2.1).

BOX 2.1

Public investment management assessment (PIMA).

Public investment performance i.e. the contribution of public investment to economic growth, is intrinsically connected with the public investment/capital stock efficiency. For the IMF, the differences in efficiency between countries are the result of the different public investment management practices. Accordingly, in order to assess the quality of a country's public investment management, the IMF (2015) created the "Public Investment Management Assessment (PIMA)" framework, which evaluates 15 key institutions for planning, allocation, and implementation (the three stages) of public investment. In 2018 the IMF (2018) updated the framework and, for now, the countries are evaluated in terms of:

i) planning: fiscal principles or rules, national and sectoral planning, coordination between entities, project appraisal, and alternative infrastructure financing;

ii) allocation: multiyear budgeting, budget comprehensiveness and unity, budgeting for investment, maintenance funding, and project selection;

iii) implementation: procurement, availability of funding, portfolio management and oversight, management of project implementation, and monitoring of public assets.

The PIMA framework allows wide-ranging analysis to be applied to the stronger and weaker features of the public investment management of a determined country, enabling the identification of areas that can be improved and helping the development of future public investment projects.

The IMF (2015; p. 26) provides an example of strategic public investment planning from Brazil, saying that "The Growth Acceleration Program (PAC) introduced in 2007 is a comprehensive rolling 4-year plan that coordinates investment and PPPs made by the central government, subnational governments, and state-owned enterprises. The program includes large infrastructure projects in the areas of energy, transportation, housing, water and sanitation, environment, and health". For the IMF, the PAC enhanced the investment from 2006 to 2010 in Brazil, given that it has allowed construction of a development strategy centred on crucial sectors, with increased collaboration between the public and private sectors, and improvement of investment efficiency and transparency.

2.3 Public capital stock, private capital stock, and economic growth in Latin America and the Caribbean

As is known, and as was already explained in Chapter 1, the LAC region faced a severe debt crisis in the 1980s. This crisis led the LAC region countries into a deep depression which in turn led to a deterioration of their macroeconomic condition. Given this situation, several LAC governments decided to support the implementation of strong economic adjustment and stabilisation programmes in order to overcome the effects of this crisis. Among various effects, this led to an ever greater reduction in both the public investment and public expenditure of the LAC countries (again, see Chapter 1).

The study by Ramirez and Nazmi (2003) entitled "Public Investment and Economic Growth in Latin America: an Empirical Test" was based on the analysis of the impact of public investment on economic growth for nine LA countries (Argentina, Bolivia, Brazil, Chile, Colombia, Ecuador, Mexico, Peru, and Uruguay) during the 1983–93 period. Their conclusion was that the excessive reduction of public investment (as also of the private investment) was detrimental to the economic growth of this group of countries. The results of this study support the idea that both public and private investment spending contribute to increased economic growth. These outcomes may indicate that some of the strategies that were followed by the majority of the LAC governments (uncritical cuts in public expenditure and investment) were not the most suitable and that they should have improved some other investment-associated factors, such as investment efficiency or appraisal of investment projects.

Although some time has passed since the 1980s, it seems that the LAC region continues to suffer from the consequences of the investment cuts of that same decade, as well as from the low investment levels that were witnessed in the following decades. One aspect that was greatly affected by the lack of public investment was the state of the regional public capital, particularly infrastructures. As was already mentioned, for the IMF (Faruqee, 2016), the weak state of the regional infrastructure can be problematic for the LAC's competitiveness and growth prospects and they state that there is a need for the LAC governments to concentrate their efforts on this field. Despite the need for higher investment levels, the IMF agrees on the fact that it is also essential that these governments improve some of their associated investment factors (e.g. they should improve public investment management processes and practices and make an effort to maintain existing infrastructures).

The outcomes and conclusions from more recent studies (e.g. Castellani et al., 2019; Ruiz-Nuñez and Wei, 2015) also seem to support the idea that there is a lack of public investment in the LAC. Following their recommendations, in order to close the LAC infrastructure gap, regional governments should increase their investment levels. However, increased investment should be accompanied by the development of sustainable financing methods and strategies. Additionally, in some cases, the strategy can start with the improvement of the existing infrastructures, rather than by the investment in new ones (which is an example of improved public investment management).

Despite the somewhat consensual idea that increased public capital stock fosters economic growth, and although the study by Ramirez and Nazmi (2003) is, undoubtedly, one of the most cited studies on this theme for the LAC region, there is a need to extend (and update) the knowledge surrounding the impacts that public and private capital stock has had on the economic growth of this region.

New databases [e.g. the "Investment and Capital Stock Dataset" by the IMF (2017)], new econometric techniques and new ideas have emerged after the publication of the Ramirez and Nazmi (2003) study. Thus in this chapter, we will empirically expand the analysis of the relationship between public capital stock, private capital stock, and economic growth for the LAC region for a time span ranging from 1970 until 2014. As we previously explained, the inclusion of private capital stock is mainly due to the fact that the authors whose studies are centred on the effects of public capital on growth usually include private capital in their estimations as a way to compare the effects from both of them on growth. Alternatively given their interconnection, they can be used to explore the effects of public capital on private capital, which can be a channel for enhancing or diminishing a country's economic output (e.g. Nguyen and Trinh, 2018; Dreger and Reimers, 2014).

To achieve the goals of this analysis, we collected annual data from the "Investment and Capital Stock Dataset" by the IMF (2017) on gross domestic product (current prices) in billions of national currency (*y*)—which will be our proxy for economic growth; general government capital stock (current cost) in billions of national currency (*kpub*); and private capital stock (current cost) in billions of national currency (*kpriv*). These will be our proxies for public capital stock and private capital stock, respectively—from 1970 until 2014 for 30 LAC countries; namely, Antigua and Barbuda, Argentina, Bahamas, Barbados, Belize, Bolivia, Brazil, Chile, Colombia, Costa Rica, Dominica, Dominican Republic, Ecuador, El Salvador, Grenada, Guatemala, Haiti, Honduras, Mexico, Nicaragua, Panama, Paraguay, Peru, St. Kitts and Nevis, St. Lucia, St. Vincent and the Grenadines, Suriname, Trinidad and Tobago, Uruguay, and Venezuela. Countries and time horizon were chosen according to the data availability. In Table 2.1, we give the name, definition, and sources of the variables included in the analysis.

TABLE 2.1 Description of variables.

Variable	Definition	Source
y	Gross domestic product (current prices), in billions of national currency	Investment and Capital Stock Dataset (FMI)
p	Total population, the total number of people	World Development Indicators (WB)
kpub	General government capital stock (current cost), in billions of national currency	Investment and Capital Stock Dataset (FMI)
kpriv	Private capital stock (current cost), in billions of national currency	Investment and Capital Stock Dataset (FMI)

As can be seen from Table 2.1, annual data was also collected for the variable total population as the total number of people (*p*) from the World Development Indicators of the World Bank, and subsequently used to convert *y*, *kpub*, and *kpriv* variables into their per capita values (*ypc*, *kpubpc*, and *kprivpc*) in order to remove possible distortions that can be produced by population variations.

Recurring to the use of the Stata 15.0 and E-Views 10 statistical software, the empirical analysis of this chapter will be based on the panel vector autoregression (PVAR) developed by Holtz-Eakin et al. (1988), using the estimator proposed by Love and Zicchino (2006), and on the use of the dynamic ordinary least square (DOLS) and the fully modified ordinary least squares (FMOLS), which were extended to the panel framework by Pedroni (2001a, b).

Although Romp and De Haan (2007) state that most authors used Cobb-Douglas production function approaches to investigate the effects of public capital on growth, approaches based on the vector autoregression (VAR) and vector error correction (VECM) models have been gaining some considerable attention in most recent years (e.g. De Jong et al., 2018). This is because the production function approach has been accused of not addressing the possibility of reverse causation and violating the marginal productivity theory, which has led many authors to start looking for different estimation methods. According to De Jong et al. (2018), the VAR approach has the advantages of not imposing causal relationships, allowing tests for the existence of causal relationships in any direction (overcomes the reverse causation problem), and enabling indirect links between the variables.

When we work with macro-panels (such as ours), problems related to the presence of cointegration between the variables and the presence of endogeneity may arise. One way to overcome these issues is by using the PVAR, which is an estimator that can deal with these two phenomena (Abrigo and Love, 2016). Nevertheless, the PVAR requires that all variables be stationary of the same order. For this very reason, it is often necessary to resort to the transformation of variables into their first differences. The only problem is that, if we do this, we lose the capacity to analyse the long-run relationships. Because of this fact, we will use the PDOLS and PFMOLS to explore the long-run impacts that the public capital stock and the private capital stock had on the economic growth of the LAC countries from our panel. We use these two different estimators because we want to see whether the results are oversensitive to the estimation technique. The specification of the PVAR model is described in the following equation:

$$Z_{it} = T_0 + T_1 Z_{it-1} + f_i + d_{c,t} + \varepsilon_t \tag{2.1}$$

where Z_{it} represents the vector of the variables in our analysis (dlypc, dlkpubpc, and dlkprivpc), T_0 denotes the vector of constants, $T_1 Z_{it-1}$ denotes the polynomial matrix, f_i denotes the fixed effects, $d_{c,\,t}$ denotes the time fixed effects, and ε_t is the random error term. Regarding the long-run relationship between economic growth, public capital stock, and private capital stock, it is given by the following equation:

$$lypc_{it} = \alpha_i + \beta_i lkpubpc_{it} + \gamma_i lkpubpc_{it} + \varepsilon_{it} \tag{2.2}$$

where the α_i represents the intercept, β_i and γ_i represent the elasticities of public capital stock and private capital stock, respectively, and ε_{it} represents the error term. The prefix "*L*" denotes natural logarithms. For a better understanding of the mathematical expressions behind the PDOLS and FMOLS estimators (see Pedroni, 2001a, b).

As is known, before the estimation of the PVAR and the PDOLS and PFMOLS, we had to conduct a series of preliminary and specification tests in order to confirm the requirements of the models and thus guarantee the trustworthiness of the achieved outcomes. Nevertheless, we should say that as this chapter is not primarily aimed at being too technical, we focus our analysis mainly on the results of the models. Although we give an explanation of the preliminary and specification tests results, their respective tables and specific information (e.g. null hypothesis), as well as the "**How to do**" part, will be displayed in the appendix of this chapter (Appendix).

The first step of the estimation was to transform the variables *ypc*, *kpubpc*, and *kprivpc* into their natural logarithms (*lypc*, *lkpubpc*, and *lkprivpc*) and first differences (*dlypc*, *dlkpubpc*, and *dlkprivpc*) and see their respective descriptive statistics (Table 2.A1). Then, in order to see if a correlation exists across countries between our series, we computed

Pesaran's cross-sectional dependence test (Pesaran, 2004). The results of this test revealed the presence of cross-sectional dependence in all variables, whether in natural logarithms or in first differences (Table 2.A2), which means that there is, in fact, a correlation between the series across countries. The reason for this interdependency may well be associated with the common shocks that our crosses (the 30 LAC countries) share. If we do pay no attention to the presence of such a phenomenon, it may lead to inconsistent and incorrect conclusions in the econometric approach (Eberhardt and Teal, 2011).

The following step of the analysis was to test if collinearity and multicollinearity could be a problem for our estimations, through the computation of the correlation matrix and of the variance inflation factor (VIF) statistics (Belsley et al., 1980). Although the results point to a relatively high degree of collinearity between *lypc*, *lkpubpc*, and *lkprivpc* (the expected outcome gave their close relationship), the variables show a low degree of collinearity when in differences. Moreover, as the VIF and mean VIF values were both low for the variables in natural logarithms and first differences, we can conclude that multicollinearity does not raise concerns for our analysis and so we can proceed with our estimation without further concerns (see Table 2.A3).

To investigate the stationarity of the variables i.e. their order of integration, the cross-sectionally augmented IPS (CIPS) test (Pesaran, 2007) was carried out. This test, which is also called the second-generation unit root test, was used due to the presence of cross-sectional dependence found in all variables, given that it is robust to this same characteristic. The results of this test clearly indicated that all variables in natural logarithms (except for *lkpub* with trend) were integrated of order one and that in first differences, all variables were stationary, with and without trend (see Table 2.A4). This last result i.e. the fact that all variables are stationary in first differences, is extremely important given that it is a required condition so that we can estimate the PVAR.

Before the estimation of the PVAR, the presence of random or fixed effects should also be checked. To this end, we conducted the Hausman test (Hausman, 1978) on the three possible PVAR model specifications, with *dlypc*, *dlkpubpc*, and *dlkprivpc*, as the dependent variable (see Table 2.A5). The presence of fixed effects was detected in both specifications with *dlypc* and *dlkprivpc* as the dependent variables. Due to the confirmed presence of fixed effects, correlation problems between the regressors can arise, and so, in order to overcome this problem, the PVAR estimation will be conducted using the "Hermelet procedure" (Arellano and Bover, 1995), which allows these effects to be removed.

The last preliminary test related to the PVAR estimation is the test of lag-order selection criteria (see Table 2.A6). The procedure of this test is the following: after passing the Hansen J test (Hansen, 1982), the optimal lag length should be the one that minimises the Bayesian information criterion (MBIC), the Akaike information criterion (MAIC), and the Quinn information criterion (MQIC) (Andrews and Lu, 2001). According to the results of this test, a first-order PVAR should be estimated.

As we also want to use the PDOLS and PFMOLS to access the long-run elasticities of public capital stock and private capital stock in economic growth, we also need to compute a cointegration test [in this case the cointegration test of Westerlund (2007), which is robust to cross-sectional dependence] to guarantee that there is a cointegrating relationship between the *lypc*, *lkpubpc*, and *lkprivpc* variables. The results of cointegration test of Westerlund (2007) were unanimous, indicating that the variables are cointegrated (see Table 2.A7). The other condition is that the variables should be $I(1)$ i.e. integrated of the first order, a condition that we already confirmed previously when we conducted the CIPS test.

After carrying out all these preliminary and specification tests, the PVAR model was estimated, using one lag, the "gmmstyle" option (Holtz-Eakin et al., 1988), and with the first four lags of the regressors as instruments. In Table 2.2, we give the results from the estimation of the first-order PVAR.

The results from the PVAR model (Table 2.2) seem to indicate that economic growth (*dlypc*) is indeed a driving force for the increase in the LAC capital stock (*dlkpubpc* and *dlkprivpc*). A 1% increase in economic growth (per capita) leads to an increase of 0.46962% in public capital stock (per capita), and an increase of 0.31004% in private capital stock (per capita). Moreover, from the outcomes of Table 2.2, we can also see that an increase in private capital stock also has a positive effect on the economic growth of this group of countries. A 1% increase in private capital stock (per capita)

TABLE 2.2 PVAR model results.

Variables	*dlypc*	*dlkpubpc*	*dlkprivpc*
dlypc	−0.0252587	0.4696266***	0.3100442***
dlkpubpc	−0.1166286***	−0.2240945***	−0.1387485***
dlkprivpc	0.4073608***	0.0871552	0.089169**

*** and ** denote statistical significance at the 1%, and 5% levels, respectively.

leads to an increase of 0.40736% in the economic growth of these countries (per capita). Regarding the effect from private capital stock the public capital stock, although it appears to be positive (a 1% increase in private capital stock (per capita) leads to an increase of 0.08715% in public capital stock), there is a lack of statistical significance, which means that there is no guarantee that this effect will occur. Conversely, looking at the effects of the public capital stock on both economic growth and private capital stock, we see that they appear to be negative. A 1% increase in general government capital stock (per capita) seems to lead to a decrease of 0.11662% in the case of economic growth (per capita) and of 0.13874% in the case of private capital stock.

After the PVAR estimation, it is also important to check its stability by computing the eigenvalue. If all eigenvalues are within the unit circle, the stability condition is confirmed and, therefore, the PVAR model is stable. The graph of eigenvalues is shown in Table 2.A8. There it can be seen that the stability condition is fulfilled (all eigenvalues are within the unit circle). Moreover, this result is a reinforcement of the conclusion that our variables are all stationary (see e.g. Lütkepohl, 2005).

With the stability of the PVAR model confirmed, we can now proceed to the performance of the Granger causality test (Abrigo and Love, 2016). This test will allow us to identify and explore the possible causal relationships between the variables. In Table 2.3, we give the results of this test.

Looking at Table 2.3, it can be concluded that a bidirectional causality seems to exist between economic growth (*dlypc*) and public capital stock (*dlkpubpc*), bidirectional causality between economic growth (*dlypc*) and private capital stock (*dlkprivpc*), and unidirectional causality running from public capital stock (*dlkpubpc*) to private capital stock (*dlkprivpc*). All of these causal relationships were found at the 1% level of significance. Furthermore, the analysis of the exogeneity blocks (ALL) seem to confirm the presence of endogeneity. In Fig. 2.3, we show the signal and the direction of the causal relationships that were found between the variables, allowing a more concise interpretation of the outcomes that were found in the Granger causality test.

The outcomes from the Granger causality test, together with the results from the PVAR estimation, point to the presence of bidirectional causality between economic growth (*dlypc*) and public capital stock (*dlkpubpc*) in this group of LAC countries, with a positive sign when the causality runs from economic growth to public capital stock, and a negative sign when it runs from public capital stock to economic growth. Additionally, evidence was also found of the existence of bidirectional causality between economic growth (*dlypc*) and private capital stock (*dlkprivpc*), with a positive sign in both directions. Finally, we were also able to uncover a unidirectional causality which runs from public capital stock (*dlkpubpc*) to private capital stock (*dlkprivpc*). As in the case of the causality that runs from public capital stock to economic growth, this last one also has a negative sign.

TABLE 2.3 Granger causality test.

	dlypc	*dlkpubpc*	*dlkprivpc*
dlypc does not cause	–	124.094***	113.287***
dlkpubpc does not cause	15.803***	–	15.633***
dlkprivpc does not cause	88.626***	2.375	–
ALL	89.039***	200.053***	124.625***

*** denotes statistical significance of 1%; H0: the absence of causality i.e. if we reject H0, it means that there is a causality between the variables.

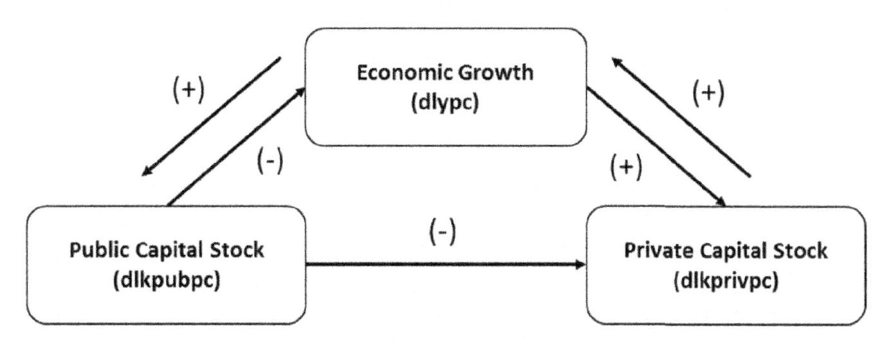

FIG. 2.3 Summary of the causalities according to the Granger causality (created by the authors). The causality signals were based on the coefficients of the PVAR estimation (Table 2.2); the arrows denote a 1% significance level.

In sum, from the previous results, we can say that economic growth seems to be a contributor to increases in both the public and private capital stocks in this group of LAC countries. Conversely, the results also indicate that private capital stock also appears to positively contribute to LAC countries' economic growth. Regarding public capital stock, it can be seen that its effect on growth was far from being desirable, with this variable showing a detrimental effect on the economic output of these countries. Moreover, public capital stock also seems to have an adverse effect on LAC private capital stock, which opens the possibility for the existence of a crowding-out effect from the public capital on the private capital in the LAC region. Before we proceed, it is important to note that all these inferences and conclusions are only related to the short-run analysis.

In addition to the assessment of the Granger causalities, when estimating a PVAR, it is also advisable to compute the so-called forecast error variance decomposition (FEVD) and the impulse response functions (IRFs). Regarding the FEVD, this is a test that allows an evaluation of the percentage that a variable explains of the forecast error variance of another variable which has been faced with a shock i.e. it allows an evaluation of the time needed by a variable to attain equilibrium and the influence of each variable for that same purpose. Concerning IRFs, they enable an analysis of the performance of one variable, the response variable, when faced with a shock (or innovation) in another variable, also known as the impulse variable. Moreover, IRFs also allow us to ascertain the time that the response variable needs to return to equilibrium. In Table 2.4, we give the results from the FEVD estimation.

Looking at the outcomes from Table 2.4, it can be seen that, in the first period, economic growth (*dlypc*), public capital stock (*dlkpubpc*), and private capital stock (*dlkprivpc*) are mainly self-explanatory i.e. their forecast error variance is mostly explained by shocks to themselves (100%, 88.86%, and 96.93%, respectively). However, as time moves forward, the percentage of the forecast error variance explained by shocks to themselves decreases. In contrast, the percentage of the forecast error variance explained by shocks in the other variables increases. In the 10th period, 91.95% of the *dlypc* forecast error variance is explained by itself, 0.4% by *dlkpubpc*, and 7.65% by *dlkprivpc*. Concerning *dlkpubpc*, 76.31% of its forecast error variance is explained by itself, 22.09% by *dlypc* and 1.6% by *dlkprivpc*. In the 10th period, it can be seen that 83.33% of the forecast error variance of *dlkprivpc* is explained by itself, 11.70% is explained by *dlypc* and 4.97% by the *dlkpubpc*.

Overall, the FEVD results seem to acknowledge the idea that private capital stock exerts a more significant influence on the growth of LAC countries when compared with the public capital stock, and that economic growth has a considerable influence on the variance of both the public and private capital stocks. Regarding the two types of capital, the FEVD results show that the private capital stock forecast error variance is more influenced by shocks in the public capital stock than the public capital stock forecast error variance is influenced by shocks in the private capital stock. To complete the PVAR analysis the outcomes from the IRF estimation can be seen in Fig. 2.4.

TABLE 2.4 Forecast error variance decomposition (FEVD).

		Impulse variables		
Response variables	**Forecast horizon**	*dlypc*	*dlkpubpc*	*dlkprivpc*
dlypc	1	1	0	0
	2	0.9201382	0.0035039	0.0763579
	5	0.9194797	0.004039	0.0764813
	10	0.9194771	0.004039	0.0764838
dlkpubpc	1	0.1113593	0.8886407	0
	2	0.2162131	0.7807284	0.0030585
	5	0.2208831	0.7631297	0.0159872
	10	0.220901	0.7631051	0.0159939
dlkprivpc	1	0.0007584	0.0299305	0.9693111
	2	0.1129587	0.0505504	0.8364909
	5	0.1169415	0.0497397	0.8333188
	10	0.116957	0.0497396	0.8333033

FEVD followed the Cholesky decomposition and was performed using 1000 Monte Carlo simulations for 10 periods.

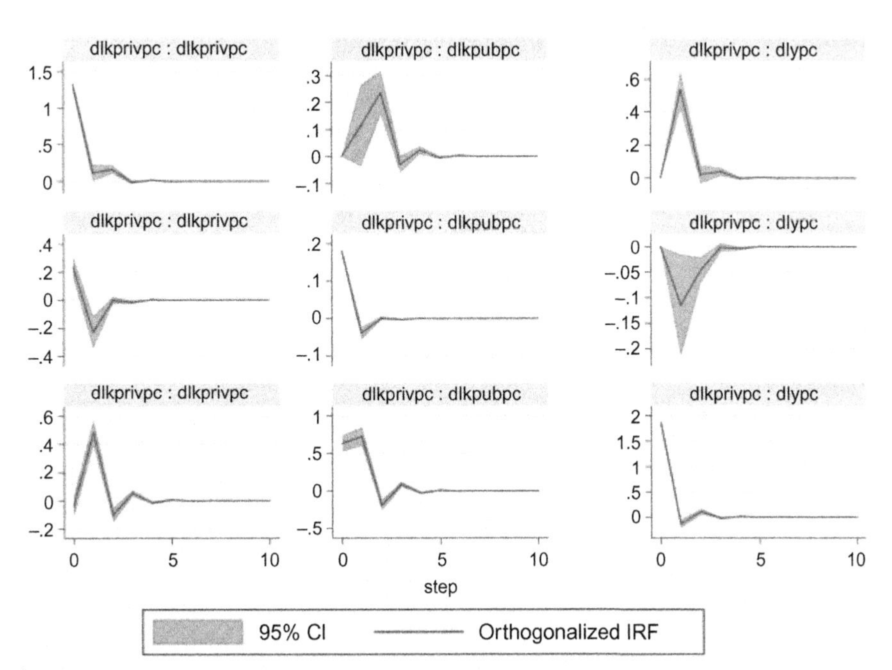

FIG. 2.4 Impulse response functions (IRFs) (created by the authors).

First, from Fig. 2.4, it can be seen that all variables converge to equilibrium after a shock, which once again supports the stationarity of the variables that were used in the model (*dlypc*, dlkpubpc, and *dlkprivpc*). Second, concerning the responses of the variables to the impulses, we see that both private capital (*dlkprivpc*) and public capital (*dlkpubpc*) stocks respond positively to an impulse in economic growth (*dlypc*); both economic growth (*dlypc*) and private capital stock (*dlkprivpc*) respond negatively to an impulse in public capital stock (*dlkpubpc*); and both economic growth (*dlypc*) and public capital stock (*dlkpubpc*) respond positively to an impulse in private capital stock (*dlkprivpc*).

In sum, the conclusions from the FEVD and the IRFs highlight the effects which were found in the PVAR model and the subsequent Granger causality test. This is, in the short run, economic growth seems to have positive effects on both private and public capital stocks, private capital stock seems to have a positive effect on economic growth, and public capital stock appears to affect both economic growth and private capital stock adversely.

As previously stated, in addition to the PVAR estimation, the PDOLS and PFMOLS estimators were also used in order to assess the effects of the public capital stock and private capital stock on the long-run economic growth of these LAC countries. As stressed above, the use of these two different estimators is linked with the possibility of the estimated parameters being oversensitive to the estimation technique. The results from the PDOLS and PFMOLS estimation are presented in Table 2.5.

The outcomes from the PDOLS and PFMOLS displayed in Table 2.5 seem to be unanimous in indicating that both public and private capital stock have a positive impact on these LAC countries' long-run economic growth, with the coefficients which were achieved in both models [following Farhani (2013) they can be considered as long-run elasticities] being very similar. Moreover, it is important to stress that in both estimation techniques, the impact of private capital stock on economic growth is slightly more significant than that of public capital stock.

TABLE 2.5 PDOLS and PFMOLS results.

	Dependent variable: lypc	
Independent variables	**PDOLS**	**PFMOLS**
lkpubpc	0.405942***	0.413921***
lkprivpc	0.610754***	0.600292***

*** denotes statistical significance of 1%.

Now that the proposed analysis is completed, with the relationship between public capital stock, private capital stock, and economic growth in LAC having been empirically studied, in the next section (Section 2.4) we will extend the assumptions that can be drawn from the results which were found in the PVAR, PDOLS, and PFMOLS estimations. In Section 2.4, further to the discussion surrounding the justifications for the achieved outcomes, we also draw some conclusions in order to be able to help LAC policymakers in the development of growth-promoting policies (especially those linked to public investment).

2.4 Conclusion

In the empirical analysis conducted in this chapter, PVAR, PDOLS, and PFMOLS methodologies were used in order to reveal the short- and long-run relationships between public capital, private capital, and economic growth in the LAC region, with the use of annual data from 1970 to 2014 for a group of 30 countries from this region.

Starting with the PVAR model results, the overall indication was that economic growth could indeed contribute to increasing both public and private capital stocks in this region. This result is far from unexpected, given that an enhancement in a country's economic output is usually expected to lead to a rise in the economy's degree of investment i.e. to an increase in both public and private capital investment. As an example, we can cite the study by Blomstrom et al. (1996) who had already reached a similar conclusion, such as increases in economic growth enhance capital formation rates.

Conversely, increases in both types of capital stocks (public and private) were also expected to reveal positive effects on growth, as the neoclassic growth models postulate (e.g. Solow, 1956). However, following the outcomes from the PVAR estimation, in the short run, only private capital stock seems to affect the economic growth of these countries positively. This fact could be possibly linked with the theory of Agenor and Moreno-Dodson (2006) that, in the short run, public capital stock may have harmful effects on growth if it produces a crowding-out effect on private investment. This effect was clearly identified in our estimation, with public capital showing a negative unidirectional causal relation with private capital (see Fig. 2.3). This result seems to point to the possibility that, in the LAC, the public and private capital act as substitutes and compete with each other, rather than acting as complements (Erden and Holcombe, 2006).

Given the fact that an increase in public capital should be able to raise the returns for private capital (Aschauer, 1989), it might be thought that increases in public capital would have a different effect on private capital (a crowding-in effect) and would contribute to an increase in the economic growth of these countries'. Nevertheless, if we read the conclusions from some previous studies regarding the relationship between public and private investment (e.g. Bahal et al., 2018; Presbitero, 2016; Cavallo and Daude, 2011) we can find a set of critical factors that can easily lead to this relationship not being as linearly positive as one might think. Overall, the conclusions from these studies state that the quality and strength of institutions, the ease of a country's access to finance, and the degree of absorptive capacity can influence the effect that public capital can have on private capital. Basically, countries with weak institutions, with greater difficulty in accessing financing and with worse absorptive capacity are more prone to revealing a crowding-out effect. All these last characteristics seem to be present in most countries in the LAC region, meaning that they could probably explain why we found this result.

Moreover, the insufficient level of public capital stock in the LAC, with special emphasis on the regional infrastructure shortfalls (Faruqee, 2016), and inaccurate public investment strategies (Gupta et al., 2014), can also be pointed to as reasons for the negative effect that public capital stock seems to have on economic growth. Additionally, we should also consider that this adverse effect could be exacerbated by factors which are usually linked to this group of countries for example, corruption, political instability, or "white elephants" (e.g. Pritchett, 2000). Lastly, we should clarify that public capital stock is not always centred on profit and is often aimed at increasing social welfare. It can also act in areas where it is difficult to make a profit (where there is no private interest), leading to a situation where the positive economic effects are not immediate (i.e. they may not be immediately felt in the short run).

Turning to the results of the PDOLS and PFMOLS estimations, it can be seen that in the long run, both public and private capital stock have a positive impact on the economic growth of this group of LAC countries. Concerning the turn in the public capital stock effect, following Agenor and Moreno-Dodson (2006) and Erenburg and Wohar (1995), we can say that the crowding-out effects of public capital stock are generally observed in the short run, and as we move forward in time to the medium/long term, this effect is usually suppressed. The fact that, according to the results, public capital stock seems to be able to foster economic growth in these countries in the long run leads us to believe that this situation occurred in the LAC and the public capital stock probably started to crowd-in private capital. Additionally, the alteration of the public capital stock effect on long-run growth can also be associated with the large marginal returns produced by the increases in both public capital stock levels (e.g. Fournier, 2016) and public capital efficiency (e.g. Berg et al., 2019). As previously stressed, in situations like that of the LAC, where there is a relatively low level of public capital and/or where past investments were inefficient, the large marginal returns produced by a rise in the public investment levels and efficiency usually lead to increased output growth. This is probably another reason for the positive result which was found.

As in the short run, private capital stock continues to positively influence the economic growth of these countries in the long run, still supporting the idea that higher investment rates lead to higher output levels (e.g. Solow, 1956). Moreover, it should be stressed that the fact that private capital is demonstrated to have a higher coefficient than public capital is supported by the results from past studies (e.g. Arslanalp et al., 2010). Devadas and Pennings (2018) further mention that private investment is most significantly responsible for increases in the output of developing countries. As in the case of public capital, these effects are also linked to the low levels of private capital in these countries, which produce relatively high returns for private investment.

Turning now to some of the policy implications that may arise from the results of our estimations, we think that firstly, LAC governments should continue to promote their public investment at the same time as creating or improving the conditions to encourage private investment. This is mainly due to the fact that both types of capital are proven to enhance the growth of these countries in the long run. Moreover, as already stressed, an increase in LAC public and private capital is not only essential for increased regional growth but is also crucial to boost the competitiveness of countries from the LAC region regarding their main competitors, such as the emerging Asian countries.

Nevertheless, given the adverse effects that we found from the public capital on both private capital and economic growth in the short run, we think that LAC governments should pay considerable attention to how public investment is carried out and evaluate its effects on several variables, especially the effects that this investment may generate on private capital. This degree of attention is essential to ensure that the possibility of public capital stock displacing i.e. crowding out, private capital stock does not occur. In order to produce positive effects, public investment should be planned in such a way that public and private capital act as complements and not as substitutes. Thus there is the prospect that public spending for example, on infrastructures, will be able to raise not only general social well-being but also the marginal productivity of private capital. The development of public-private partnership (PPP) structures can be one solution to increase cooperation between the public and private sectors and private participation in areas of potential interest. In cases where governments face more serious difficulty in financing public infrastructure projects for example this could be a viable solution. However, it is important that this type of contracts must be very well planned so that the investment can be sustainable and able to meet the desired quality standards. It is also a priority to ensure that it does not give rise to harmful effects for the general public (and for taxpayers), with the creation of undesirable barriers to accessing these services or with the emergence of rent-seeking situations, for example.

Furthermore, we believe that in order to guarantee that the money goes to projects of the most significant interest and to ensure the increased efficiency of these projects, it is imperative to improve the selection, evaluation, and management of public investment projects in the LAC region. An example can be the fact that governments could sometimes achieve better outcomes by investing in maintenance and upgrading of existing capital instead of moving directly to investment in new capital. Finally, due to the problems associated with the macroeconomic stability of the countries in this region, it is important that LAC governments weigh their fiscal space to ensure that no negative fiscal costs such as distortionary taxes will be associated with their investment, which may end up in a situation where investment could hamper their growth. In the most problematic situations such as cases where there is a high public-debt-to-GDP ratio, improving public financing instruments searching for new means of funding (e.g. resorting to institutional investors or PPP schemes) is of vital importance. In order to end this section, as well as this chapter, in Fig. 2.5, we display a scheme representing the ideal public capital materialisation in LAC economies.

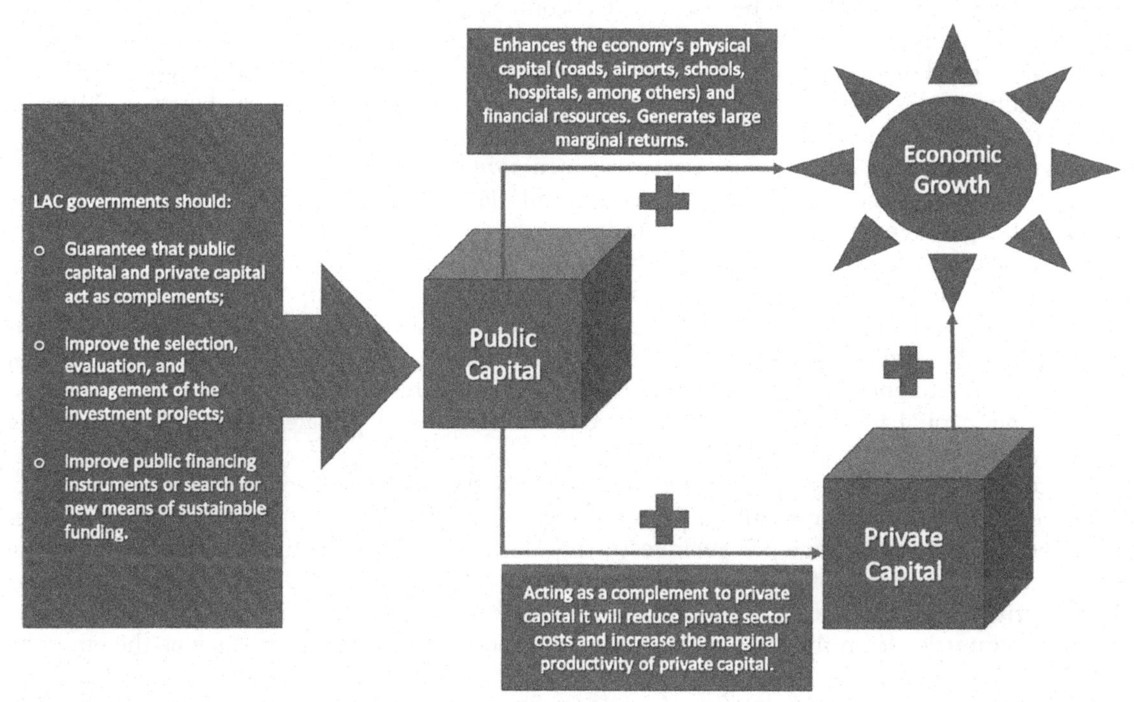

FIG. 2.5 Ideal public capital materialisation in the LAC economies (created by the authors).

Appendix

TABLE 2.A1 Descriptive statistics.

Variables	Obs.	Mean	Std. dev.	Min.	Max.
lypc	1350	−11.40378	4.7331	−31.14783	9.127953
lkpubpc	1350	−−11.65031	4.804759	−30.13868	8.777058
lkprivpc	1350	−11.01015	4.603375	−28.63935	9.17065
dlypc	1320	0.0733062	1.918062	−31.24936	28.30312
dlkpubpc	1320	0.0499681	2.088308	−31.71728	28.01527
dlkprivpc	1320	0.0696745	1.516823	−27.12892	4.560019

TABLE 2.A2 Cross section dependence (CD) test.

Variables	CD-test	Corr.	Abs. (corr.)
lypc	81.73***	0.584	0.844
lkpubpc	75.96***	0.543	0.825
lkprivpc	81.46***	0.582	0.834
dlypc	22.94***	0.166	0.240
dlkpubpc	10.72***	0.078	0.161
dlkprivpc	11.96***	0.086	0.167

The CD test has $N(0,1)$ distribution under the H0: cross-section independence, *** denotes statistical significance at the 1% level.

TABLE 2.A3 Correlation matrices and VIF statistics.

	lypc	*lkpubpc*	*lkprivpc*		*dlypc*	*dlkpubpc*	*dlkprivpc*
lypc	1.0000			*dlypc*	1.0000		
lkpubpc	0.9047	1.0000		*dlkpubpc*	0.2611	1.0000	
lkprivpc	0.9294	0.8887	1.0000	*dlkprivpc*	−0.0446	0.3509	1.0000
VIF		4.76	4.76			1.14	1.14
Mean VIF		4.76				1.14	

In the case of the VIF test, the values are lower than the typically assumed benchmarks: 10 in the case of the VIF values and 6 in the case of the mean VIF values.

TABLE 2.A4 Panel unit root test (CIPS).

		CIPS (Zt-bar)	
	Lags	**Without trend**	**With trend**
lypc	0	−0.320	5.682
	1	0.027	1.161
	2	0.624	1.325
	3	0.834	1.685
lkpubpc	0	2.693	2.126
	1	0.521	−1.718**
	2	1.675	0.292
	3	3.400	1.621
lkprivpc	0	4.144	5.156
	1	1.921	2.015
	2	1.619	2.441
	3	3.760	3.566
dlypc	0	−11.321***	−13.447***
	1	−6.819***	−8.214***
	2	−4.935***	−6.008***
	3	−3.402***	−3.504***
dlkpubpc	0	−16.663***	−17.456***
	1	−11.093***	−11.246***
	2	−7.525***	−6.489***
	3	−5.556***	−4.139***
dlkprivpc	0	−16.255***	−16.612***
	1	−10.576***	−11.548***
	2	−6.725***	−6.800***
	3	−4.338***	−3.672***

*** and ** denote statistical significance at the 1% and 5% levels, respectively; Pesaran (2007) panel unit root test (CIPS) assumes that cross-sectional dependence is in the form of a single unobserved common factor and H0: series is $I(1)$.

TABLE 2.A5 Hausman test.

	Model with dlypc as dependent	**Model with dlkpubpc as dependent**	**Model with dlkprivpc as dependent**
	FE vs. RE	FE vs. RE	FE vs. RE
Hausman test	Chi2(2) = 6.39 (0.0409)	Chi2(2) = 0.05 (0.9737)	Chi2(2) = 9.00 (0.0111)

H0: difference in coefficients not systematic (random-effects). The values in () represent the Prob > chi2, if <5%, it means that we can reject H0.

TABLE 2.A6 Lag order selection criteria.

Lag	CD	*J*	*J P* value	MBIC	MAIC	MQIC
1	0.4063907	28.44341	0.3883925	−162.3051	−25.55659	−77.13302
2	0.6168017	25.56291	0.1101798	−101.6027	−10.43709	−44.82137
3	0.7423995	8.059086	0.5282029	−55.52375	−9.940914	−27.13305

This procedure gives us the coefficient of determination (CD), Hansen's *J* statistic (*J*), and its *P*-value (*J P*-value—if *J P*-value is higher than 10%, then we cannot reject the null-hypothesis that the overidentification restrictions are valid) (Hansen, 1982), and the Bayesian information criterion (MBIC), the Akaike information criterion (MAIC), and the Quinn information criterion (MQIC) introduced by Andrews and Lu (2001); the test was conducted for first- to third-order panel VAR using the first four lags of the regressors as instruments.

TABLE 2.A7 Westerlund cointegration test.

Statistics	Value	Z value	*P*-value	Robust *P*-value
Gt	−1.974	−3.096	0.001	0.004
Ga	−14.027	−8.217	0.000	0.000
Pt	−17.313	−9.167	0.000	0.013
Pa	−41.335	−43.400	0.000	0.000

Bootstrapping regression with 800 reps. H0: No cointegration; H1 Gt and Ga test the cointegration for each country individually, and Pt and Pa test the cointegration of the panel as a whole.

TABLE 2.A8 Eigenvalue stability condition.

Eigenvalue			Graph
Real	Imaginary	Modulus	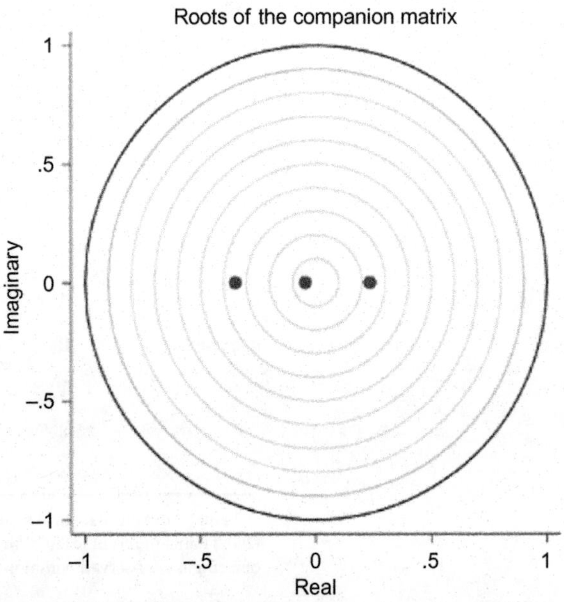
−0.3480847	0	0.3480847	
0.2333451	0	0.2333451	
−0.0454445	0	0.0454445	

All the eigenvalues are inside the unit circle, meaning PVAR satisfies stability condition.

How to do:

STATA:

```
**Transform the variables y, kpub, and kpriv into per capita values**

gen ypc=y/p
gen kpubpc=kpub/p
gen kprivpc=kpriv/p

**Transform the variables ypc, kpubpc, and kprivpc into natural logarithms**
gen lypc=ln(ypc)
gen lkpubpc=ln(kpubpc)
gen lkprivpc=ln(kprivpc)

**Transform the variables lypc, lkpubpc, and lkprivpc into first differences of logarithms**

gen dlypc=d.lypc
gen dlkpubpc=d.lkpubpc
gen dlkprivpc=d.lkprivpc

**Descriptive statistics**

sum lypc lkpubpc lkprivpc
sum dlypc dlkpubpc dlkprivpc

**Cross section dependence (CD) test**

xtcd lypc lkpubpc lkprivpc
xtcd dlypc dlkpubpc dlkprivpc

**Correlation matrices**

corr lypc lkpubpc lkprivpc
corr dlypc dlkpubpc dlkprivpc

**VIF statistics**

qui:regress lypc lkpubpc lkprivpc
estat vif
qui:regress dlypc dlkpubpc dlkprivpc
estat vif

**Panel unit root test (CIPS)**

multipurt lypc lkpubpc lkprivpc, lags(3)
multipurt dlypc dlkpubpc dlkprivpc, lags(3)

**Hausman test**

qui: xtreg dlypc dlkpubpc dlkprivpc, fe
estimate store fixed
qui: xtreg dlypc dlkpubpc dlkprivpc, re
estimate store random
hausman fixed random
qui: xtreg dlkpubpc dlypc dlkprivpc, fe
estimate store fixed
qui: xtreg dlkpubpc dlypc dlkprivpc, re
estimate store random
hausman fixed random
qui: xtreg dlkprivpc dlypc dlkpubpc, fe
estimate store fixed
```

qui: xtreg dlkprivpc dlypc dlkpubpc, re
estimate store random
hausman fixed random

Lag order selection criteria

pvarsoc dlypc dlkpubpc dlkprivpc, maxlag(3) pvaropts (instl(1/4))

Westerlund cointegration test

set matsize 800
xtwest lypc lkpubpc lkprivpc, lags(1)lrwindow(3) bootstrap(800)
xtwest lypc lkpubpc lkprivpc, lags(1)lrwindow(3) bootstrap(800) constant
xtwest lypc lkpubpc lkprivpc, lags(1)lrwindow(3) bootstrap(800) constant trend

PVAR model

pvar dlypc dlkpubpc dlkprivpc, lags(1) instl(1/4) gmmst

Granger causality test

pvargranger

Eigenvalue stability condition

pvarstable, graph

Forecast error variance decomposition (FEVD)

pvarfevd, mc(1000) st(10)

Impulse response functions (IRFs)

pvarirf, mc(1000) oirf byopt(yrescale) st(10)

EViews:

Transform the variables y, kpub, and kpriv into per capita values

Quick > Generate Series > ypc=y/p > OK
Quick > Generate Series > kpubpc=kpub/p > OK
Quick > Generate Series > kprivpc=kpriv/p > OK

Transform the variables ypc, kpubpc, and kprivpc into natural logarithms

Quick > Generate Series > lypc=log(ypc) > OK
Quick > Generate Series > lkpubpc=log(kpubpc) > OK
Quick > Generate Series > lkprivpc=log(kprivpc) > OK

PDOLS model

Quick > Estimate Equation > (Equation specification) lypc lkpubpc lkprivpc > (Estimation settings) Method: COINTREG—Cointegrating Regression > (Trend Specification) None > (Nonstationary estimation settings) Method: Dynamic OLS (DOLS) > Panel method: Grouped > Lag & lead method: Fixed > Lags: 1 > Leads: 1 > OK

PFMOLS model

Quick > Estimate Equation > (Equation specification) lypc lkpubpc lkprivpc > (Estimation settings) Method: COINTREG—Cointegrating Regression > (Trend Specification) None > (Nonstationary estimation settings) Method: Fully modified OLS (FMOLS) > Panel method: Grouped > OK

References

Abrigo, M.R.M., Love, I., 2016. Estimation of panel vector autoregression in stata. Stata J. 16 (3), 778–804. https://doi.org/10.1177/1536867X1601600314.

Agenor, P.R., Moreno-Dodson, B., 2006. Public Infrastructure and Growth: New Channels and Policy Implications. Policy Research Working Paper No. 4064, World Bank, Washington, DC, https://doi.org/10.1596/1813-9450-4064.

Andrews, D.W.K., Lu, B., 2001. Consistent model and moment selection procedures for GMM estimation with application to dynamic panel data models. J. Econ. 101 (1), 123–164. https://doi.org/10.1016/S0304-4076(00)00077-4.

Arellano, M., Bover, O., 1995. Another look at the instrumental variable estimation of error-components models. J. Econ. 68 (1), 29–51. https://doi.org/10.1016/0304-4076(94)01642-D.

Arslanalp, S., Bornhorst, F., Gupta, S., Sze, E., 2010. Public Capital and Growth. IMF Working Paper No. 10/175, International Monetary Fund, Washington, DC, https://doi.org/10.5089/9781455201860.001.

Aschauer, D.A., 1989. Does public capital crowd out private capital? J. Monet. Econ. 24 (2), 171–188. https://doi.org/10.1016/0304-3932(89)90002-0.

Bahal, G., Raissi, M., Tulin, V., 2018. Crowding-out or crowding-in? Public and private investment in India. World Dev. 109, 323–333. https://doi.org/10.1016/j.worlddev.2018.05.004.

Barbiero, F., Darvas, Z., 2014. In Sickness and in Health. Protecting and Supporting Public Investment in Europe. Bruegel Policy Contribution No. 2014/02, Bruegel, Brussels. Available at: https://www.bruegel.org/wp-content/uploads/imported/publications/pc_2014_02.pdf.

Belsley, D.A., Kuh, E., Welsch, R.E., 1980. Regression Diagnostics: Identifying Influential Data and Sources of Collinearity. Wiley, New York, https://doi.org/10.1002/0471725153.

Berg, A., Buffie, E.F., Pattillo, C., Portillo, R., Presbitero, A.F., Zanna, L.F., 2019. Some misconceptions about public investment efficiency and growth. Economica 86, 409–430. https://doi.org/10.1111/ecca.12275.

Blomstrom, M., Lipsey, R.E., Zejan, M., 1996. Is fixed investment the key to economic growth. Q. J. Econ. 111 (1), 269–276. https://doi.org/10.2307/2946665.

Castellani, F., Olarreaga, M., Panizza, U., Zhou, Y., 2019. Investment gaps in Latin America and the Caribbean. Rev. Int. Polit. Dév. https://doi.org/10.4000/poldev.2894.

Cavallo, E., Powell, A., 2019. Building Opportunities for Growth in a Challenging World. 2019 Latin American and Caribbean Macroeconomic Report, Inter-American Development Bank, Washington, DC, https://doi.org/10.18235/0001633.

Cavallo, E., Daude, C., 2011. Public investment in developing countries: a blessing or a curse? J. Comp. Econ. 39 (1), 65–81. https://doi.org/10.1016/j.jce.2010.10.001.

De Jong, J.F.M., Ferdinandusse, M., Funda, J., 2018. Public capital in the 21st century: as productive as ever? Appl. Econ. 50 (51), 5543–5560. https://doi.org/10.1080/00036846.2018.1487002.

Devadas, S., Pennings, S., 2018. Assessing the Effect of Public Capital on Growth: An Extension of the World Bank Long-Term Growth Model. Policy Research Working Paper No. 8604, World Bank, Washington, DC, https://doi.org/10.1596/1813-9450-8604.

Dreger, C., Reimers, H.E., 2014. On the Relationship Between Public and Private Investment in the Euro Area. DIW Berlin Discussion Papers No. 1365, DIW Berlin, German Institute for Economic Research, Berlin, https://doi.org/10.2139/ssrn.2403885.

Eberhardt, M., Teal, F., 2011. Econometrics for grumblers: a new look at the literature on cross-country growth empirics. J. Econ. Surv. 25 (1), 109–155. https://doi.org/10.1111/j.1467-6419.2010.00624.x.

Erden, L., Holcombe, R., 2006. The linkage between public and private investment: a co-integration analysis of a panel of developing countries. East. Econ. J. 32 (3), 479–492. Available at: https://www.jstor.org/stable/40326291.

Erenburg, S.J., Wohar, M.E., 1995. Public and private investment: are there causal linkages? J. Macroecon. 17 (1), 1–30. https://doi.org/10.1016/0164-0704(95)80001-8.

Farhani, S., 2013. Renewable energy consumption, economic growth and CO_2 emissions: evidence from selected MENA countries. Energy Econ. Lett. 1 (2), 24–41. Available at: https://ssrn.com/abstract=2294995.

Faruqee, H., 2016. Regional Economic Outlook, April 2016, Western Hemisphere Department: Managing Transitions and Risks. International Monetary Fund, Washington, DC, https://doi.org/10.5089/9781498329996.086.

Fisher, W., Turnovsky, S., 1998. Public investment, congestion, and private capital accumulation. Econ. J. 108 (447), 399–413. https://doi.org/10.1111/1468-0297.00294.

Fournier, J., 2016. The Positive Effect of Public Investment on Potential Growth. OECD Economics Department Working Papers No. 1347, OECD Publishing, Paris, https://doi.org/10.1787/15e400d4-en.

Gupta, S., Kangur, A., Papageorgiou, C., Wane, A., 2014. Efficiency-adjusted public capital and growth. World Dev. 57, 164–178. https://doi.org/10.1016/j.worlddev.2013.11.012.

Hansen, L.P., 1982. Large sample properties of generalized method of moments estimators. Econometrica 50 (4), 1029–1054. Available at: https://www.jstor.org/stable/1912775.

Hausman, J.A., 1978. Specification tests in econometrics. Econometrica 46 (6), 1251–1271. https://doi.org/10.2307/1913827.

Holtz-Eakin, D., Newey, W., Rosen, H.S., 1988. Estimating vector autoregressions with panel data. Econometrica 56 (6), 1371. Available at: https://www.jstor.org/stable/1913103.

IMF, 2015. Making Public Investment More Efficient. Policy Papers No. 15/003, International Monetary Fund, Washington, DC, https://doi.org/10.5089/9781498344630.007.

IMF, 2017. Estimating the Stock of Public Capital in 170 Countries. Fiscal Affairs Department, International Monetary Fund, Washington, DC. Available at: https://www.imf.org/external/np/fad/publicinvestment/pdf/csupdate_aug19.pdf.

IMF, 2018. Public Investment Management Assessment—Review and Update. Policy Papers No. 18/025, International Monetary Fund, Washington, DC, https://doi.org/10.5089/9781498308441.007.

IMF, 2014. Is it time for an infrastructure push? The macroeconomic effects of public investment. In: World Economic Outlook, October 2014: Legacies, Clouds, Uncertainties. Research Department, International Monetary Fund, Washington, DC, https://doi.org/10.5089/9781498331555.081 (Chapter 3).

Liaqat, Z., 2019. Does government debt crowd out capital formation? A dynamic approach using panel VAR. Econ. Lett. 178, 86–90. https://doi.org/10.1016/j.econlet.2019.03.002.

Love, I., Zicchino, L., 2006. Financial development and dynamic investment behavior: evidence from panel VAR. Q. Rev. Econ. Fin. 46 (2), 190–210. https://doi.org/10.1016/j.qref.2005.11.007.

Lütkepohl, H., 2005. New Introduction to Multiple Time Series Analysis. Springer, Berlin, https://doi.org/10.1007/978-3-540-27752-1.

Nguyen, C.T., Trinh, L.T., 2018. The impacts of public investment on private investment and economic growth. J. Asian Bus. Econ. Stud. 25 (1), 15–32. https://doi.org/10.1108/JABES-04-2018-0003.

Pedroni, P., 2001a. Fully modified OLS for heterogeneous cointegrated panels. In: Baltagi, B. (Ed.), Nonstationary Panels, Panel Cointegration, and Dynamic Panels (Advances in Econometrics). vol. 15. JAI Press, pp. 93–130, https://doi.org/10.1016/S0731-9053(00)15004-2.

Pedroni, P., 2001b. Purchasing power parity tests in cointegrated panels. Rev. Econ. Stat. 83 (4), 727–731. https://doi.org/10.1162/003465301753237803.

Perotti, R., 2004. Estimating the Effects of Fiscal Policy in OECD Countries. IGIER Working Paper No. 276, Bocconi University, https://doi.org/10.2139/ssrn.637189.

Pesaran, M.H., 2004. General Diagnostic Tests for Cross Section Dependence in Panels. Cambridge Working Papers in Economics No. 435, Faculty of Economics, University of Cambridge, https://doi.org/10.17863/CAM.5113.

Pesaran, M.H., 2007. A simple panel unit root test in the presence of cross-section dependence. J. Appl. Econ. 22 (2), 265–312. https://doi.org/10.1002/jae.951.

Presbitero, A.F., 2016. Too much and too fast? Public investment scaling-up and absorptive capacity. J. Dev. Econ. 120, 17–31. https://doi.org/10.1016/j.jdeveco.2015.12.005.

Pritchett, L., 2000. The tyranny of concepts: CUDIE (cumulated, depreciated, investment effort) is not capital. J. Econ. Growth 5 (4), 361–384. https://doi.org/10.1023/A:1026551519329.

Ramirez, M.D., Nazmi, N., 2003. Public investment and economic growth in Latin America: an empirical test. Rev. Dev. Econ. 7 (1), 115–126. https://doi.org/10.1111/1467-9361.00179.

Romp, W., De Haan, J., 2007. Public capital and economic growth: a critical survey. Perspekt. Wirtsch. 8, 6–52. https://doi.org/10.1111/j.1468-2516.2007.00242.x.

Ruiz-Nuñez, F., Wei, Z., 2015. Infrastructure Investment Demands in Emerging Markets and Developing Economies. Policy Research Working Paper No. 7414, World Bank, Washington, DC, https://doi.org/10.1596/1813-9450-7414.

Solow, R.M., 1956. A contribution to the theory of economic growth. Q. J. Econ. 70 (1), 65–94. https://doi.org/10.2307/1884513.

Westerlund, J., 2007. Testing for error correction in panel data. Oxf. Bull. Econ. Stat. 69 (6), 709–748. https://doi.org/10.1111/j.1468-0084.2007.00477.x.

CHAPTER

3

Concentration hurts: Exploring the effects of capital stock on Latin American and Caribbean income inequality

JEL codes D63, E22, F21, O54

3.1 Introduction

Inequality is, without doubt, a central issue in today's world. The increased attention that this topic has been gaining in most recent times is primarily linked to the worrying fact that inequality has been rising in recent decades (OECD, 2015) and the results can be very harmful to the economic and social development of many countries, including the developed ones (UN, 2020). The increased division between *the rich* and *the others* has become a significant source of concern for many economists (as well as policymakers), which stresses that the gains from increased economic growth should not be primarily conducted to the small part of the population that already enjoys a considerable level of wealth. Following Cingano (2014, p. 28): *"focusing exclusively on growth and assuming that its benefits will automatically trickle down to the different segments of the population may undermine growth in the long run inasmuch as inequality actually increases"*. This is why works by author such as Thomas Piketty, especially his book "Le capital au XXIe siècle" (Piketty, 2013), have been receiving more and more attention from the general public. In sum, the idea that governments should act in the fight against inequality has been gaining a sort of consensual status, thus increasing the calls (and the pressure) for them to develop sustainable and inclusive growth policies.

Even though the LAC region has recently presented positive trends in some economic indicators—from 2000 to 2014 the region registered an average output growth of 3% per year (OECD, 2016)—it is clear that the region still needs to make an effort in progress linked to the design and materialisation of structural policies that can lead to a sustainable growth path. As we have already seen in Chapter 1, the 2000–14 growth was mostly based on the "commodity boom"; with the end of this boom, the LAC saw its growth shrink again. These policies should also tackle their persistent socio-economic problems (which may prevent the achievement of the previously announced objective). One of these "problems" is undoubtedly the high level of inequality. Although in Chapter 1 of the "World Social Report 2020" (UN, 2020) it is stressed that income inequality has declined since 2000 in the LAC, the report also describes that income inequality has been rising again in the leading countries from the region (e.g. Argentina, Brazil, Mexico) in most recent years. Additionally, it is important to note that even with the registered decrease in the new millennium, the LAC, jointly with Africa, continues to stand among the regions with the highest income inequality (UN, 2020).

As seen in Fig. 3.1, the share of the pretax national income that goes to the richest 10% in LA has always been above 50% of the total national income since 1980, and almost always above the world average (with only a few periods of exception). Regarding the top 1%, it can be seen that in LA since the 1980s, the 1% richest have received more than 25% of the pretax national income. In this case, the LA average was always above the world average.

Given the previously mentioned facts and the theme of Part II of this book, we believe that in addition to the study of the relationship that capital stock has with economic growth in the LAC (Chapter 2), the analysis of the impacts that LAC capital stock has on regional income inequality is also extremely important. This is, as we know, because income

https://doi.org/10.1016/B978-0-12-824429-6.00011-5

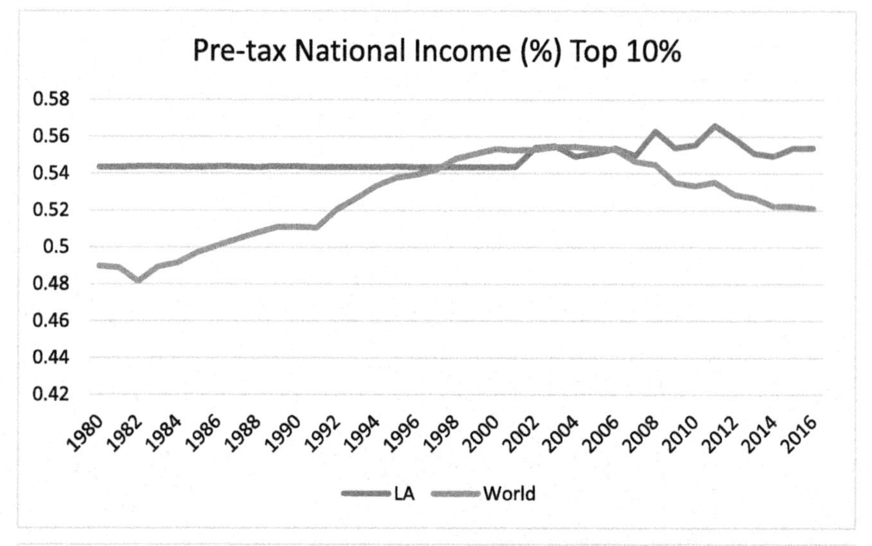

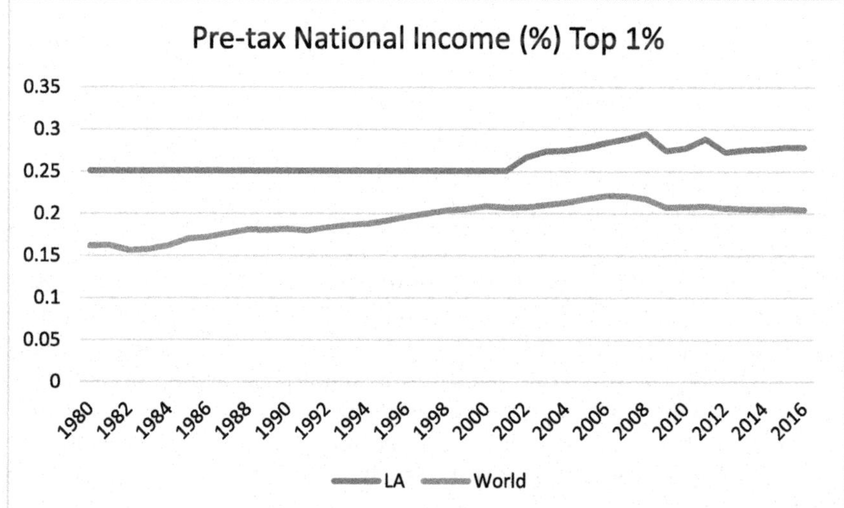

FIG. 3.1 Pretax national income (%) of the top 10% and top 1%, LA vs. world (1980–2016). These graphs were created by the authors and were based on the data from the "World Inequality Database": https://wid.world/data/.

inequality continues to be a serious problem for this region. So, it is imperative to guarantee that future policies regarding the physical capital development of the region will be not focused only on promoting growth but also on the reduction of regional inequality levels.

3.2 Public capital stock and income inequality

There are diverse tools that governments could use to control income inequality levels in their countries for example through revisions on their fiscal policy, increased minimum wages, or through the control of interest rates. Still, in general, there is also the belief that governments can tackle income inequality through their spending (e.g. Anderson et al., 2017). Although most authors centre their analysis on the impacts that government spending on education, health, and social welfare have on income inequality (e.g. Martínez-Vázquez et al., 2012), we believe that there are more types of government spending which may impact income inequality and whose effects should be thoroughly analysed.

As is known, capital stock represents the available physical capital of an economy at a given moment, and it is calculated by the value of new investments minus depreciation. As the public component of capital stock i.e. public capital stock, can be directly related to the government's investment on economic and social public infrastructures (roads, highways, railroads, airports, ports, hospitals, schools, among others), it becomes increasingly interesting (and pertinent) to investigate the effects that this variable can have on income inequality. Nevertheless, we should stress that there is a scarcity of literature that directly addresses the effects of the public capital stock on income inequality, with most of it being focused on the relationship between this variable and economic growth (e.g. De Jong et al., 2018; Romp and De Haan, 2007), which was precisely the theme of Chapter 2. However, if we consider public capital as a form of government spending (public investment), or as a representation of the public provision of infrastructure, the number of studies from which we can retrieve valuable information significantly increases, allowing some light to be shed on the relationship between these two variables.

In general, public investment is seen as an important tool in tackling inequality, with the outcomes of most studies showing that raising public investment levels can generate equal income distribution (e.g. Bom and Goti, 2018; Furceri and Li, 2017). However, as in the case of the relationship that was analysed in Chapter 2, the magnitude of this effect can be influenced by several factors such as the country's investment efficiency, the way that it finances public investment, and its degree of economic slack. Public investment, as already stated, can take several forms. One of these forms is public physical capital investment, with the government spending on the construction of roads, railways, bridges, schools, hospitals, sanitation and water systems, telecommunications and energy systems, among others. Looking at the literature centred on the effects of public infrastructure, (and from infrastructure in general) on income inequality, the general conclusion seems very similar to that of the relationship we emphasised earlier (i.e. public investment-income inequality): infrastructure development tends to reduce income inequality. For a useful review of the literature surrounding the effects of infrastructure development on income distribution and economic growth, see Calderón and Servén (2014).

Notwithstanding, despite this conclusion, it is important to state that there are also some authors whose results seem to support the opposite idea (e.g. Turnovsky, 2015; Chatterjee and Turnovsky, 2012; Artadi and Sala-i-Martin, 2003). The main problem with the literature devoted to this theme is, indeed, the lack of knowledge regarding the channels through which infrastructure can affect income distribution. However, we can find some authors who have offered some explanations on this issue.

One aspect that can lead to investment in infrastructure succeeding in reducing income inequality is the fact that these investments can be a precious help for connecting the poorest/rural areas to the richer areas. These areas have more thriving economic activity and are able to reduce production and transportation costs, facilitate information flow, and increase access to further productive opportunities (Calderón and Servén, 2004, 2014; Calderón and Chong, 2004; Estache, 2003; Lopez, 2003). However, it is also important to note that if these investments are channelled to the already rich and developed areas, they can lead to an increase in inequality (Lopez, 2003). Moreover, the literature also points out the positive effects that increased investment in physical and social infrastructure can have on human capital as a way of enhancing productivity, earnings, and social welfare (e.g. Calderón and Servén, 2014; Agenor and Moreno-Dodson, 2006). According to Pi and Zhou (2012), an increased supply of public infrastructure can also raise the marginal productivity of both skilled and unskilled labour, consequently raising their earnings. If the sector which is most intensive in public infrastructure services is the one which uses unskilled labour, the skilled-unskilled wage inequality will be reduced due to the capital shift from the skilled to the unskilled sector, because this shift can lead to a decrease in the wage rate of skilled labour, and to an increase in the wage rate of unskilled labour. However, Pi and Zhou (2012) also state that if the sector which is more public infrastructure intensive is one of the skilled labour, then the effect can be the opposite (Box 3.1).

As in the analysis of Chapter 2, we will also use private capital in the analysis that we will conduct in this chapter. As Easterly and Servén (2003) have stressed, due to pressures associated with fiscal consolidation, many countries have increasingly reduced their public investment in infrastructure, leading to an insufficient provision of it. One alternative that the governments found to resolve this situation was to increase private participation in the infrastructure sector. Due to this fact, it may also be interesting to analyse the possible effects that the private provision of infrastructure can have on income distribution.

Following Estache et al. (2002), after privatisation, it is possible that formerly public companies become profitable, due to the downsizing strategy from the new private providers. The effect on income distribution depends on the number of lower-income workers in the infrastructure sector, and on the compensation that laid-off workers receive during the downsizing process. Apart from this effect, following Estache et al. (2000), if the fiscal resources that come from privatisation are used to improve the quality and efficiency of public services, they can lead to a reduction in income inequality. However, Estache et al. (2002) also advise that due to market effects, private providers will probably eliminate subsidies, charge higher connection fees, and be unwilling to invest in the poorest/most undeveloped areas, which ultimately can lead to infrastructure services becoming too expensive for lower-income groups, thus increasing the gap between the poor and the rich. This idea follows that of Ferreira (1995), who also points out the fact that credit constraints faced by the poor eventually inhibit them from using the private substitutes for infrastructure. In contrast, the rich can complement the public provision of infrastructure with private alternatives. In the end, as stressed by Calderón and Servén (2014), the outcome of the increased privatisation of infrastructure is extremely dependent on the design of the reforms of the infrastructure sector involving private participation.

BOX 3.1

Income inequality in LAC: The role of infrastructure.

Following the "World Social Report 2020" (UN, 2020), income inequality has followed a decreasing trend in the LAC region since 2000, at least until now, in that there are some signs that it could be rising again. Nevertheless, the report stresses the case of Brazil, which was capable of an impressive reduction in its income inequality levels from 1995 to 2015. It states that, in addition to the declining disparities in labour earnings (through the rise in education levels and the minimum wage, and through the transition of workers from informal to formal employment), social policies (with reinforced fiscal redistribution) also played a major role during the period of 1995–2015. The "Previdência Social Rural," "Benefício de Prestação Continuada," and the "Bolsa Familia" are some of the social programmes applied in Brazil which were highlighted, and which were also fundamental in reducing the country's income inequality levels. However, the report also stresses that the future does not look so bright, given that the new President Jair Bolsonaro supports the idea of a powerful cut in public spending (which includes the social protection programmes).

As can be perceived from the Brazilian case, it seems that public spending is indeed important in reducing income inequality. However, given the theme of Part II of this book, it should be asked: what is known regarding the role of infrastructure investment in income inequality in LAC? Well, the report from the IDB called "Building Opportunities for Growth in a Challenging World" (Cavallo and Powell, 2019) clearly states that in addition to the fact that low infrastructure investments can have enormous positive impacts on the LAC countries' growth, the failure to invest in infrastructure can hurt the poor strata of the population to a significant magnitude. According to the report's calculations, the failure to invest in infrastructure can lead *"households in the poorest two quintiles of the income distribution"* to *"lose 11 percentage points of real income over a 10-year period"* (Cavallo and Powell, 2019, pp. 85–86). In another report from the IDB called "From Structures to Services: The Path to Better Infrastructure in Latin America and the Caribbean" (Cavallo et al., 2020), the relationship between infrastructure and income inequality in LAC is once again analysed, with the report concluding that *"increasing efficiency in infrastructure would likely raise aggregate output, support the high-productivity sectors in the economy, and reduce income inequality"* (Cavallo et al., 2020, p. 285) because according to the report's calculations *"improving efficiency in infrastructure benefits low-income households more than high-income households"* (Cavallo et al., 2020, p. 283). Overall, the report states that the *"quality of infrastructure in Latin America and the Caribbean conspires against the region's aspirations of joining the ranks of upper-income countries. In that context, upgrading infrastructure and improving services is not an option; it is a necessity"* (Cavallo et al., 2020, p. 304).

3.3 Public capital stock, private capital stock, and income inequality in Latin America and the Caribbean

In recent decades, public capital stock as a share of output has been declining worldwide, a fact which contributes to the expanding gap between the infrastructure levels of developing (e.g. LAC) and developed countries (IMF, 2014). In addition to this circumstance, it seems that the lack of new infrastructure investments in the LAC region as well as the lack of the maintenance of their existing physical capital have given rise to an "infrastructure gap" which, according to a vast number of authors, could be very harmful to the development, economic sustainability, and competitiveness of this region (e.g. Faruqee, 2016; Lardé and Sánchez, 2014; Perrotti, 2011).

As already explained in Chapter 1, capital stock represents the available physical capital of an economy at a given moment and is calculated as the value of new investments minus depreciation. As the public component of capital stock i.e. public capital stock, is directly related to the government's investment in economic and social public infrastructures, it is of increasing interest to try to analyse the effects that the state of the LAC public capital stock has had on various macroeconomic aspects of the region. As economic growth is an indicator which shows (to a great extent) if a country is on the right development path and is a tool that the countries can use to achieve various macroeconomic objectives, in Chapter 2 we centred our analysis on the relationship that LAC capital stock has had with the region's growth. However, there are more standards that countries/regions need to attain in order to consider them developed. In the case of the LAC, we have the issue of its high-income inequality levels which, if nothing is done, can prevent this region from starting on a desired sustainable development path (OECD, 2019).

Given these statements, we feel that there are a vast number of reasons to study the effects of public capital stock on income inequality. Still, the primary reason is linked with our belief that the LAC's development, and the convergence of the region's living standards with those of the advanced economies, will depend greatly on the public investment strategies that LAC governments follow, particularly concerning their infrastructure (e.g. Cavallo et al., 2020; Cavallo and Powell, 2019). As previously stated in Section 3.2, in our analysis, we will also use private capital stock, given that due to the fiscal pressure that many countries (including those in the LAC) face or have faced, private participation in the infrastructure sector has significantly increased.

In order to conduct our analysis, annual data from 1995 to 2017 were collected for 18 countries from the LAC region, namely Argentina, Bolivia, Brazil, Chile, Colombia, Costa Rica, Dominican Republic, Ecuador, El Salvador, Guatemala, Honduras, Mexico, Nicaragua, Panama, Paraguay, Peru, Uruguay, and Venezuela. As in Chapter 2, the countries and time horizon were chosen according to the data available. To conduct our analysis, we used the statistical software package Stata 15.0. In Table 3.1, we have provided the name, definition, and sources of the raw variables.

The dependent variable will be represented by the Gini index of disposable income (*ineq*). This variable was collected from the "Standardised World Income Inequality Database" (SWIID) and will represent our measure of income inequality. Its values range from 0% to 100%, with 0% meaning perfect equality and 100% meaning maximum inequality, see Solt (2020) for supplementary information on the construction of this variable.

Regarding the interest variables, there will be two, each one in a different model. General government capital stock or public capital stock (*kpub*) will be the interest variable in Model I and private capital stock (*kpriv*) will be the interest variable in Model II. Public capital stock (*kpub*) and private capital stock (*kpriv*) were both retrieved from the "Investment and Capital Stock Dataset" (IMF, 2017). It is also important to stress that the variable consumer prices index (*cpi*)—retrieved from the CEPALSTAT—was used to adjust the variables public capital stock (*kpub*), private capital stock (*kpriv*), and gross domestic product (*y*), to the effects from price changes. This means it was used to transform these variables into their real or constant values, and that both the public and private capital stock variables i.e. "*kpub*" and "*kpriv*," were then transformed into percentages of GDP.

Concerning the control variables, we can say that they are some of the numerous commonly used variables in income inequality regressions. However, it is important to stress that among the full range of variables that could be used, the control variables that were chosen were those for which a considerable amount of data was available. These were: (1) gross domestic product (*y*) from the IMF "Investment and Capital Stock Dataset," (2) human development index (*hdi*) from the United Nations "Human Development Reports," (3) trade-in percentage of the gross domestic product (*trd*) from the World Bank "World Development Indicators," (4) tax revenue as a percentage of gross domestic product (*tr*) from CEPALSTAT, and (5) unemployment rate in percentage of the total labour force (*unp*) from the World Bank "World Development Indicators".

About the gross domestic product (*y*), we can say that the relationship between this variable and income inequality has been at the centre of the researchers' interest for the past decades, with the nexus between growth and income inequality being the subject of a vast number of past studies (e.g. Yang and Greaney, 2017; Rubin and Segal, 2015). For example, Tsounta and Osueke (2014) found that, after policy measures, economic growth was the main reason for the decrease in the LA income inequality. Concerning the human development index (*hdi*), we should clarify that

TABLE 3.1 Description of variables.

Variable	Definition	Source
ineq	Gini index	Standardized World Income Inequality Database (SWIID)
kpub	General government capital stock (current cost), in billions of national currency	Investment and Capital Stock Dataset (FMI)
kpriv	Private capital stock (current cost), in billions of national currency	Investment and Capital Stock Dataset (FMI)
y	Gross domestic product (current prices), in billions of national currency	Investment and Capital Stock Dataset (FMI)
hdi	Human development index	Human Development Reports (UNDP)
trd	Trade (% of GDP)	World Development Indicators (WB)
tr	Tax revenue (% GDP)	CEPALSTAT
unp	Unemployment, total (% of total labour force)	World Development Indicators (WB
cpi	Annual consumer prices indices general level (Base Index 2010 = 100)	CEPALSTAT

most of the authors focused their investigations on the effects of education on income inequality (e.g. Coady and Dizioli, 2018). However, as most education variables are highly lacking in data, we decided to use the HDI, which, in addition to taking education into account, also incorporates data on population health and standard of living. Theyson and Heller (2015) for example analysed the relationship between development, proxied by HDI, and income inequality and concluded that human development could have different effects on income inequality, depending on the development stage of the countries.

Regarding trade (*trd*), we must say that the vast literature which addresses its relationship with income inequality has found mixed results (e.g. Cerdeiro and Komaromi, 2017; Meschi and Vivarelli, 2009). However, Cerdeiro and Komaromi (2017), who carried out a study to be included in an IMF report on trade integration in the LAC, found that trade tends to have the effect of reducing income inequality. When it comes to tax revenue (*tr*), there is a wide range of studies which are focused on the effects of tax policies on income inequality (e.g. Martorano, 2018; Balseven and Tugcu, 2017). Given the LAC's high levels of income inequality, the literature on this theme is composed by several studies applied to the region. Martorano (2018) for example found that the low levels of tax revenue from LA countries are an obstacle to the promotion of equality in this region.

Moreover, we should state the results from a study by Balseven and Tugcu (2017) which indicate that tax revenues contribute to the decrease in income inequality in developing economies. Finally, there is also an extensive literature which addresses the relationship between the unemployment rate (*unp*) and income inequality (e.g. Sheng, 2011; Helpman et al., 2010; Cysne, 2009; Mocan, 1999). The general conclusion is that the unemployment rate has an augmenting effect on income inequality. Gasparini and Lustig (2011) for example stated that unemployment could have contributed to rising inequality in Argentina due to its indirect effect on wages. Moreover, Hacibedel et al. (2019) concluded that policies to support employment are an essential tool for reducing inequality regardless of whether the countries are facing "good" or "bad" economic conjunctures.

Turn now to the empirical analysis conducted in this chapter, it was based on a panel autoregressive distributed lag (PARDL) model in the form of an unrestricted error correction model (UECM). There are many reasons for using this methodology, among them being: PARDL allows us to identify the short- and long-run impacts of the explanatory variables on the dependent variable; it deals properly with cointegration, and allows the inclusion of $I(0)$, $I(1)$, and fractionally integrated variables in the same estimation; additionally, PARDL is robust when there are signals of endogeneity and gives consistent estimates with a small/moderate number of observations.

The PARDL specifications of our two models, Model I with public capital stock as the interest variable and Model II with private capital as the interest variable, can be described, respectively, by the following equations:

$$\begin{aligned} lineq_{it} = {} & \alpha_{1i} + \beta_{1i1} lineq_{it-1} + \beta_{1i2} lkpub_{it} + \beta_{1i3} lkpub_{it-1} + \beta_{1i4} ly_{it} + \beta_{1i5} ly_{it-1} + \beta_{1i6} lhdi_{it} + \beta_{1i7} lhdi_{it-1} + \beta_{1i8} ltrd_{it} \\ & + \beta_{1i9} ltrd_{it-1} + \beta_{1i10} ltr_{it} + \beta_{1i11} ltr_{it-1} + \beta_{1i12} lunp_{it} + \beta_{1i13} lunp_{it-1} + \varepsilon_{1it} \end{aligned} \tag{3.1}$$

$$\begin{aligned} lineq_{it} = {} & \alpha_{2i} + \beta_{2i1} lineq_{it-1} + \beta_{2i2} lkpriv_{it} + \beta_{2i3} lkpriv_{it-1} + \beta_{2i4} ly_{it} + \beta_{2i5} ly_{it-1} + \beta_{2i6} lhdi_{it} + \beta_{2i7} lhdi_{it-1} \\ & + \beta_{2i8} ltrd_{it} + \beta_{2i9} ltrd_{it-1} + \beta_{2i10} ltr_{it} + \beta_{2i11} ltr_{it-1} + \beta_{2i12} lunp_{it} + \beta_{2i13} lunp_{it-1} + \varepsilon_{3it} \end{aligned} \tag{3.2}$$

In order to obtain the dynamic relationships between the variables and the dynamic general UECM form of the PARDL model, Eqs. (3.1), (3.2) can be reparameterised into Eqs. (3.3), (3.4), respectively.

$$\begin{aligned} dlineq_{it} = {} & \alpha_{3i} + \beta_{3i1} dlkpub_{it} + \beta_{3i2} dly_{it} + \beta_{3i3} dlhdi_{it} + \beta_{3i4} dltrd_{it} + \beta_{3i5} dltr_{it} + \beta_{3i6} dlunp_{it} + \gamma_{3i1} lineq_{it-1} + \gamma_{3i2} lkpub_{it-1} \\ & + \gamma_{3i3} ly_{it-1} + \gamma_{3i4} lhdi_{it-1} + \gamma_{3i5} ltrd_{it-1} + \gamma_{3i6} ltr_{it-1} + \gamma_{3i7} lunp_{it-1} + \varepsilon_{3it} \end{aligned} \tag{3.3}$$

$$\begin{aligned} dlineq_{it} = {} & \alpha_{4i} + \beta_{4i1} dlkpriv_{it} + \beta_{4i2} dly_{it} + \beta_{4i3} dlhdi_{it} + \beta_{4i4} dltrd_{it} + \beta_{4i5} dltr_{it} + \beta_{4i6} dlunp_{it} + \gamma_{4i1} lineq_{it-1} + \gamma_{4i2} lkpriv_{it-1} \\ & + \gamma_{4i3} ly_{it-1} + \gamma_{4i4} lhdi_{it-1} + \gamma_{4i5} ltrd_{it-1} + \gamma_{4i6} ltr_{it-1} + \gamma_{4i7} lunp_{it-1} + \varepsilon_{4it} \end{aligned} \tag{3.4}$$

In Eqs. (3.3), (3.4), α_i represents the intercept, β_{ik} and γ_{ik} represent the estimated parameters, and ε_{it} denotes the error term. In both equations, the variables are presented in natural logarithms, with the prefix "*l*," and first differences, with the prefix "*d*."

As in the case of the models from Chapter 2, before the estimation of the PARDL model, it is necessary to compute a range of preliminary tests and specification tests. To understand the characteristics of our series and cross sections, we applied the following preliminary tests: the correlation matrix; the variance inflation factor (VIF) (Belsley et al., 1980);

the cross-sectional dependence test (Pesaran, 2004); and the second-generation unit root test (CIPS) (Pesaran, 2007). The specification tests conducted were as follows: the Hausman test (Hausman, 1978) to confront the random effects (RE) and fixed effects (FE) models; the Hausman test (Hausman, 1978) to confront the mean group (MG), the pooled mean group (PMG), and the pooled estimators; the modified Wald test (Greene, 2002); the Pesaran test of cross-sectional independence (Pesaran, 2004); and the Wooldridge test (Wooldridge, 2002). Again, as in (Chapter 2), in the appendix of this chapter, we present the preliminary and specification tests tables jointly with the How to do part.

Starting with the analysis of the correlation matrices and VIFs, we concluded that both collinearity and multicollinearity were far from being a concern in the estimation of our models, given the low correlation and VIF (and mean VIF) values (see Table 3A.1). In Table 3A.2, we provide the descriptive statistics of the variables in natural logarithms and first differences, as well as the results from the cross-sectional dependence test, which endorsed the presence of cross-sectional dependence in all the variables, either in natural logarithms or in first differences. As in Chapter 2, this outcome seems to point to the existence of interdependency between our variables across countries, maybe due to the mutual shocks that our countries share.

Before advancing, it is appropriate to explain that the variables "*ineq*," "*kpub*," "*kpriv*," "*y*," and "*trd*" have fewer observations (see the descriptive statistics in Table 3A.2) because there was a lack of observations for the Gini index of disposable income (*ineq*) in the cases of the Dominican Republic in 2017, Guatemala in 2015, 2016, and 2017, Mexico in 2017, Nicaragua in 2015, 2016, and 2017, and Venezuela in 2016 and 2017. Venezuela also has a shortage of data for public capital stock (*kpub*) and private capital stock (*kpriv*) in 2016 and 2017, and for trade in the percentage of the gross domestic product (*trd*) in 2015, 2016, and 2017. Despite this fact, the statistical software Stata 15 assumes the panel to be "strongly balanced," given that the lack of data only occurs at the end of the series. This means that we can continue to carry out the analysis without further concerns.

Now, in order to investigate the stationarity of the variables, we applied the cross-sectionally augmented IPS (CIPS) test. As in Chapter 2, the use of this test is linked with the fact that it is robust when cross-sectional dependence is detected in the variables. From the outcomes of the CIPS test, displayed in Table 3A.3, it can be seen that all variables in natural logarithms are $I(1)$ i.e. they are integrated of order one, and that they are all stationary in first differences, except DLUNP with the trend. As PARDL allows the inclusion of $I(0)$, $I(1)$, and fractionally integrated variables in the same estimation, there would only be a concern if any of the variables were $I(2)$, which was not the case.

Turning now to the specification tests, it should be noted that we first conducted the Hausman test between the RE and FE in order to see if the countries' individual effects must be considered. In Table 3A.4, it can be seen that the null hypothesis of this test—the difference in coefficients is not systematic, or that the RE are the most suitable—is rejected for all specifications (with the standard specification and with the *sigmamore* and *sigmaless* options) for the case of Model I and Model II. The conclusion is that the FE are the most suitable specification i.e. we should account for the individual effects. The use of the *sigmamore* and *sigmaless* options was in order to correct the fact that in the standard specification "*the covariance matrix has not been positively defined*." Nevertheless, this can also be seen as a robustness test to the standard Hausman test result.

The next step of the estimation was to compare the MG, the PMG, and the pooled estimators in order to test the slope heterogeneity of the parameters. For a more in-depth discussion on the MG and PMG estimators, see Pesaran et al. (1999). To compute this test, the Stata commands "xtdcce2" (Ditzen, 2018) and "Hausman" were jointly used (see How to do in Appendix). In Table 3A.5, we display the results from the Hausman tests between these estimators, which indicated the pooled estimator as preferable for both models, suggesting that the panel is homogeneous. Therefore, the estimation will proceed with the FE specification rather than with the MG and PMG specifications.

The remaining specification tests conducted before the estimation of the models were the modified Wald test, to test for group-wise heteroscedasticity, the Pesaran test of cross-sectional independence, to test for the presence of contemporaneous correlation among cross-sections, and the Wooldridge test, to test for the presence of serial correlation. In Table 3A.6, the results from these tests can be seen, as well as their respective null hypotheses. From the outcomes shown in Table 3A.6 it can be seen that all null hypotheses are rejected at the 1% level for all models, meaning that heteroscedasticity, contemporaneous correlation, and first-order autocorrelation, are all present in Model I and Model II. Due to this fact, in order to deal with the presence of these phenomena (i.e. heteroskedasticity, contemporaneous correlation, first-order autocorrelation, and cross-sectional dependence), we decided to use the Driscoll and Kraay (1998) estimator to perform the analysis of the two models, given that it produces standard errors robust to the disturbances being cross-sectionally dependent, heteroskedastic, and autocorrelated. In Table 3.2, the results from the estimation of Model I and Model II with the DK-FE estimator are presented.

We should clarify that the long-run elasticities are not provided in Table 3.2, because they had to be calculated through the application of a ratio between the long-run coefficients of the variables and the "lineq" coefficient, both

TABLE 3.2 P-ARDL estimation results.

Dependent variable: *dlineq*	Model I	Model II
Constant	0.3749395***	0.3804234***
dlkpub	0.0145526**	–
dlkpriv	–	0.0205358**
dly	−0.0137031***	−0.0104574*
dlhdi	0.0546026	0.0514217
dltrd	−0.0005006	−0.0001897
dltr	−0.0038807	−0.0028675
dlunp	0.0088814***	0.0089465**
lineq (−1)	−0.0869482***	−0.0874074***
lkpub (−1)	0.0012425	–
lkpriv (−1)	–	−0.0012596
ly (−1)	−0.0029274**	−0.0024863*
lhdi (−1)	−0.065256***	−0.0675616***
ltrd (−1)	−0.008963***	−0.0087956***
ltr (−1)	−0.0142447***	−0.0137585***
lunp (−1)	0.0094884***	0.0104827***
Diagnostic statistics		
N	385	385
R^2	0.3525	0.3599
F	F(13, 21) = 63.56***	F(13, 21) = 109.37***

***, **, and * denote statistical significance at 1%, 5%, and 10% level, respectively.

lagged once, and then we had to multiply this ratio by −1. Table 3.3 presents the long-run elasticities, the short-run impacts, and the adjustment speed of the three models.

Following the rule of parsimony (e.g. Santiago et al., 2020), after the first estimation of our models, we decided to remove the variables that did not produce any statistically significant coefficients in the short and long run. Thus, we removed the variables human development index (*hdi*), trade (*trd*), and tax revenue (*tr*) from the short run in both models, and the variables public capital stock (*kpub*), and private capital stock (*kpriv*), from the long run in Model I and Model II, respectively. Now, we can replace the specifications from Eqs. (3.3), (3.4) with

$$dlineq_{it} = \alpha_{5i} + \beta_{5i1} dlkpub_{it} + \beta_{5i2} dly_{it} + \beta_{5i3} dlunp_{it} + \gamma_{5i1} lineq_{it-1} + \gamma_{5i2} ly_{it-1} + \gamma_{5i3} lhdi_{it-1} + \gamma_{5i4} ltrd_{it-1} + \gamma_{5i5} ltr_{it-1} + \gamma_{5i6} lunp_{it-1} + \varepsilon_{5it} \quad (3.5)$$

$$dlineq_{it} = \alpha_{6i} + \beta_{6i1} dlkpriv_{it} + \beta_{6i2} dly_{it} + \beta_{6i3} dlunp_{it} + \gamma_{6i1} lineq_{it-1} + \gamma_{6i2} ly_{it-1} + \gamma_{6i3} lhdi_{it-1} + \gamma_{6i4} ltrd_{it-1} + \gamma_{6i5} ltr_{it-1} + \gamma_{6i6} lunp_{it-1} + \varepsilon_{6it} \quad (3.6)$$

Eqs. (3.5), (3.6) stand for the most parsimonious specifications that we have reached. All specification tests were redone to ensure that all assumptions remained the same (see Tables 3A.7, 3A.8, and 3A.9, in the Appendix). The results from most parsimonious versions of Model I and Model II are presented in Table 3.4.

Once again, we should note that Table 3.4 does not provide the long-run elasticities; as in the previous case (Table 3.2), they had to be calculated. The long-run elasticities are provided in Table 3.5, together with the short-run impacts and with the adjustment speed of the model i.e. the error correction mechanism (ECM).

As can be seen from the outcomes of Tables 3.3 and 3.5, the results from the nonparsimonious and parsimonious models are very similar, with minor differences in the coefficient values and the statistical significances of some of the variables. Overall, the results from both specifications seem to indicate that, in the short run, both gross domestic product (y) and the unemployment rate (*unp*) have a statistically significant effect on income inequality (*ineq*). However, while the gross domestic product (*y*) seems to contribute to reducing income inequality (*ineq*) in this group of LAC countries, the unemployment rate (*unp*) seems to present an inverse effect, showing signs of having an enhancing impact on income inequality (*ineq*). Regarding our interest variables, it can be seen that, in the short run, both public capital stock (*kpub*)—in Model I—and private capital stock (*kpriv*)—in Model II—also seem to contribute to the increase in income inequality (*ineq*) in the countries in our sample. Additionally, it can be seen that capital stocks seem to be the main drivers of income inequality (*ineq*) among the variables which were included in the models in the short run. Finally, comparing both models (Model I and Model II), we realise that the enhancing effect of private capital stock (*kpriv*) on income inequality (*ineq*) is higher than that of public capital stock (*kpub*).

Concerning the long run, the results from Tables 3.3 and 3.5 seem to indicate that gross domestic product (*y*), the human development index (*hdi*), trade (*trd*), and tax revenue (*tr*) all contribute to decreasing income inequality (*ineq*) in these countries. Among these variables, the human development index (*hdi*) seems to contribute the most to reducing income inequality (*ineq*). In contrast, the unemployment rate (*unp*) continues to promote income inequality (*ineq*) in the long run. Although the unemployment rate (*unp*) has a similar effect in the short and long run, the magnitude of its effect can be seen to be considerably larger in the long run. Finally, conversely to what was found in the short run, in the long run, neither public capital stock (*kpub*)—in Model I—nor private capital stock (*kpriv*)—in Model II—seem to demonstrate a statistically significant effect on income inequality (*ineq*)—see Tables 3.2 and 3.3. Because of this, and as was previously stressed, these variables were not included in the most parsimonious models (Tables 3.4 and 3.5).

Regarding the ECM terms in Model I and Model II, represented by the variable "*lineq*" lagged once, it can be seen that they are all negative and statistically significant at the 1% level [both in the parsimonious specification (Table 3.5) and in the nonparsimonious specification (Table 3.3)], which is a sign of the presence of cointegration/long memory in the variables. Moreover, from this result, we can also consider that when a parameter is statistically significant in these models, it is identical to testing for Granger causality (e.g. Jouini, 2015). Concerning the magnitude of the ECM coefficients, they can be said to indicate that the speed at which the dependent variable returns to equilibrium after variations in the independent variables is relatively low/moderate for both models; i.e. when the models are faced with shocks they require a substantial amount of time to return to equilibrium.

TABLE 3.3 Elasticities, short-run impacts and adjustment speed.

Dependent variable: *dlineq*	**Model I**	**Model II**
Short-run impacts		
dlkpub	0.0145526**	–
dlkpriv	–	0.0205358**
dly	−0.0137031***	−0.0104574*
dlhdi	0.0546026	0.0514217
dltrd	−0.0005006	−0.0001897
dltr	−0.0038807	−0.0028675
dlunp	0.0088814***	0.0089465**
Long-run (computed) elasticities		
lkpub (−1)	0.0142906	–
lkpriv (−1)	–	−0.0144103
ly (−1)	−0.033668**	−0.0284454*
lhdi (−1)	−0.7505163***	−0.7729507***
ltrd (−1)	−0.1030845***	−0.1006281***
ltr (−1)	−0.1638301***	−0.1574071***
lunp (−1)	0.1091275***	0.119929***
Speed of adjustment		
ECM	−0.0869482***	−0.0874074***

***, **, and * denote statistical significance at 1%, 5%, and 10% level, respectively; the ECM denotes the coefficient of the variable "*lineq*" lagged once.

TABLE 3.4 P-ARDL estimation results (parsimonious).

Dependent variable: *dlineq*	Model I	Model II
Constant	0.3690458***	0.3704958***
dlkpub	0.0118132***	–
dlkpriv	–	0.0167744***
dly	−0.0131713***	−0.0102333***
dlunp	0.0097863***	0.0095263***
lineq (−1)	−0.0869753***	−0.0881925***
ly (−1)	−0.0023486**	−0.00204*
lhdi (−1)	−0.0730341***	−0.073642***
ltrd (−1)	−0.0092784***	−0.0089951***
ltr (−1)	−0.0123148***	−0.0127744***
lunp (−1)	0.0099225***	0.0102706***
Diagnostic statistics		
N	386	386
R^2	0.3510	0.3572
F	$F(9, 21) = 39.12$***	$F(9, 21) = 46.77$***

***, **, and * denote statistical significance at 1%, 5%, and 10% level, respectively.

TABLE 3.5 Elasticities, short-run impacts, and speed of adjustment (parsimonious).

Dependent variable: *dlineq*	Model I	Model II
Short-run impacts		
dlkpub	0.0118132***	–
dlkpriv	–	0.0167744***
dly	−0.0131713***	−0.0102333***
dlunp	0.0097863***	0.0095263***
Long-run (computed) elasticities		
ly (−1)	−0.0270034**	−0.0231307**
lhdi (−1)	−0.8397108***	−0.8350141***
ltrd (−1)	−0.1066784***	−0.1019935***
ltr (−1)	−0.1415892***	−0.1448471***
lunp (−1)	0.1140838***	0.1164562***
Speed of adjustment		
ECM	−0.0869753***	−0.0881925***

*** and ** denote statistical significance at 1% and 5% level, respectively; ECM denotes the coefficient of the variable "*lineq*" lagged once.

Before proceeding to Section 3.4, we should bear in mind that when researchers analyse regions such as the LAC, they should not ignore the possible existence of several political and economic shocks which could influence the results of their estimations. Following this assumption, in order to test the robustness of the previous results and conclusions, a set of dummy variables was added to both models, in both the parsimonious and the nonparsimonious specifications, in order to control for the shocks which may have affected the income inequality levels of these countries in several ways.

This method consists of identifying events that may have produced peaks/breaks of large magnitude in the income inequality of the countries in our sample, followed by residuals analysis to allow us to confirm the existence of such shocks. Finally, we incorporate dummies in the regressions with the aim of correcting the peaks/breaks identified (e.g. Santiago et al., 2020; Fuinhas et al., 2017). In this case, the dummies which were added in order to deal with the detected outliers were the following: BRA2016; GTM2013; GTM2014; PRY2004; URY2010; URY2011; URY2012, see Table 3.6 for a comprehensive description of the detected outliers.

The results of Model I and Model II with the correction of outliers are presented in Tables 3.7 and 3.8, nonparsimonious and parsimonious, respectively.

Looking at the results of Tables 3.7 and 3.8, it can be concluded that the coefficients from all the dummies which were introduced in the models are all statistically significant at 1% level, thus proving the suitability of their inclusion. As in the previous case, the long-run elasticities had to be calculated. Tables 3.9 and 3.10 display the long-run elasticities, the short-run impacts, and the adjustment speed of the two models corrected for outliers, again, nonparsimonious and parsimonious, respectively.

With the inclusion of dummies (Tables 3.9 and 3.10), it can be seen that the overall results remain similar to the ones from the models without the correction of shocks (Tables 3.3 and 3.5), which means that the previous inferences remain true, with public capital stock (*kpub*) and private capital stock (*kpriv*) continuing to show an enhancing effect on income inequality (*ineq*) in the short run and an absence of a statistically significant effect in the long run. Moreover, we also note that the gross domestic product (*y*) continues to contribute to reducing income inequality (*ineq*) in the short and long run, and the unemployment rate (*unp*) continues to promote income inequality (*ineq*) in the short and long run. The human development index (*hdi*), trade (*trd*), and tax revenue (*tr*) still show a decreasing effect on income inequality (*ineq*) in the long run. Finally, the ECM terms of the models all continue negative and statistically significant at 1% level. However, they suffered a decrease in their magnitude, which means that with the inclusion of the dummy variables, the adjustment speed of the models became slightly slower.

As discussed in Chapter 2, after the conclusion of the empirical analysis, we will now move our attention to the discussion of the results and their respective findings in Section 3.4. Additionally, we will also present some policy implications associated with our results in order to help LAC governments with the development of strategies to decrease this region's income inequality levels.

TABLE 3.6 Detected outliers.

Dummies	Description
BRA2016	Corrects the peak observed in Brazil in 2016. This peak could be explained by the effects of the Brazilian crisis, which started in mid-2014, with the deceleration of the Chinese economy and the fall in commodity prices and culminated with the impeachment of Dilma Rousseff in 2016. This unfavourable situation generated a set of nefarious effects on Brazilian macroeconomic stability, particularly income inequality, with a rise in unemployment and a decline in real wages
GTM2013, GTM2014	Corrects the breaks observed in Guatemala in 2013 and 2014, respectively. These breaks could probably be linked to the tax reforms adopted in 2012 by the Guatemalan government in order to improve its revenues and its public social spending, and which increased the progressivity of the country's tax system
PRY2004	Corrects the break observed in Paraguay in 2004. This break could be possibly connected with the fact that, after some years of decline and stagnation, Paraguay registered a recovery in 2003 and 2004 (in part due to high commodity prices). At the same time, in 2003, the Paraguayan government also introduced a set of welfare programmes which, combined with the country's economic recuperation, could have contributed to influencing its income inequality levels greatly
URY2010, URY2011, and URY2012	Corrects the breaks observed in Uruguay in 2010, 2011, and 2012. These breaks could be linked with the election of José Mujica as President of Uruguay in 2010. Although income inequality began to fall around 2007 when José Mujica rose to power, one of his biggest flags was the fight against inequalities and wealth concentration. The measures which were taken under his presidency for example the rise in minimum wage and the expansion of social spending, certainly affected Uruguay's income gap

TABLE 3.7 P-ARDL estimation results (corrected for outliers).

Dependent variable: *dlineq*	Model I	Model II
Constant	0.3622899***	0.3685722***
dlkpub	0.0128172*	–
dlkpriv	–	0.0200936***
dly	−0.0151064***	−0.0113837**
dlhdi	0.1489848	0.1544508
dltrd	0.000386	0.0006801
dltr	−0.0052934	−0.0043075
dlunp	0.0061857**	0.0062728
lineq (−1)	−0.0778069***	−0.0804609***
lkpub (−1)	0.0024813	–
lkpriv (−1)	–	0.0005425
ly (−1)	−0.0040874**	−0.0034539**
lhdi (−1)	−0.044427***	−0.050532***
ltrd (−1)	−0.0081788**	−0.0080467**
ltr (−1)	−0.0177203***	−0.0170059***
lunp (−1)	0.0060525**	0.0071748***
BRA2016	0.0260843***	0.0260914***
GTM2013	−0.0213843***	−0.0212918***
GTM2014	−0.026532***	−0.0267326***
PRY2004	−0.0210444***	−0.0206358***
URY2010	−0.0246049***	−0.0231927***
URY2011	−0.0312096***	−0.0310709***
URY2012	−0.0349603***	−0.0353811***
Diagnostic statistics		
N	385	385
R^2	0.4913	0.4969
F	$F(20, 21) =$ 1337.34***	$F(20, 21) =$ 1117.34***

***, **, and * denote statistical significance at 1%, 5%, and 10% level, respectively.

TABLE 3.8 P-ARDL estimation results (parsimonious) (corrected for outliers).

Dependent variable: *dlineq*	Model I	Model II
Constant	0.3503321***	0.3519591***
dlkpub	0.0094165**	–
dlkpriv	–	0.0144921***
dly	−0.0125846***	−0.0096596***
dlunp	0.0071428***	0.006892**
lineq (−1)	−0.0770527***	−0.0782553***

TABLE 3.8 P-ARDL estimation results (parsimonious) (corrected for outliers)—cont'd

Dependent variable: *dlineq*	Model I	Model II
ly (−1)	−0.0030031**	−0.0027128**
lhdi (−1)	−0.0608617***	−0.0615741***
ltrd (−1)	−0.0090063***	−0.0087275***
ltr (−1)	−0.0143732***	−0.0148788***
lunp (−1)	0.0067394***	0.0070916***
BRA2016	0.0260935***	0.0260335***
GTM2013	−0.0214255***	−0.0215789***
GTM2014	−0.0208854***	−0.0210336***
PRY2004	−0.0206415***	−0.0207195***
URY2010	−0.0265102***	−0.0256541***
URY2011	−0.0305581***	−0.0303302***
URY2012	−0.0342909***	−0.0343981***
Diagnostic statistics		
N	386	386
R^2	0.4845	0.4898
F	$F(16, 21) = 490.76$***	$F(16, 21) = 473.34$***

*** and **denote statistical significance at 1% and 5% level, respectively.

TABLE 3.9 Elasticities, short-run impacts and adjustment speed (corrected for outliers).

Dependent variable: *dlineq*	Model I	Model II
Short-run impacts		
dlkpub	0.0128172*	–
dlkpriv	–	0.0200936***
dly	−0.0151064***	−0.0113837**
dlhdi	0.1489848	0.1544508
dltrd	0.000386	0.0006801
dltr	−0.0052934	−0.0043075
dlunp	0.0061857**	0.0062728
Long-run (computed) elasticities		
lkpub (−1)	0.0318911	–
lkpriv (−1)	–	0.0067419
ly (−1)	−0.0525327***	−0.0429259**
lhdi (−1)	−0.5709912**	−0.6280314***
ltrd (−1)	−0.1051168**	−0.1000079***
ltr (−1)	−0.2277474***	−0.2113565***
lunp (−1)	0.0777883**	0.0891718***
Speed of adjustment		
ECM	−0.0778069***	−0.0804609***

***, **, and * denote statistical significance at 1%, 5%, and 10% level, respectively; ECM denotes the coefficient of the variable "*lineq*" lagged once.

TABLE 3.10 Elasticities, short-run impacts, and speed of adjustment (parsimonious) (corrected for outliers).

Dependent variable: *dlineq*	Model I	Model II
Short-run impacts		
dlkpub	0.0094165**	–
dlkpriv	–	0.0144921***
dly	−0.0125846***	−0.0096596***
dlunp	0.0071428***	0.006892**
Long-run (computed) elasticities		
ly (−1)	−0.0389748***	−0.0346666**
lhdi (−1)	−0.7898707***	−0.7868353***
ltrd (−1)	−0.1168845***	−0.1115265***
ltr (−1)	−0.1865375***	−0.1901311***
lunp (−1)	0.0874654***	0.0906217***
Speed of adjustment		
ECM	−0.0770527***	−0.0782553***

*** and ** denote statistical significance at 1% and 5% level, respectively; ECM denotes the coefficient of the variable "*lineq*" lagged once.

3.4 Conclusion

In this chapter, we tried to uncover the effects that Latin American and Caribbean (LAC) capital stock had on the income inequality levels of 18 countries in the region between 1995 and 2017. In order to achieve our goals, two models were built: Model I with public capital stock as the interest variable, and Model II with private capital stock as the interest variable. The empirical analysis conducted in this chapter was based on the use of a PARDL model.

Looking at the results of our analysis, we can start by referring that economic growth seems to be a powerful tool to reduce the LAC income inequality levels, given that gross domestic product is shown to have the effect of depressing income inequality in all estimated models, both in the short and long run. This result suggests that the economic performance of these countries can indeed influence their income inequality levels and, more precisely, it suggests that LAC countries are probably making advances in the promotion of inclusive growth policies, so that the positive effects of their economic performance can benefit all population layers i.e. they are trying to combine their growth-enhancing policies with measures focused on the promotion of a more equitable society. In our view, this strategy should continue to be followed by these countries' governments in order to guarantee that the gains from their growth will not be channelled only to the highest strata of the population. Additionally, we should also stress that this result seems to be in line with the ones from previous studies such as that by Tsounta and Osueke (2014), who used a sample similar to ours.

Conversely to economic growth, unemployment (proxied by the unemployment rate) was shown to have an augmenting effect on these countries' income inequality levels, both in the short and long run, indicating that increased unemployment is indeed a factor that contributes to widening the income gap in these countries, with a persistent effect which extends over time. This means that, in order to tackle income inequality, LAC governments should concentrate a substantial part of their efforts on the fight against unemployment for example with policies to encourage job creation and measures that guarantee increased job opportunities for all. Primarily, the conclusion that can be derived from this observed outcome is that fighting against unemployment is also fighting against income inequality. As in the previous case, this result is also in accordance with past literature findings (e.g. Hacibedel et al., 2019).

According to the results from our models, we also saw that the human development index (which is a variable that incorporates information related to population health, education, and standard of living) is capable of contributing to the reduction of the LAC countries' income inequality levels in the long run. This outcome reflects the general view that policies aimed at the improvement of the standard of living of populations contribute to an equal society, with public investment in education and health or with the development of social policies, namely social protection programmes (e.g. Martinez-Vázquez et al., 2012). Therefore, we suggest that LAC governments should continue to invest in the well-being of their populations, especially of the lower-income groups, in order to achieve greater social cohesion and a more equal income distribution.

Along the same lines, tax revenues are also shown to have a decreasing effect on income inequality in the long run, which means that taxation can in fact generate a redistributive effect in these countries if properly done and hence contribute to alleviating the income gap. One example of this could be the promotion of their social welfare and social protection programmes with their tax revenues. However, as most of these countries have low levels of tax revenues (see e.g. Martorano, 2018), it could be essential to improve their tax schemes with for example more progressive taxation. In addition to providing higher revenues, this will also help these governments to support their public expenditure policies, namely the ones focused on the reduction of income inequality. This outcome also seems to be validated by some of the previous literature (e.g. Martorano, 2018; Balseven and Tugcu, 2017).

Concerning the effects of trade on the income inequality levels of the LAC countries, we must say that while the literature has found mixed results, our outcomes seem to support the view that trade does indeed have the effect of reducing income inequality in the long run (e.g. Cerdeiro and Komaromi, 2017). We should point out the fact that although the LAC region saw income inequality levels increase in the 1980s and 1990s, which coincided with the increasing integration of the region in the global economy, this trend was reversed with the beginning of the new millennium due to the stabilisation of the trade liberalisation process in the region and the policy reforms which were put into place in order to promote growth and control inequality (Székely and Sámano-Robles, 2014; Cornia, 2011). Following these considerations, we suggest that LAC countries should continue their integration process, given the positive effects that this can have on their economic output (e.g. Santiago et al., 2020), at the same time as they continue to develop policies aimed at extending the gains from trade to the general population.

Finally, turning our attention to the interest variables and to the central theme of our analysis, from the outcomes of all of our models we saw that both public capital stock and private capital stock are demonstrated to have an enhancing effect on income inequality i.e. both variables seem to contribute to the deterioration of the income distribution in these countries, with the effect from private capital stock being slightly higher than that of public capital stock. However, we should mention that these effects were only observed in the short run. In the long run, none of these variables showed a statistically significant effect on income inequality. We should also mention that these results held when we corrected the models for the presence of outliers, suggesting that even with this correction i.e. with the inclusion of dummy variables, the previous inferences remained accurate.

In sum, our findings revealed that these countries' physical capital investment strategies must be rethought. First, concerning their public investment strategies, it can be seen that, contrary to what is intended, public capital stock fails to contribute to the progress of these countries in terms of equality, with the outcomes from our models showing that public capital stock has contributed to an increase in these countries' income inequality levels in the short run. This result probably means that public investment in physical capital (e.g. roads, railways, bridges, schools, hospitals, sanitation and water systems, telecommunications and energy systems and public transportation, among others) is being made in the already wealthiest areas, where there is evidence of a certain economic dynamism, rather than being channelled to the poorest/undeveloped areas (see e.g. Lopez, 2003).

Similarly to public capital stock, private capital stock can also be seen to demonstrate an enhancing effect on income inequality in the short run, with the justifications for this result probably matching those of the previous case. However, in the case of private capital, we should also account for the fact that private interest is majorly driven by profit and therefore it is natural that, in the absence of government incentives, they invest in areas where higher profits are guaranteed (normally in the most developed areas). In addition to this, we must also consider the possible barriers that the private control of energy, infrastructure, and transport services for example can generate to the poorest groups of the population. Private enterprises usually charge higher prices for their services compared to public enterprises, which could lead to a reduction in access to and affordability of these services by the most disadvantaged strata of the population (these additional assumptions can help to explain why the magnitude of the effect from private capital on income inequality is greater than that of the public capital).

Now, regarding the absence of a statistically significant effect from both variables i.e. public and private capital stock, on income inequality, in the long run, we can say that it may be due to the fact that these countries' governments may try to correct the detected negative impact over time through investing in less developed areas and through the creation of incentives for the private sector to also invest in them. However, the level of investment does not seem to be large enough to have a reducing effect on income inequality in the long run. In general, the results seem to point to problems related to the lack of investment. Looking at some of the data available in the World Development Indicators Database of the World Bank ["Access to electricity, rural (% of rural population)" (EG.ELC.ACCS.RU.ZS) and "Access to electricity, urban (% of urban population)" (EG.ELC.ACCS.UR.ZS) for the LAC aggregate], we see that although the progresses in some fields for example in the electricity coverage (in 1995 only 62.9% of the rural population of the LAC had access to electricity, which contrasts with the value from 2017, where this percentage achieved the 91%), some economic indicators continue to point the rural areas of this region as the areas where people need to struggle more to get out of poverty or, in other words, where the risk of falling into poverty is greater. For example, the LAC Equity Lab and the World Bank's $1.90-a-day (2011 PPP prices) International Poverty Line for the LAC aggregate show that while in the region's rural areas the poverty rate in 2017 was at 16.2%, in urban areas this same rate was only at 1%.

According to these outcomes, it seems that the LAC governments should rethink their physical capital investment strategies, given that it seems that the investment is being concentrated in the regions where there is already a certain level of development/wealth. If no changes are made in this field, continuing to ignore the areas where investment in physical capital is really necessary i.e. in the poorest/undeveloped areas, the cohesion of these countries will continue to be threatened. As stressed in the book by Fay et al. (2017, p. 25) book focusing on infrastructure in the LAC: "*Connecting rural communities to the "outside world" is essential for inclusive development*". In this sense, the region's governments should improve the management and the selection criteria of public investments in order to develop the poorest/rural areas, linking them to the richer areas where there is a more thriving economic activity, allowing an increased convergence of income. Given the low degree of economic slack that many of these countries face, it could also be important for these governments to create incentives so that private initiatives also invest in these areas, as otherwise, this is unlikely to happen. Still, private initiatives should be intensively examined/discussed by public entities to guarantee that the low-income lawyers of the population are not neglected. Lastly, investment in the areas where physical capital is scarce is also required to be sufficient to allow this investment to have the desired effect on income inequality in the long run and also to guarantee the maintenance of existing capital so that the impact of new investments does not vanish over time. As Pérez (2020, p. 14) states: "*Infrastructure planning with a long-term and territorially balanced perspective will allow effective support to be provided for industrial transformation, as well as enabling better adaptation to economic changes and new social and environmental concerns that emerge as development progresses*". Finally, in Fig. 3.2, we display a scheme representing the ideal scenario so that physical capital stock can decrease income inequality in the LAC.

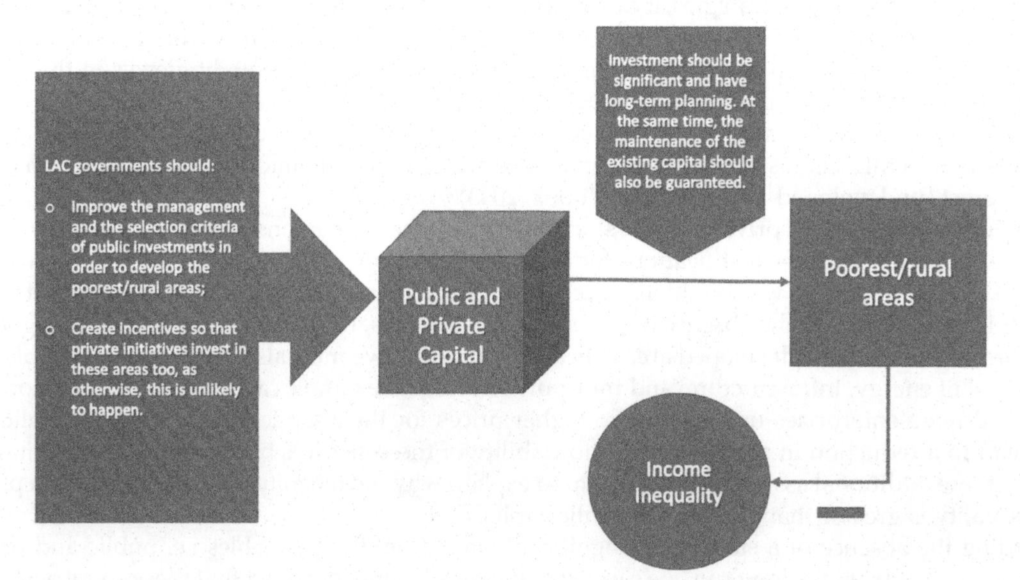

FIG. 3.2 Ideal scenario for physical capital stock to decrease income inequality in the LAC (created by the authors).

Appendix

TABLE 3A.1 Correlation matrices and VIF statistics.

			Model I				
	lineq	***lkpub***	***ly***	***lhdi***	***ltrd***	***ltr***	***lunp***
lineq	1.0000						
lkpub	−0.1155	1.0000					
ly	0.0842	−0.2221	1.0000				
lhdi	−0.4848	−0.4016	0.3306	1.0000			
ltrd	0.0772	−0.0070	−0.2605	−0.2374	1.0000		
ltr	−0.3300	0.1344	−0.0546	0.1739	0.0059	1.0000	
lunp	−0.1554	0.0848	0.4369	0.3069	−0.4691	0.0228	1.0000
VIF		1.42	1.38	1.57	1.32	1.10	1.64
Mean VIF		1.40					
	dlineq	***dlkpub***	***dly***	***dlhdi***	***dltrd***	***dltr***	***dlunp***
dlineq	1.0000						
dlkpub	0.1099	1.0000					
dly	−0.1873	−0.4097	1.0000				
dlhdi	0.0050	−0.2932	0.2392	1.0000			
dltrd	0.0633	−0.0653	0.0702	0.0791	1.0000		
dltr	0.0168	−0.0503	−0.0634	0.1182	0.2035	1.0000	
dlunp	0.1913	0.2520	−0.2359	−0.1109	−0.2072	−0.1900	1.0000
VIF		1.40	1.31	1.12	1.08	1.11	1.21
Mean VIF		1.20					
			Model II				
	lineq	**lkpriv**	**ly**	**lhdi**	**ltrd**	**ltr**	**lunp**
lineq	1.0000						
lkpriv	0.3135	1.0000					
ly	0.0842	0.0527	1.0000				
lhdi	−0.4848	−0.1141	0.3306	1.0000			
ltrd	0.0772	−0.0033	−0.2605	−0.2374	1.0000		
ltr	−0.3300	−0.3703	−0.0546	0.1739	0.0059	1.0000	
lunp	−0.1554	0.1173	0.4369	0.3069	−0.4691	0.0228	1.0000
VIF		1.24	1.33	1.24	1.31	1.24	1.56
Mean VIF		1.32					

Continued

TABLE 3A.1 Correlation matrices and VIF statistics.—cont'd

	dlineq	*dlkpriv*	*dly*	*dlhdi*	*dltrd*	*dltr*	*dlunp*
dlineq	1.0000						
dlkpriv	0.1783	1.0000					
dly	−0.1873	−0.4433	1.0000				
dlhdi	0.0050	−0.2772	0.2392	1.0000			
dltrd	0.0633	−0.0842	0.0702	0.0791	1.0000		
dltr	0.0168	−0.0767	−0.0634	0.1182	0.2035	1.0000	
dlunp	0.1913	0.2677	−0.2359	−0.1109	−0.2072	−0.1900	1.0000
VIF		1.47	1.37	1.10	1.08	1.12	1.22
Mean VIF		1.23					

In the case of the VIF test, the values are lower than the typically assumed benchmarks: 10 in the case of the VIF values, and 6 in the case of the mean VIF values.

TABLE 3A.2 Descriptive statistics and cross-section dependence (CD) test.

	Descriptive statistics					Cross-sectional dependence test		
Variables	**Obs.**	**Mean**	**Std. dev.**	**Min.**	**Max.**	**CD-test**	**Corr.**	**Abs. (corr.)**
lineq	404	3.829967	0.088778	3.580737	3.972177	38.21***	0.660	0.813
lkpub	412	4.155507	0.5826732	3.021907	6.256092	6.94***	0.121	0.501
lkpriv	412	5.120085	0.2258767	4.414647	5.824798	5.21***	0.091	0.464
ly	412	7.106889	2.821871	2.535461	13.44719	46.79***	0.808	0.868
lhdi	414	−0.374639	0.1064339	−0.6792443	−0.1707883	56.73***	0.979	0.979
ltrd	411	4.061701	0.4562949	2.74955	5.116187	15.54***	0.264	0.462
ltr	414	2.53745	0.2501715	1.713798	3.095849	27.21***	0.460	0.523
lunp	414	1.772423	0.4910779	0.696641	3.0214	14.48***	0.248	0.467
dlineq	386	−0.0051319	0.0095045	−0.0376389	0.0218825	17.69***	0.310	0.369
dlkpub	394	−0.0043109	0.0756566	−0.2453785	0.8519788	8.66***	0.151	0.266
dlkpriv	394	0.0023594	0.0742963	−0.2536368	0.8357296	12.51***	0.220	0.277
dly	394	0.0327024	0.077779	−0.7740245	0.2634878	16.82***	0.295	0.315
dlhdi	396	0.0072372	0.0058224	−0.0116808	0.0428978	6.51***	0.115	0.220
dltrd	393	0.0027417	0.0946325	−0.3370762	0.6474607	20.87***	0.363	0.377
dltr	396	0.011132	0.0804589	−0.7910475	0.2923148	6.04***	0.104	0.207
dlunp	396	−0.0084556	0.130235	−0.4741598	0.4782575	11.38***	0.199	0.247

The CD test has $N(0,1)$ distribution under the H0: cross-sectional independence; *** denotes statistical significance at the 1% level.

TABLE 3A.3 Panel unit root test (CIPS).

	CIPS (Zt-bar)	
	Without trend	**With trend**
lineq	−0.232	−1.774**
lkpub	1.417	1.090
lkpriv	3.586	3.638
ly	1.402	1.218
lhdi	−0.503	1.136
ltrd	−0.946	0.601
ltr	−1.042	0.022
lunp	0.182	1.710
dlineq	−3.439***	-1.741**
dlkpub	−3.362***	−1.729**
dlkpriv	−3.004***	−2.142**
dly	−5.570***	−4.414***
dlhdi	−5.152***	−3.649***
dltrd	−4.883***	−3.218***
dltr	−6.688***	−5.486***
dlunp	−3.558***	−0.897

*** and ** denote statistical significance at the 1% and 5% level, respectively; the Pesaran (2007) panel unit root test (CIPS) assumes that cross-sectional dependence is in the form of a single unobserved common factor and H0: series is $I(1)$.

TABLE 3A.4 Hausman test (FE vs. RE).

	Model I	**Model II**
	FE vs. RE	FE vs. RE
Hausman test	Chi2(13) = 86.39***	Chi2(13) = 81.90***
Hausman test (with *sigmamore*)	Chi2(13) = 81.81***	Chi2(13) = 73.37***
Hausman test (with *sigmaless*)	Chi2(13) = 101.29***	Chi2(13) = 88.26***

*** denotes significance at the 1% level; H0: difference in coefficients not systematic (random-effects).

TABLE 3A.5 Hausman test (MG vs. PMG vs. Pooled).

	Model I	**Model II**
Hausman test	MG vs. PMG	MG vs. PMG
	Chi2(13) = 60.14***	Chi2(13) = 31.81***
	PMG vs. Pooled	PMG vs. Pooled
	Chi2(13) = 7.39	Chi2(13) = 0.75
	MG vs. Pooled	MG vs. Pooled
	Chi2(13) = 15.54	Chi2(13) = −1.14

*** denotes statistically significant at 1%; H0: difference in coefficients not systematic or that: (1) the pooled mean group (PMG) is the most suitable (when MG vs. PMG); (2) Pooled is the most suitable (when PMG vs. Pooled and MG vs. Pooled); Negative "Chi2" values can be interpreted "*as strong evidence that we cannot reject the null hypothesis*" (see "Hausman specification test" from the Stata Manual, p. 8—available at: https://www.stata.com/manuals13/rhausman.pdf).

TABLE 3A.6 Specification tests.

	Model I	Model II
	Statistics	Statistics
Modified Wald test	Chi2 (18) = 144.96***	Chi2 (18) = 139.46***
Pesaran's test	3.051***	2.868***
Wooldridge test	F(1, 17) = 19.482***	F(1, 17) = 19.193***

*** denotes statistical significance at the 1% level; H0 of modified Wald test: sigma(*i*)^2 = sigma^2 for all *I*; H0 of Pesaran's test: residuals are not correlated; H0 of Wooldridge test: no first-order autocorrelation.

TABLE 3A.7 Hausman test (FE vs. RE) (parsimonious).

	Model I	Model II
	FE vs. RE	FE vs. RE
Hausman test	Chi2(9) = 96.54***	Chi2(9) = 97.36***
Hausman test (with *sigmamore*)	Chi2(9) = 88.82***	Chi2(9) = 85.24***
Hausman test (with *sigmaless*)	Chi2(9) = 113.52***	Chi2(9) = 107.60***

*** denotes significance at the 1% level; H0: difference in coefficients not systematic (random-effects).

TABLE 3A.8 Hausman test (MG vs. PMG vs. Pooled) (parsimonious).

	Model I	Model II
Hausman test	MG vs. PMG	MG vs. PMG
	Chi2(9) = −3.66	Chi2(9) = 3.10
	PMG vs. Pooled	PMG vs. Pooled
	Chi2(9) = 3.01	Chi2(9) = 3.35
	MG vs. Pooled	MG vs. Pooled
	Chi2(9) = 35.59***	Chi2(9) = 14.91

*** denotes statistically significant at 1%; H0: difference in coefficients not systematic or that: (1) the PMG is the most suitable (when MG vs. PMG); (2) Pooled is the most suitable (when PMG vs. Pooled and MG vs. Pooled); Negative "Chi2" values can be interpreted *"as strong evidence that we cannot reject the null hypothesis"* (see "Hausman specification test" from the Stata Manual, p. 8—available at: https://www.stata.com/manuals13/rhausman.pdf).

TABLE 3A.9 Specification tests (parsimonious).

	Model I	Model II
	Statistics	Statistics
Modified Wald test	Chi2 (18) = 144.97***	Chi2 (18) = 145.36***
Pesaran's test	2.981***	2.798***
Wooldridge test	*F*(1, 17) = 19.484***	*F*(1, 17) = 19.484***

*** denotes statistical significance at the 1% level; H0 of modified Wald test: sigma(*i*)^2 = sigma^2 for all *I*; H0 of Pesaran's test: residuals are not correlated; H0 of Wooldridge test: no first-order autocorrelation.

How to do:

STATA:

****Transform the variables y, kpub, and kpriv in nominal values (n) into real values (r)****

```
gen y_r=y_n/(cpi/100)
gen kpub_r=kpub_n/(cpi/100)
gen kpriv_r=kpriv_n/(cpi/100)
```

****Transform the variables kpub and kpriv into % of the GDP****

```
gen kpub=(kpub_r/y_r)*100
gen kpriv=(kpriv_r/y_r)*100
```

****Transform the variables into natural logarithms****

```
gen lineq=ln(ineq)
gen ly=ln(y_r)
gen lkpub=ln(kpub)
gen lkpriv=ln(kpriv)
gen lhdi=ln(hdi)
gen ltr=ln(tr)
gen ltrd=ln(trd)
gen lunp=ln(unp)
```

****Transform the variables into first differences of logarithms****

```
gen dlineq=d.lineq
gen dly=d.ly
gen dlkpub=d.lkpub
gen dlkpriv=d.lkpriv
gen dlhdi=d.lhdi
gen dltr=d.ltr
gen dltrd=d.ltrd
gen dlunp=d.lunp
```

****Transform the variables into natural logarithms lagged once****

```
gen l_lineq=l.lineq
gen l_ly=l.ly
gen l_lkpub=l.lkpub
gen l_lkpriv=l.lkpriv
gen l_lhdi=l.lhdi
gen l_ltr=l.ltr
gen l_ltrd=l.ltrd
gen l_lunp=l.lunp
```

****Descriptive statistics****

```
sum lineq lkpub lkpriv ly lhdi ltrd ltr lunp
sum dlineq dlkpub dlkpriv dly dlhdi dltrd dltr dlunp
```

****Cross section dependence (CD) test****

```
xtcd lineq lkpub lkpriv ly lhdi ltrd
xtcd ltr lunp
xtcd dlineq dlkpub dlkpriv dly dlhdi dltrd
xtcd dltr dlunp
```

Panel unit root test (CIPS)

```
multipurt lineq lkpub lkpriv ly lhdi ltrd, lags(3)
multipurt ltr lunp, lags(3)
multipurt dlineq dlkpub dlkpriv dly dlhdi dltrd, lags(3)
multipurt dltr dlunp, lags(3)
```

Correlation matrices

Model I

```
pwcorr lineq lkpub ly lhdi ltrd ltr lunp
pwcorr dlineq dlkpub dly dlhdi dltrd dltr dlunp
```

Model II

```
pwcorr lineq lkpriv ly lhdi ltrd ltr lunp
pwcorr dlineq dlkpriv dly dlhdi dltrd dltr dlunp
```

VIF statistics

Model I

```
qui: reg lineq lkpub ly lhdi ltrd ltr lunp
estat vif
qui: reg dlineq dlkpub dly dlhdi dltrd dltr dlunp
estat vif
```

Model II

```
qui: reg lineq lkpriv ly lhdi ltrd ltr lunp
estat vif
qui: reg dlineq dlkpriv dly dlhdi dltrd dltr dlunp
estat vif
```

Hausman test (FE vs. RE)

Model I

```
qui: xtreg dlineq dlkpub dly dlhdi dltrd dltr dlunp l_lineq l_lkpub l_ly l_lhdi l_ltrd l_ltr l_lunp, fe
estimates store fixed
qui: xtreg dlineq dlkpub dly dlhdi dltrd dltr dlunp l_lineq l_lkpub l_ly l_lhdi l_ltrd l_ltr l_lunp, re
estimates store random
hausman fixed random
hausman fixed random, sigmamore
hausman fixed random, sigmaless
```

Model II

```
qui: xtreg dlineq dlkpriv dly dlhdi dltrd dltr dlunp l_lineq l_lkpriv l_ly l_lhdi l_ltrd l_ltr l_lunp, fe
estimates store fixed
qui: xtreg dlineq dlkpriv dly dlhdi dltrd dltr dlunp l_lineq l_lkpriv l_ly l_lhdi l_ltrd l_ltr l_lunp, re
estimates store random
hausman fixed random
hausman fixed random, sigmamore
hausman fixed random, sigmaless
```

Hausman test (MG vs. PMG vs. Pooled)

Model I

```
qui: xtdcce2 dlineq dlkpub dly dlhdi dltrd dltr dlunp, lr(l.lineq l.lkpub l.ly l.lhdi l.ltrd l.ltr l.lunp) nocross
estimates store mg
```

```
qui: xtdcce2 dlineq dlkpub dly dlhdi dltrd dltr dlunp, lr(l.lineq l.lkpub l.ly l.lhdi l.ltrd l.ltr l.lunp) p(l.lineq l.lkpub l.ly l.lhdi l.ltrd
l.ltr l.lunp) nocross
estimates store pmg
qui: xtdcce2 dlineq dlkpub dly dlhdi dltrd dltr dlunp, lr(l.lineq l.lkpub l.ly l.lhdi l.ltrd l.ltr l.lunp) p(l.lineq l.lkpub l.ly l.lhdi
l.ltrd l.ltr l.lunp dlkpub dly dlhdi dltrd dltr dlunp) nocross
estimates store pooled
hausman mg pmg, sigmamore
hausman pmg pooled, sigmamore
hausman mg pooled, sigmamore
```

Model II

```
qui: xtdcce2 dlineq dlkpriv dly dlhdi dltrd dltr dlunp, lr(l.lineq l.lkpriv l.ly l.lhdi l.ltrd l.ltr l.lunp) nocross
estimates store mg
qui: xtdcce2 dlineq dlkpriv dly dlhdi dltrd dltr dlunp, lr(l.lineq l.lkpriv l.ly l.lhdi l.ltrd l.ltr l.lunp) p(l.lineq l.lkpriv l.ly l.lhdi
l.ltrd l.ltr l.lunp) nocross
estimates store pmg
qui: xtdcce2 dlineq dlkpriv dly dlhdi dltrd dltr dlunp, lr(l.lineq l.lkpriv l.ly l.lhdi l.ltrd l.ltr l.lunp) p(l.lineq l.lkpriv l.ly l.lhdi
l.ltrd l.ltr l.lunp dlkpriv dly dlhdi dltrd dltr dlunp) nocross
estimates store pooled
hausman mg pmg, sigmamore
hausman pmg pooled, sigmamore
hausman mg pooled, sigmamore
```

****Modified Wald test****

Model I

```
qui: xtreg dlineq dlkpub dly dlhdi dltrd dltr dlunp l_lineq l_lkpub l_ly l_lhdi l_ltrd l_ltr l_lunp, fe
xttest3
```

Model II

```
qui: xtreg dlineq dlkpriv dly dlhdi dltrd dltr dlunp l_lineq l_lkpriv l_ly l_lhdi l_ltrd l_ltr l_lunp, fe
xttest3
```

****Pesaran test of cross-sectional independence****

Model I

```
qui: xtreg dlineq dlkpub dly dlhdi dltrd dltr dlunp l_lineq l_lkpub l_ly l_lhdi l_ltrd l_ltr l_lunp, fe
xtcsd, pesaran abs
```

Model II

```
qui: xtreg dlineq dlkpriv dly dlhdi dltrd dltr dlunp l_lineq l_lkpriv l_ly l_lhdi l_ltrd l_ltr l_lunp, fe
xtcsd, pesaran abs
```

****Wooldridge test****

Model I

```
xtserial dlineq dlkpub dly dlhdi dltrd dltr dlunp l_lineq l_lkpub l_ly l_lhdi l_ltrd l_ltr l_lunp
```

Model II

```
xtserial dlineq dlkpriv dly dlhdi dltrd dltr dlunp l_lineq l_lkpriv l_ly l_lhdi l_ltrd l_ltr l_lunp
```

****PARDL estimation with the Driscoll and Kraay estimator ****

Model I

```
xtscc dlineq dlkpub dly dlhdi dltrd dltr dlunp l_lineq l_lkpub l_ly l_lhdi l_ltrd l_ltr l_lunp, fe
```

Model II

```
xtscc dlineq dlkpriv dly dlhdi dltrd dltr dlunp l_lineq l_lkpriv l_ly l_lhdi l_ltrd l_ltr l_lunp, fe
```

Long-run elasticities

Model I

```
qui:xtscc dlineq dlkpub dly dlhdi dltrd dltr dlunp l_lineq l_lkpub l_ly l_lhdi l_ltrd l_ltr l_lunp, fe
nlcom (ratio1: -_b[l_lkpub]/_b[l_lineq])
nlcom (ratio1: -_b[l_ly]/_b[l_lineq])
nlcom (ratio1: -_b[l_lhdi]/_b[l_lineq])
nlcom (ratio1: -_b[l_ltrd]/_b[l_lineq])
nlcom (ratio1: -_b[l_ltr]/_b[l_lineq])
nlcom (ratio1: -_b[l_lunp]/_b[l_lineq])
```

Model II

```
qui:xtscc dlineq dlkpriv dly dlhdi dltrd dltr dlunp l_lineq l_lkpriv l_ly l_lhdi l_ltrd l_ltr l_lunp, fe
nlcom (ratio1: -_b[l_lkpriv]/_b[l_lineq])
nlcom (ratio1: -_b[l_ly]/_b[l_lineq])
nlcom (ratio1: -_b[l_lhdi]/_b[l_lineq])
nlcom (ratio1: -_b[l_ltrd]/_b[l_lineq])
nlcom (ratio1: -_b[l_ltr]/_b[l_lineq])
nlcom (ratio1: -_b[l_lunp]/_b[l_lineq])
```

Hausman test (FE vs. RE) (parsimonious)

Model I

```
qui:xtreg dlineq dlkpub dly dlunp l_lineq l_ly l_lhdi l_ltrd l_ltr l_lunp, fe
estimates store fixed
qui:xtreg dlineq dlkpub dly dlunp l_lineq l_ly l_lhdi l_ltrd l_ltr l_lunp, re
estimates store random
hausman fixed random
hausman fixed random, sigmamore
hausman fixed random, sigmaless
```

Model II

```
qui:xtreg dlineq dlkpriv dly dlunp l_lineq l_ly l_lhdi l_ltrd l_ltr l_lunp, fe
estimates store fixed
qui:xtreg dlineq dlkpriv dly dlunp l_lineq l_ly l_lhdi l_ltrd l_ltr l_lunp, re
estimates store random
hausman fixed random
hausman fixed random, sigmamore
hausman fixed random, sigmaless
```

Hausman test (MG vs. PMG vs. Pooled) (parsimonious)

Model I

```
qui: xtdcce2 dlineq dlkpub dly dlunp, lr(l.lneqi l.ly l.lhdi l.ltrd l.ltr l.lunp) nocross
estimates store mg
qui: xtdcce2 dlineq dlkpub dly dlunp, lr(l.lineq l.ly l.lhdi l.ltrd l.ltr l.lunp) p(l.lineq l.ly l.lhdi l.ltrd l.ltr l.lunp) nocross
estimates store pmg
qui: xtdcce2 dlineq dlkpub dly dlunp, lr(l.lineq l.ly l.lhdi l.ltrd l.ltr l.lunp) p(l.lineq l.ly l.lhdi l.ltrd l.ltr l.lunp dlkpub dly dlunp) nocross
estimates store pooled
hausman mg pmg, sigmamore
hausman pmg pooled, sigmamore
hausman mg pooled, sigmamore
```

Model II

```
qui: xtdcce2 dlineq dlkpriv dly dlunp, lr(l.lineq l.ly l.lhdi l.ltrd l.ltr l.lunp) nocross
estimates store mg
qui: xtdcce2 dlineq dlkpriv dly dlunp, lr(l.lineq l.ly l.lhdi l.ltrd l.ltr l.lunp) p(l.lineq l.ly l.lhdi l.ltrd l.ltr l.lunp) nocross
estimates store pmg
qui: xtdcce2 dlineq dlkpriv dly dlunp, lr(l.lineq l.ly l.lhdi l.ltrd l.ltr l.lunp) p(l.lineq l.ly l.lhdi l.ltrd l.ltr l.lunp dlkpriv dly dlunp) nocross
estimates store pooled
hausman mg pmg, sigmamore
hausman pmg pooled, sigmamore
hausman mg pooled, sigmamore
```

****Modified Wald test (parsimonious)****

Model I

```
qui:xtreg dlineq dlkpub dly dlunp l_lineq l_ly l_lhdi l_ltrd l_ltr l_lunp, fe
xttest3
```

Model II

```
qui:xtreg dlineq dlkpriv dly dlunp l_lineq l_ly l_lhdi l_ltrd l_ltr l_lunp, fe
xttest3
```

****Pesaran test of cross-sectional independence (parsimonious)****

Model I

```
qui:xtreg dlineq dlkpub dly dlunp l_lineq l_ly l_lhdi l_ltrd l_ltr l_lunp, fe
xtcsd, pesaran abs
```

Model II

```
qui:xtreg dlineq dlkpriv dly dlunp l_lineq l_ly l_lhdi l_ltrd l_ltr l_lunp, fe
xtcsd, pesaran abs
```

****Wooldridge test (parsimonious)****

Model I

```
xtserial dlineq dlkpub dly dlunp l_lineq l_ly l_lhdi l_ltrd l_ltr l_lunp
```

Model II

```
xtserial dlineq dlkpriv dly dlunp l_lineq l_ly l_lhdi l_ltrd l_ltr l_lunp
```

****PARDL estimation with the Driscoll and Kraay estimator (parsimonious)****

Model I

```
xtscc dlineq dlkpub dly dlunp l_lineq l_ly l_lhdi l_ltrd l_ltr l_lunp, fe
```

Model II

```
xtscc dlineq dlkpriv dly dlunp l_lineq l_ly l_lhdi l_ltrd l_ltr l_lunp, fe
```

****Long-run elasticities (parsimonious)****

Model I

```
qui: xtscc dlineq dlkpub dly dlunp l_lineq l_ly l_lhdi l_ltrd l_ltr l_lunp, fe
nlcom (ratio1: -_b[l_ly]/_b[l_lineq])
nlcom (ratio1: -_b[l_lhdi]/_b[l_lineq])
nlcom (ratio1: -_b[l_ltrd]/_b[l_lineq])
nlcom (ratio1: -_b[l_ltr]/_b[l_lineq])
nlcom (ratio1: -_b[l_lunp]/_b[l_lineq])
```

Model II

```
xtscc dlineq dlkpriv dly dlunp l_lineq l_ly l_lhdi l_ltrd l_ltr l_lunp, fe
nlcom (ratio1: -_b[l_ly]/_b[l_lineq])
nlcom (ratio1: -_b[l_lhdi]/_b[l_lineq])
nlcom (ratio1: -_b[l_ltrd]/_b[l_lineq])
nlcom (ratio1: -_b[l_ltr]/_b[l_lineq])
nlcom (ratio1: -_b[l_lunp]/_b[l_lineq])
```

**** Residuals analysis****

Model I

```
qui:xtreg dlineq dlkpub dly dlhdi dltrd dltr dlunp l_lineq l_lkpub l_ly l_lhdi l_ltrd l_ltr l_lunp, fe
predict double resid, e
summarize resid
xtcd resid
xtline resid, overlay
```

Model II

```
qui:xtreg dlineq dlkpriv dly dlhdi dltrd dltr dlunp l_lineq l_lkpriv l_ly l_lhdi l_ltrd l_ltr l_lunp, fe
predict double resid, e
summarize resid
xtcd resid
xtline resid, overlay
```

**** Residuals analysis (parsimonious)****

Model I

```
qui:xtreg dlineq dlkpub dly dlunp l_lineq l_ly l_lhdi l_ltrd l_ltr l_lunp, fe
predict double resid, e
summarize resid
xtcd resid
xtline resid, overlay
```

Model II

```
qui:xtreg dlineq dlkpriv dly dlunp l_lineq l_ly l_lhdi l_ltrd l_ltr l_lunp, fe
predict double resid, e
summarize resid
xtcd resid
xtline resid, overlay
```

****Generate dummy variables****

```
gen bra2016=0
replace bra2016=1 if id==3 & year==2016
gen gtm2013=0
replace gtm2013=1 if id==10 & year==2013
gen gtm2014=0
replace gtm2014=1 if id==10 & year==2014
gen pry2004=0
replace pry2004=1 if id==15 & year==2004
gen ury2010=0
replace ury2010=1 if id==17 & year==2010
gen ury2011=0
replace ury2011=1 if id==17 & year==2011
gen ury2012=0
replace ury2012=1 if id==17 & year==2012
```

PARDL estimation with the Driscoll and Kraay estimator (corrected for outliers)

Model I

```
xtscc dlineq dlkpub dly dlhdi dltrd dltr dlunp l_lineq l_lkpub l_ly l_lhdi l_ltrd l_ltr l_lunp bra2016 gtm2013 gtm2014 pry2004 ury2010 ury2011 ury2012, fe
```

Model II

```
xtscc dlineq dlkpriv dly dlhdi dltrd dltr dlunp l_lineq l_lkpriv l_ly l_lhdi l_ltrd l_ltr l_lunp bra2016 gtm2013 gtm2014 pry2004 ury2010 ury2011 ury2012, fe
```

Long-run elasticities (corrected for outliers)

Model I

```
qui:xtscc dlineq dlkpub dly dlhdi dltrd dltr dlunp l_lineq l_lkpub l_ly l_lhdi l_ltrd l_ltr l_lunp bra2016 gtm2013 gtm2014 pry2004 ury2010 ury2011 ury2012, fe
nlcom (ratio1: -_b[l_lkpub]/_b[l_lineq])
nlcom (ratio1: -_b[l_ly]/_b[l_lineq])
nlcom (ratio1: -_b[l_lhdi]/_b[l_lineq])
nlcom (ratio1: -_b[l_ltrd]/_b[l_lineq])
nlcom (ratio1: -_b[l_ltr]/_b[l_lineq])
nlcom (ratio1: -_b[l_lunp]/_b[l_lineq])
```

Model II

```
qui:xtscc dlineq dlkpriv dly dlhdi dltrd dltr dlunp l_lineq l_lkpriv l_ly l_lhdi l_ltrd l_ltr l_lunp bra2016 gtm2013 gtm2014 pry2004 ury2010 ury2011 ury2012, fe
nlcom (ratio1: -_b[l_lkpriv]/_b[l_lineq])
nlcom (ratio1: -_b[l_ly]/_b[l_lineq])
nlcom (ratio1: -_b[l_lhdi]/_b[l_lineq])
nlcom (ratio1: -_b[l_ltrd]/_b[l_lineq])
nlcom (ratio1: -_b[l_ltr]/_b[l_lineq])
nlcom (ratio1: -_b[l_lunp]/_b[l_lineq])
```

PARDL estimation with the Driscoll and Kraay estimator (parsimonious) (corrected for outliers)

Model I

```
xtscc dlineq dlkpub dly dlunp l_lineq l_ly l_lhdi l_ltrd l_ltr l_lunp bra2016 gtm2013 gtm2014 pry2004 ury2010 ury2011 ury2012, fe
```

Model II

```
xtscc dlineq dlkpriv dly dlunp l_lineq l_ly l_lhdi l_ltrd l_ltr l_lunp bra2016 gtm2013 gtm2014 pry2004 ury2010 ury2011 ury2012, fe
```

Long-run elasticities (parsimonious) (corrected for outliers)

Model I

```
qui:xtscc dlineq dlkpub dly dlunp l_lineq l_ly l_lhdi l_ltrd l_ltr l_lunp bra2016 gtm2013 gtm2014 pry2004 ury2010 ury2011 ury2012, fe
nlcom (ratio1: -_b[l_ly]/_b[l_lineq])
nlcom (ratio1: -_b[l_lhdi]/_b[l_lineq])
nlcom (ratio1: -_b[l_ltrd]/_b[l_lineq])
nlcom (ratio1: -_b[l_ltr]/_b[l_lineq])
nlcom (ratio1: -_b[l_lunp]/_b[l_lineq])
```

Model II

```
qui:xtscc dlineq dlkpriv dly dlunp l_lineq l_ly l_lhdi l_ltrd l_ltr l_lunp bra2016 gtm2013 gtm2014 pry2004 ury2010 ury2011 ury2012, fe
nlcom (ratio1: -_b[l_ly]/_b[l_lineq])
nlcom (ratio1: -_b[l_lhdi]/_b[l_lineq])
nlcom (ratio1: -_b[l_ltrd]/_b[l_lineq])
nlcom (ratio1: -_b[l_ltr]/_b[l_lineq])
nlcom (ratio1: -_b[l_lunp]/_b[l_lineq])
```

References

Agenor, P.R., Moreno-Dodson, B., 2006. Public Infrastructure and Growth: New Channels and Policy Implications. Policy Research Working Paper No. 4064, World Bank, Washington, DC, https://doi.org/10.1596/1813-9450-4064.

Anderson, E., Jalles D'Orey, M.A., Duvendack, M., Esposito, L., 2017. Does government spending affect income inequality? A meta-regression analysis. J. Econ. Surv. 31 (4), 961–987. https://doi.org/10.1111/joes.12173.

Artadi, E., Sala-i-Martin, X., 2003. The Economic Tragedy of the XXth Century Growth in Africa. NBER Working Papers No. 9865, National Bureau of Economic Research, Inc., https://doi.org/10.3386/w9865.

Balseven, H., Tugcu, C.T., 2017. Analyzing the effects of fiscal policy on income distribution: a comparison between developed and developing countries. Int. J. Econ. Financ. Issues 7 (2), 377–383. Available at: https://www.econjournals.com/index.php/ijefi/article/view/4235.

Belsley, D.A., Kuh, E., Welsch, R.E., 1980. Regression Diagnostics: Identifying Influential Data and Sources of Collinearity. Wiley, New York, https://doi.org/10.1002/0471725153.

Bom, P.R.D., Goti, A., 2018. Public capital and the labor income share. Sustainability 10 (11), 3895. https://doi.org/10.3390/su10113895.

Calderón, C., Chong, A., 2004. Volume and quality of infrastructure and the distribution of income: An empirical investigation. Rev. Income Wealth 50, 87–106. https://doi.org/10.1111/j.0034-6586.2004.00113.x.

Calderón, C., Servén, L., 2004. The Effects Of Infrastructure Development On Growth And Income Distribution. Policy Research Working Paper No. 3400, World Bank, Washington, DC, https://doi.org/10.1596/1813-9450-3400.

Calderón, C., Servén, L., 2014. Infrastructure, Growth, and Inequality: An Overview. Policy Research Working Paper No. 7034, World Bank, Washington, DC, https://doi.org/10.1596/1813-9450-7034.

Cavallo, E., Powell, A., 2019. Building opportunities for growth in a challenging world. In: 2019 Latin American and Caribbean Macroeconomic Report. Inter-American Development Bank, Washington, DC, https://doi.org/10.18235/0001633.

Cavallo, E., Powell, A., Serebrisky, T., 2020. From Structures to Services: The Path to Better Infrastructure in Latin America and the Caribbean. Inter-American Development Bank, Washington, DC, https://doi.org/10.18235/0002506.

Cerdeiro, D., Komaromi, A., 2017. Trade and Income in the Long Run: Are There Really Gains, and are They Widely Shared? IMF Working Papers No. 17/231, International Monetary Fund, Washington, DC, https://doi.org/10.5089/9781484324851.001.

Chatterjee, S., Turnovsky, S.J., 2012. Infrastructure and inequality. Eur. Econ. Rev. 56 (8), 1730–1745. https://doi.org/10.1016/j.euroecorev.2012.08.003.

Cingano, F., 2014. Trends in Income Inequality and its Impact on Economic Growth. OECD Social, Employment and Migration Working Papers No. 163, OECD Publishing, https://doi.org/10.1787/5jxrjncwxv6j-en.

Coady, D., Dizioli, A., 2018. Income inequality and education revisited: persistence, endogeneity and heterogeneity. Appl. Econ. 50 (25), 2747–2761. https://doi.org/10.1080/00036846.2017.1406659.

Cornia, G., 2011. Economic Integration, Inequality and Growth: Latin America vs. the European Economies in Transition. UN Department of Economic and Social Affairs (DESA) Working Papers No. 101, UN, New York, https://doi.org/10.18356/a6a4730a-en.

Cysne, R.P., 2009. On the positive correlation between income inequality and unemployment. Rev. Econ. Stat. 91 (1), 218–226. Available at: https://www.jstor.org/stable/25651330.

De Jong, J.F.M., Ferdinandusse, M., Funda, J., 2018. Public capital in the 21st century: as productive as ever? Appl. Econ. 50 (51), 5543–5560. https://doi.org/10.1080/00036846.2018.1487002.

Ditzen, J., 2018. Estimating dynamic common-correlated effects in stata. Stata J. 18 (3), 585–617. https://doi.org/10.1177/1536867X1801800306.

Driscoll, J.C., Kraay, A.C., 1998. Consistent covariance matrix estimation with spatially dependent panel data. Rev. Econ. Stat. 80 (4), 549–560. Available at: https://www.jstor.org/stable/2646837.

Easterly, W., Servén, L., 2003. The Limits of Stabilization: Infrastructure, Public Deficits, and Growth in Latin America. Latin American Development Forum, World Bank, Washington, DC, https://doi.org/10.1596/978-0-8213-5489-6.

Estache, A., 2003. On Latin America's Infrastructure Privatization and its Distributional Effects. Washington, DC, Mimeo, World Bank, https://doi.org/10.2139/ssrn.411942.

Estache, A., Foster, V., Wodon, Q., 2002. Accounting for Poverty in Infrastructure Reform: Learning from Latin America's Experience. World Bank Institute Development Studies, World Bank, Washington, DC, https://doi.org/10.1596/0-8213-5039-0.

Estache, A., Gomez-Lobo, A., Leipziger, D., 2000. Utility Privatization and the Needs of the Poor in Latin America—Have We Learned Enough to get it Right? Policy Research Working Paper No. 2407, World Bank, Washington, DC, https://doi.org/10.1596/1813-9450-2407.

Faruqee, H., 2016. Regional Economic Outlook, April 2016, Western Hemisphere Department: Managing Transitions and Risks. International Monetary Fund, Washington, DC, https://doi.org/10.5089/9781498329996.086.

Fay, M., Andres, L.A., Fox, C., Narloch, U., Straub, S., Slawson, M., 2017. Rethinking Infrastructure in Latin America and the Caribbean: Spending Better to Achieve More. Directions in Development-Infrastructure. World Bank, Washington, DC, https://doi.org/10.1596/978-1-4648-1101-2.

Ferreira, F.H.G., 1995. Roads to Equality: Wealth Distribution Dynamics With Public-Private Capital Complementarity. LSE STICERD Research Paper No. TE286. Available at: https://ssrn.com/abstract=1160952.

Fuinhas, J.A., Marques, A.C., Koengkan, M., 2017. Are renewable energy policies upsetting carbon dioxide emissions? The case of Latin America countries. Environ. Sci. Pollut. Res. 24 (17), 15044–15054. https://doi.org/10.1007/s11356-017-9109-z.

Furceri, D., Li, B.G., 2017. The Macroeconomic (and Distributional) Effects of Public Investment in Developing Economies. IMF Working Papers No. 17/217, International Monetary Fund, Washington, DC, https://doi.org/10.5089/9781484320709.001.

Gasparini, L., Lustig, N., 2011. The Rise and Fall of Income Inequality in Latin America, in The Oxford Handbook of Latin American Economics. Oxford University Press, Oxford, https://doi.org/10.1093/oxfordhb/9780199571048.013.0027.

Greene, W.H., 2002. Econometric Analysis, fifth ed. Prentice Hall, Upper Saddle River, ISBN: 978-0130661890.

Hacibedel, B., Mandon, P., Muthoora, P., Pouokam, N., 2019. Inequality in Good and Bad Times: A Cross-Country Approach. IMF Working Papers No. 19/20, International Monetary Fund, Washington, DC, https://doi.org/10.5089/9781484392911.001.

Hausman, J.A., 1978. Specification tests in econometrics. Econometrica 46 (6), 1251–1271. https://doi.org/10.2307/1913827.
Helpman, E., Oleg, I., Redding, S., 2010. Inequality and unemployment in a global economy. Econometrica 78 (4), 1239–1283. https://doi.org/10.3982/ECTA8640.
IMF, 2014. Is it time for an infrastructure push? The macroeconomic effects of public investment. In: World Economic Outlook, October 2014: Legacies, Clouds, Uncertainties. Research Department, International Monetary Fund, Washington, DC, https://doi.org/10.5089/9781498331555.081 (Chapter 3).
IMF, 2017. Estimating the Stock of Public Capital in 170 Countries. Fiscal Affairs Department, International Monetary Fund, Washington, DC. Available at: https://www.imf.org/external/np/fad/publicinvestment/pdf/csupdate_aug19.pdf.
Jouini, J., 2015. Economic growth and remittances in Tunisia: bi-directional causal links. J. Policy Model 37 (2), 355–373. https://doi.org/10.1016/j.jpolmod.2015.01.015.
Lardé, J., Sánchez, R., 2014. The Economic Infrastructure Gap and Investment in Latin America. FAL Bulletin No. 332, Economic Commission for Latin America and the Caribbean (ECLAC), Santiago. Available at: http://hdl.handle.net/11362/37381.
Lopez, H., 2003. Macroeconomics and Inequality. World Bank, Washington, DC. Available at: http://documents.worldbank.org/curated/en/292721468319775386/Macroeconomics-and-inequality.
Martínez-Vázquez, J., Vulovic, V., Moreno-Dodson, B., 2012. The impact of tax and expenditure policies on income distribution: evidence from a large panel of countries. Hacienda Publ. Espanola 200 (1), 95–130. https://doi.org/10.2139/ssrn.2188608.
Martorano, B., 2018. Taxation and inequality in developing countries: lessons from the recent experience of Latin America. J. Int. Dev. 30 (2), 256–273. https://doi.org/10.1002/jid.3350.
Meschi, E., Vivarelli, M., 2009. Trade and income inequality in developing countries. World Dev. 37 (2), 287–302. https://doi.org/10.1016/j.worlddev.2008.06.002.
Mocan, H.N., 1999. Structural unemployment, cyclical unemployment, and income inequality. Rev. Econ. Stat. 5, 122–134. https://doi.org/10.1162/003465399767923872.
OECD, 2015. In It Together: Why Less Inequality Benefits All. OECD Publishing, Paris, https://doi.org/10.1787/9789264235120-en.
OECD, 2016. Promoting Productivity for Inclusive Growth in Latin America. Better Policies. OECD Publishing, Paris, https://doi.org/10.1787/9789264258389-en.
OECD, 2019. Latin American Economic Outlook 2019: Development in Transition. OECD Publishing, Paris, https://doi.org/10.1787/g2g9ff18-en.
Pérez, G., 2020. Rural Roads: Key Routes for Production, Connectivity and Territorial Development. FAL Bulletin No. 377, Economic Commission for Latin America and the Caribbean (ECLAC), Santiago. Available at: http://hdl.handle.net/11362/45865.
Perrotti, D., 2011. The Economic Infrastructure Gap in Latin America and the Caribbean. FAL Bulletin No. 293, Economic Commission for Latin America and the Caribbean (ECLAC), Santiago. Available at: http://hdl.handle.net/11362/36339.
Pesaran, M.H., 2004. General Diagnostic Tests for Cross Section Dependence in Panels. Cambridge Working Papers in Economics No. 435, Faculty of Economics, University of Cambridge. https://doi.org/10.17863/CAM.5113.
Pesaran, M.H., 2007. A simple panel unit root test in the presence of cross-section dependence. J. Appl. Econ. 22 (2), 265–312. https://doi.org/10.1002/jae.951.
Pesaran, M., Shin, Y., Smith, R., 1999. Pooled mean group estimation of dynamic heterogeneous panels. J. Am. Stat. Assoc. 94 (446), 621–634. https://doi.org/10.1080/01621459.1999.10474156.
Pi, J., Zhou, Y., 2012. Public infrastructure provision and skilled-unskilled wage inequality in developing countries. Labour Econ. 19 (6), 881–887. https://doi.org/10.1016/j.labeco.2012.08.007.
Piketty, T., 2013. Le capital au XXIe siècle. Editions du Seuil, Paris, ISBN: 978-2021082289.
Romp, W., De Haan, J., 2007. Public capital and economic growth: a critical survey. Perspekt. Wirtsch. 8, 6–52. https://doi.org/10.1111/j.1468-2516.2007.00242.x.
Rubin, A., Segal, D., 2015. The effects of economic growth on income inequality in the US. J. Macroecon. 45, 258–273. https://doi.org/10.1016/j.jmacro.2015.05.007.
Santiago, R., Fuinhas, J.A., Marques, A.C., 2020. The impact of globalization and economic freedom on economic growth: the case of the Latin America and Caribbean countries. Econ. Chang. Restruct. 53, 61–85. https://doi.org/10.1007/s10644-018-9239-4.
Sheng, Y., 2011. Unemployment and Income Inequality: A Puzzling Finding From the US in 1941–2010., https://doi.org/10.2139/ssrn.2020744.
Solt, F., 2020. Measuring income inequality across countries and over time: the standardized world income inequality database. Soc. Sci. Q. 101, 1183–1199. https://doi.org/10.1111/ssqu.12795.
Székely, M., Sámano-Robles, C., 2014. Trade and income distribution in Latin America: is there anything new to say? In: Cornia, A. (Ed.), Falling Inequality in Latin America. Policy Changes and Lessons. Oxford University Press, Oxford, https://doi.org/10.1093/acprof:oso/9780198701804.003.0011.
Theyson, K., Heller, L., 2015. Development and income inequality: a new specification of the Kuznets hypothesis. J. Dev. Areas 49, 103–118. https://doi.org/10.1353/jda.2015.0153.
Tsounta, E., Osueke, A., 2014. What is Behind Latin America's Declining Income Inequality? IMF Working Papers No. 14/124, International Monetary Fund, Washington, DC, https://doi.org/10.5089/9781498378581.001.
Turnovsky, S.J., 2015. Economic growth and inequality: the role of public investment. J. Econ. Dyn. Control. 61, 204–221. https://doi.org/10.1016/j.jedc.2015.09.009.
UN, 2020. Inequality: Where We Stand Today, in World Social Report 2020: Inequality in a Rapidly Changing World. UN, New York, https://doi.org/10.18356/0de44c44-en.
Wooldridge, J.M., 2002. Econometric Analysis of Cross Section and Panel Data. The MIT Press Cambridge, MA, ISBN: 978-0262232586.
Yang, Y., Greaney, T.M., 2017. Economic growth and income inequality in the Asia-Pacific region: a comparative study of China, Japan, South Korea, and the United States. J. Asian Econ. 48, 6–22. https://doi.org/10.1016/j.asieco.2016.10.008.

CHAPTER

4

The downward trend in the energy intensity of Latin America and the Caribbean: Is the region's physical capital contributing to this tendency?

JEL codes Q43, E22, F21, O54

4.1 Introduction

As is known, the fight against environmental degradation and climate change have become central points in the agendas of the governments worldwide in the last few decades. The increased worries related to the adverse effects that these issues can produce on people's lives, assets, on economies and ecosystems, as well as the idea that there is no "planet B", have led governments and international institutions to search for more sustainable development strategies. An example of this is the well-known "Sustainable Development Goals (SDGs)" included in the "Transforming our world: the 2030 Agenda for Sustainable Development" (see UN General Assembly, 2015). These goals focused on the promotion of social, economic, and environmental sustainability until 2030 and were adopted by the United Nations Member States in 2015.

However, although the efforts that are being (and have been) made in order to achieve a more sustainable and environmentally friendly economic organisation, the truth is that worldwide production still needs to use plentiful resources such as energy, materials, and land to keep up with worldwide consumption. With the increase in the world population and the worldwide living standards, this could lead to higher levels of resource depletion (Balatsky et al., 2015). Following the UNEP (2011), the worldwide per capita resource is expected to double in 2050 compared to 2000, from 8 to 16 tonnes. So, in order to achieve sustainable consumption and production, the world needs to make an effort to promote efficient use of resources; that is, it needs *"to create more with less, delivering greater value with less input, using resources in a sustainable way and minimising their impacts on the environment"* (European Commission, 2011, p. 3).

Among the various types of resources used in production, energy is, undoubtedly, one of the most important, especially given the fact that it is present in the various stages of this process. Although the fact that energy use is essential for countries' development, we should not disregard the fact that unsustainable consumption of it could lead to serious environmental problems and to energy security concerns which could ultimately lead countries into extremely difficult environmental, social, and economic situations. As Koengkan et al. (2019, p. 44) state *"the increased worries with energy demand, energy security, and climate change have led the governments to raise their concerns regarding how energy is used"*. Apart from the improvement of the energy structure with constant support in the deployment of renewables, another important action for the increased sustainable energy consumption is to improve energy efficiency that is use less energy to produce the same output.

Looking at the Latin America and the Caribbean (LAC) region, it can be seen that regional energy consumption has more than tripled since the 1970s (Balza et al., 2016). This trend can be justified by the fact that these countries needed to expand their energy use in order to support their growth strategies, with the industrial and transport sectors accounting for *"about three-quarters of the overall growth in total final consumption since 1971"* (Balza et al., 2016, p. 15). Despite these assumptions, the report "Lights on? Energy Needs in Latin America and the Caribbean to 2040" also states that LAC energy intensity has been in decline for the past four decades (see Fig. 4.1). This may well be a sign that the LAC region has been making progress with its energy efficiency, given that *"energy intensity is a measure that is often used to*

https://doi.org/10.1016/B978-0-12-824429-6.00004-8

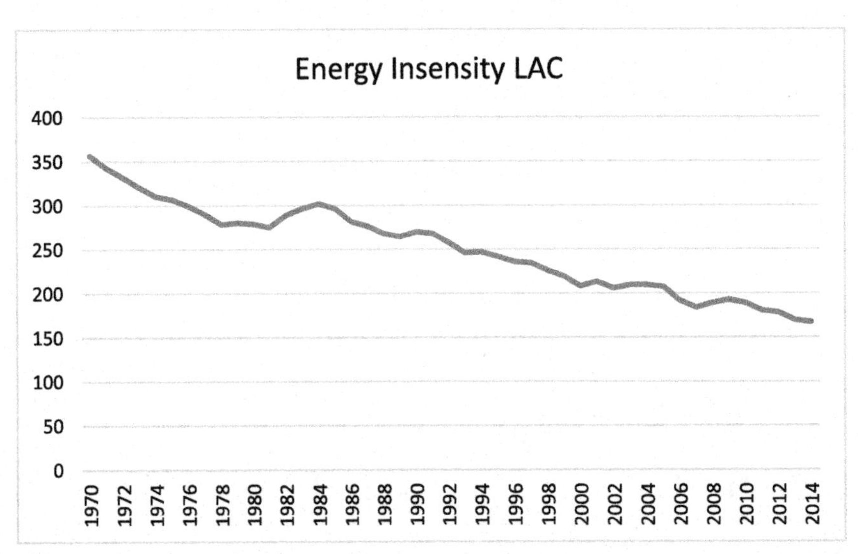

FIG. 4.1 Mean LAC energy intensity (EI_LAC) from 1970 to 2014 (in thousands of barrels of oil equivalent per billions of constant 2011 international dollars); this graph was created by the authors using data on primary energy consumption (in thousands of barrels of oil equivalent) from CEPALSTAT (https://estadisticas.cepal.org/cepalstat/Portada.html) and on the gross domestic product (in billions of constant 2011 international dollars) from the "Investment and Capital Stock Dataset" by the IMF (2017) for Argentina, Barbados, Bolivia, Brazil, Chile, Colombia, Costa Rica, the Dominican Republic, Ecuador, El Salvador, Grenada, Guatemala, Haiti, Honduras, Mexico, Nicaragua, Panama, Paraguay, Peru, Uruguay, and Venezuela.

assess the energy efficiency of a particular economy" (Martínez et al., 2019, p. 2), representing the capacity to convert energy into monetary output.

Nevertheless, given the fact that capital and energy are inherently connected (buildings, vehicles, machines, tools, and other types of physical capital require energy to produce goods and services), it is of especial interest to analyse if LAC public and private capital stocks are among the factors that have contributed to the decreasing trend of the region's energy intensity. Conversely, new and more energy-efficient physical capital investments may be needed in this region, especially considering the previously mentioned facts (Chapter 1) that the LAC seems to suffer from a lack of new physical capital investments, as also from a lack of maintenance of the existing capital (e.g. Faruqee, 2016; Lardé and Sánchez, 2014; Perrotti, 2011; Fay and Morrison, 2007).

In view of all this, in this chapter, we will try to empirically investigate the relationship between regional physical capital (public and private) and regional energy intensity in order to help LAC region policymakers with the development of physical capital investment strategies which allow the region to overcome its concerns about energy demand, energy security, and the environment.

4.2 Physical capital and energy intensity

Following Patterson (1996, p. 377), energy intensity can be defined as "*using less energy to produce the same amount of services or useful output*". The energy intensity ratio, computed through the ratio between the economy's energy use and the respective gross domestic product, is frequently used as an energy efficiency indicator, with its analysis allowing the evaluation of the energy efficiency progress of a given economy (Martínez et al., 2019). The smaller the energy intensity ratio, the lower the energy intensity (i.e. higher the energy efficiency) would be. Overall, the objective is to achieve a state where energy consumption is stable or falling, and the gross domestic product is growing (i.e. the desirable state of decoupling between energy consumption and economic growth).

The increase in the economic activity in the LAC countries, as well as the change in their economic structure (with a rise in its industrialisation), is often associated with the increase in their energy consumption pattern (Pablo-Romero and De Jesús, 2016). Hence, with the increased pressure on the regional environment, the analysis of the energy intensity determinants of this region can be very useful for regional policymakers, especially taking into account that their results can help in the design of appropriate energy policies to supplant the environmental worries and energy concerns that the LAC region currently faces. Moreover, the fact that the growth of the LAC countries seems to be still very "energy dependent" (Belke et al., 2011) makes developing policies aimed at improving energy efficiency (and raising energy consumption productivity) even more suitable compared to the more modest energy conservation policies that simply induce a decrease in energy consumption.

According to the report by Balza et al. (2016) called "Lights on? Energy Needs in Latin America and the Caribbean to 2040" (see Box 4.1), the energy intensity trend in LAC has been declining in the last four decades. Following the report, the observed trend could be a sign that the region is making some efforts to improve its energy consumption productivity. In a decomposition analysis, Jimenez and Mercado (2014) point out per capita income, petroleum prices, fuel-energy mix, and economic growth as the fundamental determinants of regional energy intensity.

In the case of OPEC (The Organisation of the Petroleum Exporting Countries), Samargandi (2019) found that trade openness and renewable energy seem to reduce energy intensity in these countries, while energy prices seem to contribute to increasing energy intensity. Moreover, the author also found that domestic technological innovation plays an insignificant role in the reduction of the energy intensity in the OPEC countries. For the EU (European Union), Filipović et al. (2015) found that energy prices and gross domestic product negatively affected energy intensity (contributed to its reduction) and energy consumption positively affected energy intensity (contributed to an increase).

Apart from the previously cited determinants, one factor that should have its relationship with concepts such as energy efficiency or energy intensity further investigated is physical capital. As we previously have mentioned, capital must use energy to produce the goods and services that we need, at the same time as energy is also used to produce

BOX 4.1

Lights on? Energy needs in Latin America and the Caribbean to 2040.

The report "Lights on? Energy Needs in Latin America and the Caribbean to 2040" (Balza et al., 2016), provides *"a basis for understanding future energy demand in LAC countries"* (Balza et al., 2016, p. 5), giving us summarised analyses of regional energy consumption and energy intensity trends.

Looking at the report, it can be seen that regional energy use underwent a substantial expansion in recent decades, from 248 million tonnes of oil equivalent in 1971 to 848 million tonnes of oil equivalent in 2013, with fossil fuels representing 74.4% of the overall energy mix in 2013 (68.9% in 1971). Nevertheless, among the various energy sources, natural gas registered a considerable increase in the mix from 1971 to 2013, passing from 11% to 23%, which means that some effort is being made to promote low-carbon sources. Moreover, the demand for renewable energy also registered an increase from 1971 to 2013 [the renewables share in the primary energy mix declined but, following the report, it was due to the switch to more modern energy sources (nevertheless, they still dominate the LAC electricity matrix)].

Regarding the weighting of the LAC countries in regional energy consumption, the report states that 57.1% of the LAC total energy use is accounted for by Mexico and Brazil. Among the major end-sectors, the transport sector registers the largest share of energy use (34.3% in 2013), followed by the industrial sector (33.2% in 2013). As could be expected, both the transport and industrial shares increased compared to the values from 1971.

Another important factor to note in the report is that it indicates that the increase on energy consumption on the LAC was closely related to the economic development of the region, and with the intensification of the regional economic activity in recent decades. The existence of this stable relationship between energy consumption and growth means that the region is still far from the desired decoupling state and that the adoption of more efficient technologies is needed to make significant advances in this issue.

Regarding energy intensity, the report stresses that the LAC region is one of the least energy-intensive regions in the world, with the regional energy intensity rate showing a decline in the past four decades. It also notes that factors such as per capita income, petroleum prices, fuel-energy mix, and GDP growth are all central determinants in LAC energy intensity. The most surprising fact is that the decreasing trend in LAC energy intensity occurred without the adoption of considerable energy-saving programmes in the region.

Looking to the future, the report unsurprisingly forecasts that LAC energy consumption will continue to grow, keeping up with the regional economic growth and with the improvement in the region's living standards. Concerning energy intensity, the prediction is that it will also continue its downward tendency, with technological progress and the adoption of more energy-efficient technologies being the prime reason for the LAC region to continue with this trend.

capital (Kennedy, 2020). Given this assumption, any strategy to reduce energy intensity should also account for the role that physical capital has on energy consumption. Further, looking at the previous literature, it can be seen that there is a lack of studies that directly address the capital stock-energy intensity relationship (Koengkan et al., 2019). This fact can probably be associated with the difficulty in measuring capital stock, which has now been solved by the publication of the "Investment and Capital Stock Dataset" by the IMF (2017).

Nevertheless, according to Martínez et al. (2019), beyond improving the energy sector extraction and conversion techniques, and transferring parts of the manufacturing production to other countries, one factor that can greatly contribute to decreasing the energy intensity of a certain economy is improving the efficiency of the materials used in their respective productive system. In the same sense, Voigt et al. (2014) also consider newer and more efficient capital equipment as an essential factor for countries to be able to decrease their energy intensity. The increased energy efficiency of buildings and equipment is especially important in energy-intensive sectors such as manufacturing (Moynihan and Barringer, 2017), given the considerable weight that these sectors have in the production of greenhouse gas emissions (O'Rielly and Jeswieta, 2014). Moreover, the International Energy Agency (IEA, 2018) reinforces the idea that there is a need to increase investment in the energy efficiency of energy-using equipment in households, firms, and governments in order to achieve the desirable clean energy transition. Similarly, the OECD (2015) states that investment in clean energy infrastructure (in which upgrading energy efficiency is included) will need to rise if green growth and development objectives are to be attained.

In this sense, the public sector, which acts as a legislator and as a heavy energy-consuming sector (with buildings like schools, universities, hospitals, among others, needing 24-hour load) should set the example, given that energy efficiency investments could lead to large positive returns for governments, energy consumers, and primarily for the environment (Singh et al., 2009). Moreover, according to Bertoldi et al. (2010), one of the major benefits of public green procurement is that, through leading by example, the public sector can show the private sector the importance of investing in energy efficiency.

As expected, one of the most important factors for reducing the energy intensity of physical capital is technological progress or innovation, due to the creation and development of greener and more energy-efficient technologies (Malaczewski, 2018; Murad et al., 2019). However, this can represent a difficulty in the case of developing countries, due to their low R&D investment level and their low absorptive capacity (Goñi and Maloney, 2017; Burns, 2009). Following the UN (2018), there are additional obstacles that countries may face in the search for energy efficiency. They include the lack of a suitable regulatory framework, which undermines investments, a lack of effective national institutions to promote and develop energy efficiency projects, low international assistance, unfavourable financial environment, with financial institutions associating energy efficiency projects with higher risk levels, insufficient incentives from energy prices and a low degree of awareness regarding its importance. The lack of knowledge concerning the benefits from energy efficiency projects, the lack of tax incentives, and low-interest loans for energy efficiency projects, and the low priority of energy efficiency for people at the core of business decision-making, jointly with the lack (or high cost) of capital and lack of government incentives, make it more difficult for companies to invest in upgrading the energy efficiency of their buildings, plants, and equipment.

4.3 Public capital stock, private capital stock, and energy intensity in LAC

As expressed in the previous chapters (Chapters 1–3), there is a sort of consensual view regarding the fact that the LAC region will have to increase its investment in physical capital, especially in infrastructures, in order to support its economic growth and development in the following decades (Faruqee, 2016; Perrotti, 2011). However, as capital needs energy, in order to foster the region's sustainable development and to overcome its energy demand and energy security concerns, we believe that these investments should also be focused on promoting energy efficiency. In our view, investment in energy efficiency is essential for the LAC, given that "*the scale and speed of economic development as well as population growth will continue to drive LAC's primary energy use over the coming decades*" (Balza et al., 2016, p. 38). So, in this picture, energy efficiency investments could help lower regional environmental damage and pollution, reduce energy costs, and decrease energy dependence on foreign suppliers (Yuan et al., 2015; Ozturk, 2013; Dean and McMullen, 2007) (Box 4.2).

Despite the improvements in the LAC energy intensity in recent decades, we believe that there is still some work to do. As Ravillard et al. (2019) state, despite the decreasing trend in regional energy intensity, the LAC industry sector still has a relatively high energy intensity level when compared with other regions, which could mean that the physical capital used in the LAC industrial production (e.g. machinery and equipment) may need to be upgraded. Additionally, doubts regarding the LAC physical capital energy efficiency and its contribution to the region's energy intensity trend

BOX 4.2

Towards greater energy efficiency in Latin America and the Caribbean.

In the 2019 report "Towards Greater Energy Efficiency in Latin America and the Caribbean: Progress and Policies" (Ravillard et al., 2019) from the Inter-American Development Bank (IDB), the authors tried to give an overview of what has been accomplished in LAC in terms of energy intensity and energy efficiency in the last decades, as well as give some suggestions regarding strategies for further improvements.

Through a conceptual framework based on four steps (law and regulation, types of incentives, targets, governance and support), the authors tried to assess the evolution of the LAC energy efficiency policy. Their analysis led them to conclude that the area which had seen the greatest progress was the area of law and regulation. Conversely, in terms of governance and support and terms of targets, the region still has much room for improvement. The authors stress Brazil, Mexico, Colombia, Chile, and Uruguay as the most advanced LAC countries in terms of energy efficiency policy. In contrast, those in the Caribbean and Central America still need substantial improvement in this area.

Regarding policy recommendations, the authors divide them into two, more precisely into national and regional initiatives. At the national level, they start by suggesting accelerating the implementation of energy efficiency laws. These laws should be consistent and centralised and should be accompanied by quantifiable national targets. In the types of incentives, the authors suggest the implementation of mandatory codes with a higher energy consumption coverage percentage, an increase in the market instruments directed to energy efficiency (e.g. more obligation schemes), increased use of financial incentives (particularly to the private sector), promotion of energy audits, and setting the labelling of appliances as mandatory. Regarding targets, the authors stress that energy efficiency programmes should be extended (or reinforced) in the private sector, that the most energy-intensive sectors (e.g. the mining sector) should also have targets, that the incentives for the residential sector to embrace energy efficiency should be increased, that the instruments chosen should allow increased flexibility in the sector or target group response and, finally, that targets should be primarily focused on improving energy efficiency. Concerning governance and support, the authors state that more technical assistance is needed for those (households, companies) that want to invest in energy efficiency projects and that governments should encourage auctions, given that it is an excellent instrument for promoting energy savings, and that so far only Brazil has used it.

At the regional level, the authors suggest the creation of more regional agencies and frameworks focused on promoting energy efficiency, the establishment of energy audits for big corporations in all the region's countries, regional synchronisation regarding labelling and the setting of minimum energy standards, and reinforcing dialogue in LAC countries on energy issues. Nevertheless, the authors also present two possible effects that can cause the policy recommendations to have undesired results: the free-rider effect and the rebound effect. If policymakers do not account for these effects in energy efficiency policy design, they can lead to the overestimation of its effectiveness.

To see some specific examples of energy efficiency initiatives in the LAC region (namely in Bolivia, Chile, Costa Rica, Ecuador, Guyana, Honduras, Nicaragua, Panama, Paraguay, Uruguay) see Ravillard et al. (2019, pp. 64–65).

increase if we look at the book by Araújo et al. (2016), which states that the low total factor productivity (TFP) of the LAC generally undermines investment in new and more efficient equipment and infrastructure in this region. In Fig. 4.2, we present the evolution of public capital stock and private capital stock (in % of the GDP) in LAC from 1970 to 2014, which reinforces Araújo et al.'s (2016) idea: from the 1980s, LAC public capital stock and private capital stock did not increase considerably. Moreover, given that technological progress is a fundamental factor in supporting the development of energy-efficient capital, another problem may arise for the LAC region due to its lack of absorptive capacity (Koengkan, 2018). Finally, the role of governments and institutions can also be a promoter, or a barrier, to energy efficiency in the LAC. Following Ravillard et al. (2019), despite efforts having been made in the region, there is still room for further improvement for example with the enforcement of energy efficiency laws and regulations, the creation of incentives to support energy efficiency measures, the creation of energy efficiency targets for the various economic sectors, and increased government support in this issue.

Given the problems associated with the state of LAC physical capital, the accentuated increase in the region's energy consumption in recent decades and the increased concern over the region's energy demand and energy security, studying the relationship between LAC capital stock and the region's energy intensity is especially important in that it may

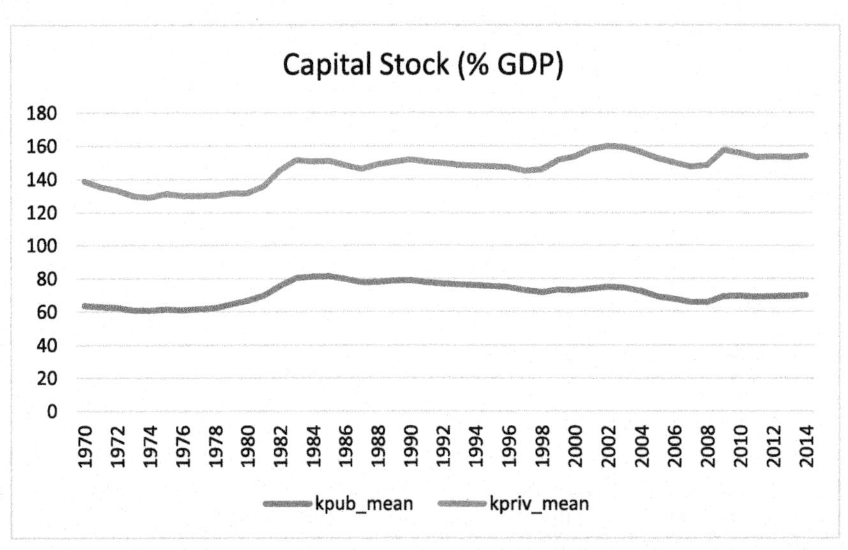

FIG. 4.2 Mean of LAC public capital stock (kpub_mean) and private capital stock (kpriv_mean) from 1970 to 2014 as % of the GDP; This graph was created by the authors using data on public capital stock, private capital stock, and GDP (all in billions of constant 2011 international dollars) from the "Investment and Capital Stock Dataset" by the IMF (2017) for Argentina, Barbados, Bolivia, Brazil, Chile, Colombia, Costa Rica, the Dominican Republic, Ecuador, El Salvador, Grenada, Guatemala, Haiti, Honduras, Mexico, Nicaragua, Panama, Paraguay, Peru, Uruguay, and Venezuela.

offer some hints regarding whether more energy-efficient capital stock is needed in the LAC. Therefore, after studying its relationship with economic growth (Chapter 2) and income inequality (Chapter 3), in this chapter, the impacts of LAC capital stock on the region's energy intensity are analysed. As we stressed in Section 4.1, our main objective is to help LAC policymakers in developing future physical capital investments, so that they can be directed to the sustainable development of the LAC region.

To accomplish the analysis, we collected annual frequency data from 1970 to 2014 for 21 LAC countries, namely: Argentina, Barbados, Bolivia, Brazil, Chile, Colombia, Costa Rica, the Dominican Republic, Ecuador, El Salvador, Grenada, Guatemala, Haiti, Honduras, Mexico, Nicaragua, Panama, Paraguay, Peru, Uruguay, and Venezuela. As in the cases of Chapters 2 and 3, the availability of data was the critical standard for setting the time span and for choosing the countries to include in the analysis. As in Chapter 3, the statistical software package Stata 15 was used to compute the statistical testing and to estimate the econometric models. In Table 4.1, we displayed the name, definition, and sources of the raw variables.

The dependent variable of our model will be energy intensity (*ei*), achieved through the ratio between the LAC countries' primary energy consumption and their respective gross domestic product. As stressed in Section 4.2, the smaller this ratio, the lower the energy intensity. The energy intensity (*ei*) formula is expressed as

$$ei_{it} = \frac{pec_{it}}{y_{it}} \tag{4.1}$$

where "*pec*" represents the primary energy consumption (in thousands of barrels of oil equivalent) of country i in period t and "y" is the gross domestic product (in billions of constant 2011 international dollars) of country i in period t.

Regarding the interest variables of the model, they will be general government capital stock (or public capital stock) (*kpub*), and private capital stock (*kpriv*), which were transformed into percentages of GDP through the ratio between general government capital stock (in billions of constant 2011 international dollars) and gross domestic product (in billions of constant 2011 international dollars), and the ratio between private capital stock (in billions of constant 2011 international dollars) and gross domestic product (in billions of constant 2011 international dollars), respectively.

The remaining independent variables will be the gross domestic product per capita (*ypc*) in billions of constant 2011 international dollars, CO_2 emissions in metric tons per capita (*co2pc*), and energy (commodities) prices (*ep*), annual indices (2011 = 100), which were selected because they have already been shown to be able to exert an influence on energy use and/or energy intensity/efficiency levels (Samargandi, 2019; Deichmann et al., 2019; Ahmad et al., 2018; Sineviciene et al., 2017; Hatzigeorgiou et al., 2011). The "*ypc*" was achieved through the computation of the ratio between the gross domestic product (in billions of constant 2011 international dollars) (y) and the total population (p), whereas the change of the "*ep*" base year to 2011 was achieved through the division of all index values by the 2011 value.

TABLE 4.1 Variables description.

Variable	Definition	Source
y	Gross domestic product (in billions of constant 2011 international dollars)	Investment and Capital Stock Dataset (IMF)
pec	Primary energy consumption (in thousands of barrels of oil equivalent)	CEPALSTAT
kpub	General government capital stock (in billions of constant 2011 international dollars)	Investment and Capital Stock Dataset (IMF)
kpriv	Private capital stock (in billions of constant 2011 international dollars)	Investment and Capital Stock Dataset (IMF)
co2pc	CO_2 emissions (in metric tons per capita)	World Development Indicators (WB)
ep	Energy (commodities) prices (annual indices, 2010 = 100, real 2010 US dollars)	World Bank Commodity Price Data
p	Total population (total number of persons)	World Development Indicators (WB)

To investigate the impacts of these variables, and primarily of the "*kpub*" and "*kpriv*", on LAC energy intensity, we will use a PARDL model in the form of an unrestricted error correction model (UECM). Some of the reasons that led us to the use of such a model were, as also explained in Chapter 2, that the PARDL model allows the decomposition of the total effects of the variables into their short- and long-run components; it is suitable to deal with cointegration; it is robust in the presence of endogeneity; and it allows the inclusion of $I(0)$, $I(1)$, and fractionally integrated variables in the same estimation. The PARDL specification of our model is described by

$$\begin{aligned} lei_{it} = {} & \alpha_{2i} + \delta_{2i1}TREND_t + \beta_{2i1}lei_{it-1} + \beta_{2i2}lkpub_{it} + \beta_{2i3}lkpub_{it-1} + \beta_{2i4}lkpriv_{it} + \beta_{2i5}lkpriv_{it-1} + \beta_{2i6}lypc_{it} + \beta_{2i7}lypc_{it-1} \\ & + \beta_{2i8}lco2pc_{it} + \beta_{2i9}lco2pc_{it-1} + \beta_{2i10}lep_{it} + \beta_{2i11}lep_{it-1} + \varepsilon_{2it} \end{aligned} \tag{4.2}$$

As discussed in Chapter 2, in order to obtain the dynamic general UECM form of the PARDL model, Eq. (4.2) can be reparameterised into

$$\begin{aligned} dlei_{it} = {} & \alpha_{3i} + \delta_{3i1}TREND_t + \beta_{3i1}dlkpub_{it} + \beta_{3i2}dlkpriv_{it} + \beta_{3i3}dllypc_{it} + \beta_{3i4}dlco2pc_{it} + \beta_{3i5}dlep_{it} + \gamma_{3i1}lei_{it-1} + \gamma_{3i2}lkpub_{it-1} \\ & + \gamma_{3i3}lkpriv_{it-1} + \gamma_{3i4}lypc_{it-1} + \gamma_{3i5}lco2pc_{it-1} + \gamma_{3i6}lep_{it-1} + \varepsilon_{3it} \end{aligned} \tag{4.3}$$

In Eq. (4.3), α_i represents the country-specific intercept, while δ_{ik}, β_{ik}, and γ_{ik} represent the estimated parameters and ε_{it} denotes the error term. The variables in Eq. (4.3) are presented in natural logarithms and first differences, with the prefixes "*l*" and "*d*" denoting natural logarithms and first differences, respectively. The ECM term, which personifies the model's adjustment speed, is represented by the natural logarithm of the energy intensity variable (*lei*) lagged once.

As in the case of the previous chapters, before estimation of the model, it is necessary to compute a series of preliminary tests and specification tests in order to understand the characteristics of our series and cross sections and, subsequently, to define a suitable estimator. As in the previous chapters, in the appendix of this chapter (Appendix), we will present the preliminary and specification tests tables together with the **How to do** part.

In our analysis, we applied the following preliminary tests: the cross-sectional dependence test (Pesaran, 2004); the second-generation unit root test (CIPS) (Pesaran, 2007); the second-generation cointegration test of Westerlund (2007); the correlation matrix; and the variance inflation factor (VIF) (Belsley et al., 1980). Regarding the specification tests, they were the Hausman test (Hausman, 1978) to confront the random effects (RE) and fixed effects (FE) models; the Hausman test (Hausman, 1978) to confront compare the mean group (MG), the pooled mean group (PMG), and the fixed effects (FE) estimators; the time fixed effects test; the modified Wald test (Greene, 2002); the Pesaran test for cross-sectional independence (Pesaran, 2004); the Frees test of cross-sectional independence (Frees, 1995, 2004); Friedman's test of cross-sectional independence (Friedman, 1937); the Breusch and Pagan Lagrangian multiplier (LM) test of independence (Breusch and Pagan, 1980); and the Wooldridge test (Wooldridge, 2002).

Starting with the cross-section dependence test, we can state that the null hypothesis (H0: cross-section independence) was rejected for all the variables, both in natural logarithms and first differences. As already explained in Chapters 2 and 3, this outcome means that a correlation exists between our series across countries, which probably derives from the common shocks shared by the countries included in our sample. In Table 4.A1, we give the results from the cross-section dependence test as well as the descriptive statistics of the variables. Finally, it is important to stress that the cross-section dependence test was not applied to the energy (commodities) prices (*ep*) variable because it has similar values for all the countries in our sample.

The following step was to investigate the order of integration of the variables. Given the results from the cross-section dependence test, we applied the cross-sectionally augmented IPS (CIPS) test, also called the second-generation unit root test, as it is robust to the fact of the variables being cross-sectionally dependent. From the results presented in Table 4.A2, the PARDL methodology appears to be a suitable methodology for our analysis, given that some of the variables seem to be on the borderline between the $I(0)/I(1)$ orders of integration. Nevertheless, the results of this test suggest that all variables are, at least, stationary at first differences, meaning that none of them seems to be $I(2)$. Again, the specific characteristics of the energy (commodities) prices (*ep*) variable, which makes it look more like a time series variable (common to all countries), led us to the computation of the augmented Dickey-Fuller (ADF) (Dickey and Fuller, 1981) and Kwiatkowski, Phillips, Schmidt, and Shin (KPSS) (Kwiatkowski et al., 1992) unit root tests, solely for this variable. The results from the ADF and KPSS tests in Table 4.A3 seem to indicate that the energy (commodities) prices (*ep*) variable is $I(1)$ i.e. integrated of the first order.

As previously stated, in this chapter, we also conducted the second-generation cointegration test of Westerlund to test for long-run relationships between the variables i.e. to test for cointegration. The results from this test, given in Table 4.A4, seem not to reject the null hypothesis (H0: no cointegration): the P-values are all higher than 10%. This outcome is verified both for the panel and for each country individually which, again, can be seen as a reason to use the PARDL (a more flexible econometric technique concerning the integration of variables).

The next step of our analysis was the analysis of the correlation matrix and of the VIF test. From the results in Table 4.A5, it can be concluded that the correlation between our variables does not cause significant concerns for the estimation, with the possible exception of the correlation between the natural logarithm of per capita gross domestic product (*lypc*) and the natural logarithm of per capita CO_2 emissions (*lco2pc*), and between the first differences of the natural logarithm of the per capita gross domestic product (*dlypc*) and the first differences of the natural logarithm of private capital stock (*dlkpriv*). However, due to the low VIF and mean VIF values (also in Table 4.A5), which indicate the absence of multicollinearity problems, we can proceed with the estimation without further concerns.

As when working on panel data, it is necessary to test for the presence of individual effects, the next step of the analysis was to compute the Hausman test in order to confront the random effects (RE) and fixed effects (FE) specifications. In Table 4.A6, we give the results from this test. As the null hypothesis of the test is rejected (including the *sigmamore* and *sigmaless* options), we can conclude that the fixed effects (FE) specification is the most suitable, meaning that the countries' individual effects are significant and must be taken into account. Similarly, when we work upon macro panels, the panel heterogeneity/homogeneity should also be tested by computing a Hausman test between the MG, the PMG, and the FE estimators. The MG (Pesaran and Smith, 1995) and PMG (Pesaran et al., 1999) estimators are usually used to deal with the possible slope heterogeneity of parameters (see Pesaran et al. (1999) for detailed information on these two estimators). Following the results in Table 4.A7, it can be seen that in MG vs PMG, PMG is the preferable estimator, whereas, for PMG vs FE and MG vs FE, the FE is preferable. These results imply that there is strong evidence that the slope heterogeneity of parameters is not found in our model and that we can consider our panel as homogeneous. Given these results, we will continue our estimation with the FE estimator.

After all this testing, it is still necessary to perform a set of specification tests and evaluate the presence of a group of phenomena that, if not accounted for, can lead to spurious conclusions. Therefore, before the estimation of the PARDL model from Eq. (4.3), the time fixed effects test, the modified Wald test, the Pesaran test for cross-sectional independence, the Frees test of cross-sectional independence, Friedman's test of cross-sectional independence, the Breusch and Pagan Lagrangian multiplier (LM) test of independence, and the Wooldridge test were conducted to see if time fixed effects are needed and to investigate if phenomena such as group-wise heteroscedasticity, cross-sectional dependence, and first-order autocorrelation are present in the model.

From the results presented in Table 4.A8, it can be concluded that no time fixed effects are needed in this case because the null hypothesis of the time fixed effects test (no time fixed effects are needed) is not rejected. Concerning the modified Wald test, we see that the null hypothesis (no group-wise heteroscedasticity) is clearly rejected, meaning that heteroscedasticity is present in our model. Regarding the presence of cross-sectional dependence, we can say that, although the Pesaran test of cross-sectional independence seems to support the hypothesis that the residuals are not correlated across countries (the null hypothesis is not rejected), the results from the remaining tests i.e. the Frees test of cross-sectional independence, Friedman's test of cross-sectional independence and the Breusch and Pagan Lagrangian multiplier (LM) test of independence all seem to support the hypothesis that there is cross-sectional dependence in the model, as all of them reject the null hypothesis that the residuals are not correlated. Finally, we also conclude that first-order autocorrelation is present in the model since the Wooldridge test rejects the null hypothesis of no first-order autocorrelation.

From the outcomes from all these tests, it can be assumed that group-wise heteroscedasticity, cross-sectional dependence, and first-order autocorrelation are all present in the model. Due to this assumption, to estimate the model, we need to use an estimator capable of dealing with the presence of all these phenomena. The best choice seems to be the

Driscoll and Kraay estimator (DK) (Driscoll and Kraay, 1998), given that it produces standard errors robust to the presence of heteroscedasticity, cross-sectional dependence, and first-order autocorrelation. In Table 4.2, we display the results from the estimation of the P-ARDL model with the DK-FE estimator.

As in the case of the models in Chapter 3, Table 4.A2 does not give us the long-run elasticities. To achieve them, we had to calculate a ratio between the variable's coefficients and the coefficient of the natural logarithm of energy intensity (*lei*), both lagged once, and then multiply it by −1. In Table 4.3, we display the short-run impacts and the long-run elasticities, as well as the model's adjustment speed [the error correction mechanism (ECM)].

After this first estimation of the P-ARDL model, it can be seen that the first differences of the natural logarithms of public capital stock (*dlkpub*), private capital stock (*dlkpriv*), and per capita CO_2 emissions (*dlco2pc*), and the natural logarithms of the per capita gross domestic product (*lypc*), lagged once, were all not statistically significant. Because of this, following the principle of parsimony (Santiago et al., 2020), we decided to remove these variables from the model. Consequently, Eq. (4.3) can be replaced by

$$dlei_{it} = \alpha_{4i} + \delta_{4i1}TREND_t + \beta_{4i1}dllypc_{it} + \beta_{4i2}dlep_{it} + \gamma_{4i1}lei_{it-1} + \gamma_{4i2}lkpub_{it-1} + \gamma_{4i3}lkpriv_{it-1} + \gamma_{4i4}lco2pc_{it-1} + \gamma_{4i5}lep_{it-1} + \varepsilon_{4it} \quad (4.4)$$

The next step was to re-estimate the model according to its most parsimonious specification (Eq. 4.4). To ensure that all assumptions remained the same, the specification tests were all redone (Tables 4.A9, 4.A10, and 4.A11 in Appendix). The results from the most parsimonious version of the model are presented in Table 4.4.

Once again, the long-run elasticities had to be calculated. The long-run elasticities of the parsimonious version of the model are now presented in Table 4.5, jointly with the short-run impacts and with the model's adjustment speed, represented by the ECM term.

On comparing the results presented in Table 4.3 with those in Table 4.5, it can be observed that the outcomes from both models (nonparsimonious and parsimonious) are not very different. First, it can be seen that, in the short run, only the per capita gross domestic product (*dlypc*) and the energy (commodities) prices (*dlep*) variables seem to demonstrate a statistically significant effect on energy intensity (*dlei*). However, while the per capita gross domestic product (*dlypc*) contributes to a reduction in LAC countries' energy intensity, the energy (commodities) prices (*dlep*) contribute to an increase in the energy intensity of these countries.

TABLE 4.2 PARDL estimation results.

Dependent variable: dlei	
Constant	−0.0134154
Trend	−0.0015759***
dlkpub	0.1490815
dlkpriv	−0.2467661
dlypc	−0.790175***
dlco2pc	0.031556
dlep	0.0287212*
lei (−1)	−0.0728701***
lkpub (−1)	0.017512
lkpriv (−1)	0.0313937*
lypc (−1)	−0.0116969
lco2pc (−1)	−0.0467759***
lep (−1)	0.0208794***
Diagnostic statistics	
N	924
R^2	0.1125
F	$F(12, 43) = 34.80$***

***, ** and * denote statistical significance at the 1%, 5%, and 10% levels, respectively.

TABLE 4.3 Elasticities, short-run impacts, and speed of adjustment

Dependent variable: dlei	
Short-run impacts	
dlkpub	0.1490815
dlkpriv	−0.2467661
dlypc	−0.790175***
dlco2pc	0.031556
dlep	0.0287212*
Long-run (computed) elasticities	
lkpub (−1)	0.2403187*
lkpriv (−1)	0.4308173*
lypc (−1)	−0.1605173
lco2pc (−1)	−0.6419077***
lep (−1)	0.286529***
Speed of adjustment	
ECM	−0.0728701***

*** and * denote statistical significance at the 1% and 10% levels, respectively; the ECM denotes the coefficient of the variable "*lei*" lagged once.

Second, regarding the long run, and starting with the interest variables, it can be seen that both public capital stock (*lkpub*) and private capital stock (*lkpriv*) are demonstrated to have an enhancing and statistically significant effect on the long-run energy intensity of the LAC. The statistical significances of these effects are greater in the parsimonious version of the model. Additionally, we can also see that private capital stock (*lkpriv*) shows a relatively larger effect than public capital stock (*lkpub*). Concerning the remaining variables, the energy (commodities) prices (*lep*) are also demonstrated to have an enhancing and statistically significant effect on the LAC countries' long-run energy intensity (as in the short run). In contrast, the per capita CO_2 emissions (*lco2pc*) reveal a depressing and statistically significant effect.

TABLE 4.4 PARDL estimation results (parsimonious).

Dependent variable: dlei	
Constant	0.0571923
Trend	−0.0017403***
dlypc	−0.6522905***
dlep	0.0281748*
lei (−1)	−0.0705164***
lkpub (−1)	0.0258047**
lkpriv (−1)	0.0358277**
lco2pc (−1)	−0.0510826***
lep (−1)	0.0216925***
Diagnostic statistics	
N	924
R^2	0.1097
F	$F(8, 43) = 36.05$***

***, ** and * denote statistical significance at the 1%, 5%, and 10% levels, respectively.

TABLE 4.5 Elasticities, short-run impacts, and speed of adjustment (parsimonious).

Dependent variable: dlei	
Short-run impacts	
dlypc	−0.6522905***
dlep	0.0281748*
Long-run (computed) elasticities	
lkpub (−1)	0.3659384***
lkpriv (−1)	0.5080759***
lco2pc (−1)	−0.7244074***
lep (−1)	0.3076233***
Speed of adjustment	
ECM	−0.0705164***

*** and * denote statistical significance at the 1% and 10% levels, respectively; the ECM denotes the coefficient of the variable *"lei"* lagged once.

Finally, the ECM terms from both the parsimonious and nonparsimonious specifications are negative and statistically significant at the 1% level, which means that there are signals of the presence of cointegration/long memory. This result is opposite to that of the Westerlund cointegration test in Table 4.A4. However, this contradiction is not a novelty, given that it has been already found in some past studies (e.g. Fuinhas et al., 2015, 2017). Moreover, the small ECM coefficient values from both specifications seem to indicate that the model's adjustment speed i.e. the speed at which the dependent variable returns to equilibrium after changes in the explanatory variables is relatively slow.

After assessing the impacts of public capital stock and private capital stock on the LAC countries' energy intensity through the PARDL estimation, we will now extend our analysis of the relationship between capital stock and energy intensity by using an approach based on the use of the log *t*-test (Phillips and Sul, 2007) and of the ordered-logit regression model.

First, through the log *t*-test, we will test the null hypothesis of convergence of the sample of countries in terms of energy intensity. Second, if this hypothesis is rejected i.e. the entire sample does not converge, we will test the convergence of subgroups through the clustering algorithm, which ultimately can lead to the identification of convergence clubs in the LAC region (see Schnurbus et al., 2017; Phillips and Sul, 2007 for detailed information on the mathematical expressions behind the log *t*-test and clustering algorithm). The operation of this algorithm can be described in the following steps: (1) sorting; (2) core group formation; (3) sieve individuals for club membership; (4) recursion and stopping rule; and (5) club merging. For further explanations regarding this algorithm and on how it can identify the clubs using the Stata statistical software package, see Du (2017, pp. 885–887).

Finally, if convergence clubs are identified through the clustering algorithm, one can use an ordered-logit regression model to investigate the possible influencing factors of the club formation (e.g. Bai et al., 2019; Yu et al., 2015). The ordered-logit regression model can be described by

$$\begin{aligned} y_i^* &= X_i'\beta + \varepsilon_i \\ y_i &= j, \text{if } \alpha_{j-1} \leq \alpha_j, \ j = 1,2,\ldots,J \end{aligned} \tag{4.5}$$

where y is the ordinal response variable which denotes the club to which a country belongs, y^* is the latent variable denoting the country's individual steady-state energy intensity level, X' is the vector of independent variables, β is the vector of regression coefficients, ε_i personifies the disturbance term, and the "α' s" represent unknown cut-points in the distribution of y^* (transition (threshold) parameters), estimated assuming $\alpha_0 = -\infty$ and $\alpha_j = \infty$, with J being the number of clubs. Fundamentally, y^* represents a country's tendency to belong to a certain club whereas the transition parameters α_j separate the clubs (when the y^* crosses a determined threshold α_j, the club membership changes).

Starting with the log t regression test, it should be stated that before its estimation, it is necessary to remove the cyclical component from the series. Thus, we first applied the Hodrick-Prescott filter (Hodrick and Prescott, 1997) to the natural *lei* in order to remove the trend component from this same variable. After this, the hypothesis of convergence can now be tested for the whole sample. From the results in Table 4.A12, we see that the *t* statistic from the log t regression test is below the critical value of −1.65 (the *t* statistic is −16.1778), meaning that we can reject the null hypothesis of convergence for the whole sample at the 5% significance level. Thus, the conclusion is that the countries in our sample do not converge to the same steady-state equilibrium concerning energy intensity.

As the null hypothesis of convergence for the whole sample was rejected, we will now test the hypothesis of club convergence within the sample through the use of the club clustering algorithm. Looking at the outcomes from Table 4.A13, we can conclude that there seem to exist four convergence clubs within our sample (with *t* statistics higher than −1.65), and one divergent group composed of countries that do not converge to any club (with a *t* statistic lower than −1.65). The countries included in the identified convergence clubs, and in the divergent group, are as follows:

- **Club 1:** Haiti, Honduras;
- **Club 2:** Argentina, Bolivia, Brazil, Guatemala, Nicaragua, Paraguay, Uruguay, Venezuela;
- **Club 3:** Barbados, Chile, Costa Rica, El Salvador, Grenada, Mexico, Peru;
- **Club 4:** Ecuador, Panama;
- **Divergent group:** Colombia and the Dominican Republic.

Finally, to test whether some of these clubs could not be merged, one should also apply the modified clustering algorithm method of Schnurbus et al. (2017). Based on the results from Table 4.A14, we see that the null hypothesis that the clubs can be merged is rejected for all combinations (*t* statistics are lower than −1.65), indicating that the convergence clubs are, indeed, the four that were formerly found.

After computing the mean of each club's energy intensity for each year, we can now present the descriptive statistics of the energy intensity of each one of the four convergence clubs (Table 4.6) as well as the graph of their average relative energy intensity transition paths (Fig. 4.3).

From Table 4.6 and Fig. 4.3, the highest average energy intensity level can be seen to belong to "Club 1". This club also has the highest initial energy intensity level, with the highest energy intensity between 1970 and 2014. In terms of average energy intensity level, after "Club 1" comes "Club 2," which is then followed by "Club 3" and "Club 4," respectively. It is also interesting to see that although "Club 3" and "Club 4" had a higher initial energy intensity when compared with "Club 2", they presented a robust decreasing trend of energy intensity immediately after 1970, whereas "Club 2" followed a stable energy intensity tendency between 1970 and 2014, presenting a slightly increasing trend until the 1990s and a modest decreasing trend after that period.

After identifying the clubs, we will now employ an ordered logit model to investigate whether public and private capital stocks were responsible, to some level, for the formation of the four convergence clubs. In the ordered logit model, the variable "club" will be the dependent variable, representing the club to which a country belongs. This variable can take values from 1 to 4 (number of clubs), and it is considered an ordinal variable since the clubs can be ranked by their energy intensity. As it can be seen in Table 4.6, the higher the value, the lower the energy intensity i.e. the highest energy intensity belongs to "Club 1" whereas the lowest belongs to "Club 4".

To ensure the robustness of our results, we adopted two alternative specifications for the estimation of the ordered logit model. According to previous literature, we estimated one model with the averages of the public and private capital stocks as a percentage of the GDP between 1970 and 2014 (Bai et al., 2019)—Specification 1—and another with the annual % averages of public capital stock and private capital stock over the period 1970–2014 (Matysiak and Olszewski, 2019)—Specification 2—with the variables being represented, respectively, by kpub_m and kpriv_m in the first specification, and kpub_g and kpriv_g in the second. Moreover, we should mention that the variables kpub_g and kpriv_g were computed using data on public and private capital stocks in billions of constant 2011 international dollars from the IMF (2017) "Investment and Capital Stock Dataset". Finally, the control variable chosen to include in both specifications was initial energy intensity (ei_1), which represents the countries' energy intensities in the first year of the analysis (1970). This variable was chosen following Yu et al. (2015) and in the belief that

TABLE 4.6 Descriptive statistics of the energy intensity of the convergence clubs.

Clubs	Obs	Mean	Std. dev.	Min	Max
1	90	6.364524	0.1284383	6.237211	6.674018
2	360	5.497266	0.0729173	5.356544	5.593428
3	315	4.894103	0.4226166	4.254727	5.524538
4	90	4.479531	0.6945355	3.120902	5.496629

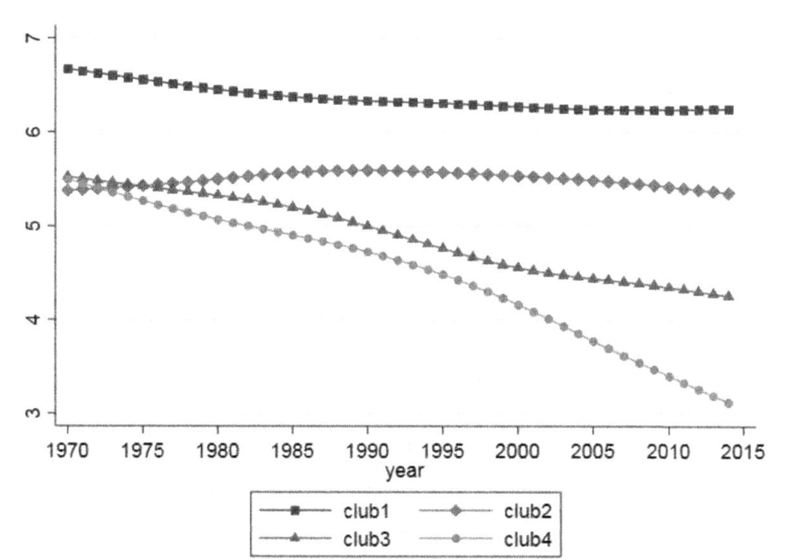

FIG. 4.3 Average relative energy intensity transition paths for different convergence clubs.

similar initial conditions are an essential convergence factor (Galor, 1996). The results from the ordered logit estimations are presented in Table 4.7.

From the results displayed in Table 4.7, it can be seen that the initial energy intensity (ei_1) is, in fact, critically determinant in convergence club membership. As it presents a negative coefficient in both specifications, it can be concluded that a positive change in the initial energy intensity reduces the likelihood that a country belongs to a low energy intensity club or, in other words, it means that a higher level of initial energy intensity increases the probability that a certain country will belong to a high energy intensity club [a result similar to that presented in a study by Yu et al. (2015)]. Regarding the interest variables (kpub_m and kpriv_m; kpub_g and kpriv_g), from the results from Table 4.7, it can be seen that none of them showed a statistically significant effect either in "Specification 1" or in "Specification 2." This means that the public and private capital stocks did not seem to be essential determinants for club membership i.e. these variables are not differentiating factors of club formation.

Similar to Chapters 2 and 3, having conducted the empirical analysis, in Section 4.4 we conclude this chapter by focusing our attention on an extended discussion of the results, as well as on the conclusions and policy implications

TABLE 4.7 Ordered logit estimation results.

Dependent variable: club			
Specification 1		**Specification 2**	
ei_1	−0.0039757**	ei_1	−0.0038458***
kpub_m	−0.0071272	kpub_g	0.1807698
kpriv_m	0.0096617	kpriv_g	0.1400534
Diagnostic statistics		Diagnostic statistics	
N	19	N	19
Pseudo R^2	0.1217	Pseudo R^2	0.0973
Log pseudolikelihood	−20.126745	Log pseudolikelihood	−20.685004
Wald chi2(3)	7.58	Wald chi2(3)	7.87
Prob > chi2	0.0556	Prob > chi2	0.0488
Brant test	1.17	Brant test	3.26

** and *** denote statistical significance at the 5% and 1% levels, respectively.

that may be derived from them. We hope that our results, and their respective inferences, would help LAC policy-makers in their quest for sustainable development in the region.

4.4 Conclusion

In this chapter, we used a PARDL model, the log t regression test method, the club clustering algorithm, and the ordered logit model to investigate the relationship between public and private capital stocks and energy intensity in a group of 21 Latin America and Caribbean (LAC) countries during the time span from 1970 to 2014.

First, the PARDL model was used to investigate the short- and long-run impacts of public and private capital stocks on energy intensity. Due to the phenomena found during the preliminary and specification testing, the model was estimated using the Driscoll and Kraay estimator with fixed effects. In our analysis, two different specifications of the PARDL were estimated: a nonparsimonious version (with the statistically significant and nonstatistically significant variables) and a parsimonious version (only with the statistically significant variables i.e. the model was re-estimated without the variables *"dlkpub"*, *"dlkpriv"*, *"dlco2pc"*, and *"lypc"*).

Overall, the results from both PARDL specifications were very similar. As in the study by Jimenez and Mercado's (2014), the results from our PARDL models also support the hypothesis that income is negatively related to energy intensity. However, according to our outcomes, this effect only seems to occur in the short run. This result seems to support the idea stressed by Deichmann et al. (2019) that the income effect on energy intensity fades when countries reach a certain income level. In addition, also in accordance with Deichmann et al. (2019), when a certain income level is reached, the effects from the development and application of energy efficiency policies become much more significant for reducing energy intensity than the income effect alone.

According to the outcomes from our PARDL models, another factor that appears to contribute to the decrease in energy intensity is CO_2 emissions, with this effect being felt only in the long run. This result seems to transmit the idea that environmental pressure (in this case, proxied by CO_2 emissions) can be a catalyst for the development of environmental and energy policies (e.g. energy efficiency policies) and the adoption of more environmentally friendly (and more energy-efficient) technologies and innovations (Khan et al., 2019, 2020). Ultimately, directly or indirectly, these actions can lead to a decrease in energy intensity. As this process takes a considerable amount of time i.e. from the moment that the problems are detected until the moment that these actions start having results, it is natural that this effect only arises in the long run.

Moreover, from the results of our PARDL models, we also saw that energy commodities prices contributed to the increase in energy intensity in both the short and long run. This result is very similar to the one that was found by Samargandi (2019) for the case of the OPEC (Organisation of the Petroleum Exporting Countries). According to Samargandi (2019, p. 8), his results are associated with the fact that *"higher energy price (crude oil) enables OPEC countries to earn more oil rent, which eventually leads to higher energy use in the domestic economies,"* thus, fostering energy intensity. In our case, as a significant portion of the LAC countries have abundant energy commodities, higher energy commodities prices can induce the LAC countries to increase their rents, leading them to higher levels of energy consumption which, eventually, can produce an increase in their energy intensity.

Finally, it can be seen that both specifications of the PARDL model indicate that public and private capital stocks did not seem to have contributed to the decreasing trend in LAC energy intensity. In fact, from the PARDL results, it can be seen that both present an enhancing effect on the energy intensity of this group of countries in the long run. This is probably a sign of the lack of investment in the LAC physical capital (Faruqee, 2016), or more properly, the lack of investment in new and more efficient capital (Araújo et al., 2016). This result indicates that there is a necessity to upgrade LAC physical capital in the public and private sectors, from equipment to infrastructure, because it still seems to be very energy intensive, which may prevent the LAC region from achieving an even lower energy intensity level. The fact that the acceleration of these economies was not accompanied by appropriate investment in new physical capital (see Fig. 4.2) may explain this effect being nearly captured in the long run, given that, over time, the effect of the lack of investment and of outdated capital on the energy intensity of this region becomes more and more noticeable and significant.

Additionally, using the log t regression test method, we found that our sample of LAC countries does not converge to the same steady-state equilibrium in terms of energy intensity. In fact, by means of the club clustering algorithm, we identified four convergence clubs and one divergent group i.e. a group composed of countries that do not converge to any club. According to our results, the composition of the convergence clubs was the following: "Club 1" was composed by Haiti and Honduras; "Club 2" by Argentina, Bolivia, Brazil, Guatemala, Nicaragua, Paraguay, Uruguay, and Venezuela; "Club 3" by Barbados, Chile, Costa Rica, El Salvador, Grenada, Mexico, and Peru; and "Club 4" by Ecuador and Panama. Regarding the so-called divergent group, it was composed by Colombia and the Dominican Republic.

Upon analysis of the descriptive statistics of the energy intensity of the four convergence clubs, it was found that they were ordered according to their energy intensity levels. Thus, "Club 1," composed by Haiti and Honduras, was the most energy-intensive club, followed by "Club 2," "Club 3," and "Club 4" (the least energy-intensive club). Then, via analysis of their average transition paths, we saw that while "Club 1" was always the most energy-intensive club from 1970 to 2014, with a relatively low downward trend at the beginning, "Club 2" had an initial energy intensity below that of "Club 3" and "Club 4." Over time, however, while the energy intensity of "Club 2" registered an upward trend (at least until the 1990s), the energy intensity from "Club 3" and "Club 4" sharply decreased immediately after the initial period. In addition, it should be noted that the decrease in the energy intensity of "Club 2" after the 1990s was very slight, with an almost constant, stagnant trend.

Given all of these findings, it is easy to perceive that there are countries in the LAC region which need to make an additional effort to reduce their energy intensity (and promote energy efficiency) when compared with others from the region. From the results of our analysis, it seems that these countries are those in "Club 1" and "Club 2," given that they present considerably higher energy intensity levels and energy transition paths, raising doubts about their efforts to reduce energy intensity. Conversely, in addition to their lower energy intensity levels, "Club 3" and "Club 4" also present a sharp, clearly decreasing trend in their energy intensity since the 1970s.

Despite the importance of the previous analysis and its respective inferences, it must not be forgotten that the main purpose of this chapter was to examine the role that public and private capital stocks play in LAC energy intensity. In accordance with this aim, one of the main objectives of the convergence analysis was to identify the LAC convergence clubs in order to be able to build an ordinal response variable that we can use in an ordered logit regression model. This variable was the dependent variable and represented the club to which a country belongs. Then, using the public and private capital stocks as the independent variables, we investigated if capital stocks are determinant factors for the formation of the energy intensity convergence clubs in the LAC i.e. to understand if public and private capital stocks are capable of decreasing (increasing) the probability of a country moving to a low-(high-)energy intensity club.

From the results of the ordered logit regression model, it can be seen that, contrary to what is found with initial energy intensity, public and private capital stocks are not determinant in forming convergence clubs i.e. their effects were not statistically significant. This result may be due to the fact that the evolution of both types of capital, and their respective effects on energy intensity, can be similar in all the clubs under analysis and, therefore, the difference of these clubs in terms of energy intensity may come from other factors. Overall, it seems that the conclusion from the ordered logit regression model supports the assumption that investment in newer and more energy-efficient capital should be made in all countries of the region, regardless of which club they are in.

Focusing on the implications from the results of the analysis conducted in this chapter, we can start by stressing that they seem to encourage the idea that LAC countries' governments should channel part of their efforts to investment in more energy-efficient physical capital. This investment focused on the improvement of the region's energy efficiency and decrease in energy intensity, should not only be made in infrastructures but also in equipment and machines used in LAC production processes, with particular attention to the most energy-intensive economic sectors. In order to achieve a suitable and satisfactory investment level in this field, LAC governments will probably need to improve their public financing instruments and, if necessary, call upon institutional investors.

Jointly with this increased investment, in the political spectrum, LAC governments should also act in order to improve their laws and their regulatory framework regarding energy efficiency. In the case such laws and regulations being absent, this is the time to move forward with creating them. As examples we can stress the establishment of energy efficiency targets for the various economic sectors (especially the most energy-intensive ones), the deployment of energy efficiency norms and regulation for equipment and appliances, and the creation and promotion of efficiency standards and energy audits for buildings/infrastructures as measures that can lead to both improvements in the energy efficiency of the physical capital and a decrease in the overall energy intensity.

Based on the outcomes of our analysis, it can be seen that it is also necessary to create incentives for the private sector to achieve the energy efficiency of their physical capital. Although improvement in public sector energy efficiency could influence the private sector to follow a similar tendency, we believe that additional measures should be developed in order to accelerate this process for example with the creation of new financing schemes and the development of financial incentives (e.g. subsidies and/or fiscal incentives) for energy efficiency investments in the private sector. Creating loans and lines of credit for investments in energy efficiency and granting tax reductions and tax credits for private investment in energy efficiency projects are some examples. Nevertheless, it should be mentioned that, in order for the private sector to move towards greater energy efficiency and consequently reduce its energy intensity, it is

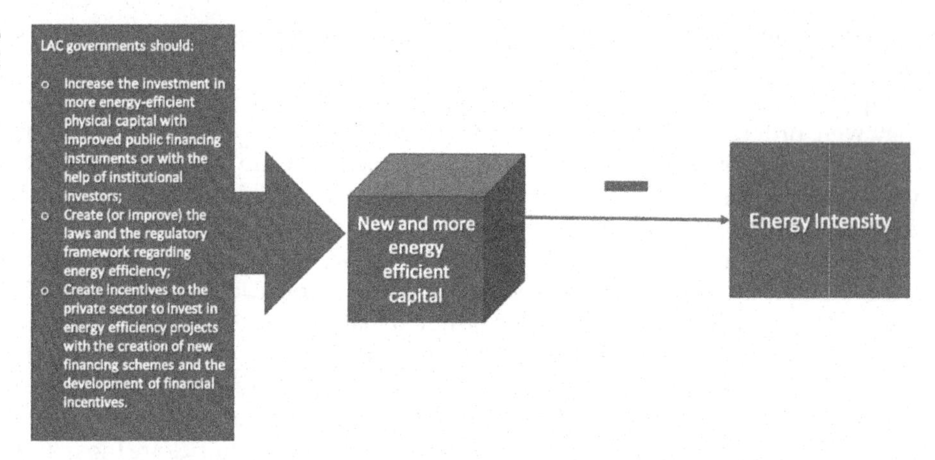

FIG. 4.4 Ways for physical capital stock to be able to reduce its contribution to LAC energy intensity (created by the authors).

desirable that the financial sector complement governments' efforts when it comes to supporting energy efficiency investment projects in the private sector.

Finally, the LAC energy transition will depend on not only the shift to renewables but also a significant increase in the region's energy efficiency. It is thus advisable that discussion of the ways in which LAC countries can attain higher levels of energy efficiency and lower levels of energy intensity should be increasingly present on the agendas of the Economic Commission for Latin America and the Caribbean (ECLAC), and the Latin American Energy Organisation (OLADE). These regional organisations can, and should, serve as a stage for the creation and promotion of energy efficiency measures for the LAC region. In fact, they can contribute to the creation of a more homogeneous energy efficiency promotion plan among the various countries in this region. In order to conclude this chapter, in Fig. 4.4, we display a scheme representing our conclusions on the ways that physical capital stock will be able to reduce its contribution to the regional energy intensity.

Appendix

TABLE 4.A1 Descriptive statistics and cross-section dependence (CD) test.

	Descriptive statistics					Cross-section dependence (CD)		
Variables	Obs	Mean	Std. dev.	Min.	Max.	CD-test	Corr	Abs(corr)
lei	945	5.246113	0.7797871	3.147825	6.898546	27.64***	0.284	0.575
lkpub	945	4.0292	0.698626	2.054651	5.52382	17.01***	0.175	0.476
lkpriv	945	4.927698	0.336838	4.141868	5.789673	11.25***	0.116	0.427
lypc	945	−11.75664	0.7540923	−13.34477	−9.048695	50.88***	0.523	0.702
lco2pc	945	0.3358615	0.9066564	−3.230116	2.041447	45.47***	0.468	0.566
lep	945	3.636263	0.6506555	2.061044	4.685315	n.a.	n.a.	n.a.
dlei	924	−0.0206048	0.1171919	−1.312275	0.8867016	2.54**	0.026	0.122
dlkpub	924	0.0040989	0.0519413	−0.1795859	0.3264704	22.01***	0.229	0.254
dlkpriv	924	0.0032835	0.0470015	−0.1878452	0.3207264	21.31***	0.222	0.242
dlkpriv	924	0.015127	0.0423243	−0.3375359	0.1506739	22.75***	0.237	0.252
dlco2pc	924	0.017039	0.1094636	−0.8105836	1.080082	3.73***	0.039	0.138
dlep	924	0.0564872	0.2550738	−0.6603057	0.9982629	n.a.	n.a.	n.a.

The CD test has $N(0,1)$ distribution under the H0: cross-sectional independence, *** and ** denote statistical significance at 1% and 5% level, respectively; n.a. denotes not applicable.

TABLE 4.A2 Panel unit root test (CIPS).

	CIPS (Zt-bar)	
	Without trend	With trend
lei	−2.215**	0.522
lkpub	1.701	0.409
lkpriv	−1.208	−1.003
lypc	−2.016**	−1.493*
lco2pc	−2.141**	1.277
lep	n.a.	n.a.
dlei	−13.011***	−12.128***
dlkpub	−8.813***	−8.096***
dlkpriv	−10.698***	−9.044***
dlkpriv	−10.575***	−8.962***
dlco2pc	−14.531***	−13.260***
dlep	n.a.	n.a.

***, **, and * denote statistical significance at 1%, 5%, 10% level, respectively; n.a. denotes not applicable; Pesaran (2007) panel unit root test (CIPS) assumes that cross-sectional dependence is in the form of a single unobserved common factor and H0: series is $I(1)$.

TABLE 4.A3 ADF and KPSS unit root tests.

	KPSS		ADF		
	Trend and Intercept	Intercept	Trend and Intercept	Intercept	None
lep	0.106475	0.476712**	−2.233689	−2.154676	1.056519
dlep	0.13579	0.163361	−6.59566***	−6.625773***	−6.429993***

*** and **denote statistical significance at 1% and 5%, respectively; KPSS has the following null hypothesis, H0: process is stationary; ADF has the following null hypothesis, H0: process has unit root (i.e. is not stationary).

TABLE 4.A4 Westerlund second-generation cointegration test.

	None			
Statistics	**Value**	***Z* value**	***P*-value**	**Robust *P*-value**
Gt	−1.801	1.790	0.963	0.626
Ga	−5.343	3.808	1.000	0.959
Pt	−6.481	1.752	0.960	0.691
Pa	−3.241	2.761	0.997	0.919
	Constant			
Statistics	**Value**	***Z* value**	***P*-value**	**Robust *P*-value**
Gt	−2.346	1.444	0.926	0.419
Ga	−9.387	3.107	0.999	0.763
Pt	−7.710	2.928	0.998	0.784
Pa	−4.536	3.728	1.000	0.956
	Constant and trend			
Statistics	**Value**	***Z* value**	***P*-value**	**Robust *P*-value**
Gt	−2.758	1.277	0.899	0.369
Ga	−9.405	4.901	1.000	0.990
Pt	−8.049	4.736	1.000	0.936
Pa	−4.914	5.469	1.000	0.990

Bootstrapping regression with 800 reps. H0: No cointegration; Gt and Ga test the cointegration for each country individually, and Pt and Pa test the cointegration of the panel as a whole.

TABLE 4.A5 Correlation matrices and VIF statistics.

	lei	*lkpub*	*lkpriv*	*lypc*	*lco2pc*	*lep*
lei	1.0000					
lkpub	−0.1621	1.0000				
lkpriv	−0.1746	0.0183	1.0000			
lypc	−0.5106	0.1810	−0.0321	1.0000		
lco2pc	−0.5807	0.2819	0.1355	0.8211	1.0000	
lep	−0.2210	0.0373	0.1034	0.1758	0.1640	1.0000
VIF		1.10	1.10	3.35	3.54	1.04
Mean VIF		2.03				
	dlei	***dlkpub***	***dlkpriv***	***dlypc***	***dlco2pc***	***dlep***
dlei	1.0000					
dlkpub	0.2146	1.0000				
dlkpriv	0.2044	0.7910	1.0000			
dlkpriv	−0.2597	−0.7613	−0.8437	1.0000		
dlco2pc	−0.0590	−0.2505	−0.2604	0.3195	1.0000	
dlep	−0.0024	−0.0800	−0.0859	0.1348	−0.0425	1.0000
VIF		2.91	4.26	3.97	1.12	1.03
Mean VIF		2.66				

In the case of the VIF test, the values are lower than the typically assumed benchmarks: 10 in the case of the VIF values, and 6 in the case of the mean VIF values.

TABLE 4.A6 Hausman test (FE vs. RE).

Hausman test	Hausman test with *sigmamore*	Hausman test with *sigmaless*
FE vs. RE	FE vs. RE	FE vs. RE
Chi2(12) = 45.77***	Chi2(9) = 42.43***	Chi2(9) = 44.06***

*** denotes significance at the 1% level; H0: difference in coefficients not systematic (random effects).

TABLE 4.A7 Hausman tests (MG vs. PMG vs. FE).

Hausman test	MG vs. PMG
	Chi2(13) = 12.26
	PMG vs. FE
	Chi2(13) = 4.94
	MG vs. FE
	Chi2(13) = 0.81

H0: difference in coefficients not systematic or that: (1) the PMG is the most suitable (when MG vs. PMG); (2) FE is the most suitable (when PMG vs. FE and FE vs. Pooled).

TABLE 4.A8 Specification tests.

	Statistics
Time fixed effects	1.13
Modified Wald test	1732.95***
Pesaran's test	0.019
Frees' test	0.185***
Friedman's test	46.863***
Breusch Pagan LM test	238.332*
Wooldridge test	64.428***

* and *** denote statistical significance at the 10% and 1% levels, respectively; H0 of time fixed effects test: dummies for all years are equal to 0 (no time fixed effects are needed); H0 of modified Wald test: sigma(*i*)ˆ2 = sigmaˆ2 for all *I*; H0 of Pesaran's, Frees', Friedman's, and Breusch-Pagan LM tests: residual are not correlated; H0 of Wooldridge test: no first-order autocorrelation.

TABLE 4.A9 Hausman test (FE vs RE) (parsimonious).

Hausman test	Hausman test with *sigmamore*	Hausman test with *sigmaless*
FE vs. RE	FE vs. RE	FE vs. RE
Chi2(8) = 44.18***	Chi2(5) = 43.79***	Chi2(5) = 45.74***

*** denotes significance at the 1% level; H0: difference in coefficients not systematic (random effects).

TABLE 4.A10 Hausman tests (MG vs. PMG vs. FE) (parsimonious).

Hausman test	MG vs. PMG
	Chi2(9) = 22.74***
	PMG vs. FE
	Chi2(9) = 1.60
	MG vs. FE
	Chi2(9) = 0.17

H0: difference in coefficients not systematic or that: (1) the PMG is the most suitable (when MG vs. PMG); (2) FE is the most suitable (when PMG vs. FE and FE vs. Pooled).

TABLE 4.A11 Specification tests—Parsimonious model.

	Statistics
Time fixed effects	1.12
Modified Wald test	1687.97***
Pesaran's test	−0.073
Frees' test	0.183***
Friedman's test	48.969***
Breusch Pagan LM test	242.808*
Wooldridge test	64.217***

* and *** denote statistical significance at the 10% and 1% levels, respectively; H0 of time fixed effects test: dummies for all years are equal to 0 (no time fixed effects are needed); H0 of modified Wald test: sigma(*i*)ˆ2 = sigmaˆ2 for all *I*; H0 of Pesaran's, Frees', Friedman's, and Breusch-Pagan LM tests: residual are not correlated; H0 of Wooldridge test: no first-order autocorrelation.

TABLE 4.A12 Log t regression test results (whole sample).

Variable	Coefficient	Standard error	*t* statistic
log(*t*)	−1.6554	0.1023	−16.1778

H0: convergence for the whole panel; if *t* statistic < −1.65, H0 is rejected at the 5% statistical significance level.

TABLE 4.A13 Club clustering algorithm results.

Clubs	Number of countries	Coefficient	*t* statistic	Countries
1	2	−1.181	−1.073	Haiti, Honduras
2	8	0.221	3.088	Argentina, Bolivia, Brazil, Guatemala, Nicaragua, Paraguay, Uruguay, Venezuela
3	7	0.107	0.805	Barbados, Chile, Costa Rica, El Salvador, Grenada, Mexico, Peru
4	2	1.251	1.602	Ecuador, Panama
Divergent group	2	−4.234	−18.893	Colombia, the Dominican Republic

H0: countries in clubs are converging; if *t* statistic < −1.65, H0 is rejected at the 5% level.

TABLE 4.A14 Club merging test.

Clubs	Coefficient	*t* statistic
1+2	−0.212	−7.467
2+3	−1.007	−34.056
3+4	−0.911	−11.541
4+5	−3.000	−29.315

H0: clubs can be merged; if *t* statistic < −1.65, H0 is rejected at the 5% level.

How to do:

STATA:

****Transform the variables kpub and kpriv into % of the GDP****

*(kpub1 and kpriv1 represent the raw capital stock variables)

```
gen kpub=(kpub1/y)*100
gen kpriv=(kpriv1/y)*100
```

Transform the variable y into per capita values

```
gen ypc = y/p
```

Create ei variable

```
gen ei = pec/y
```

Change the base year of ep from 2010 to 2011

```
*(ep2010 represents ep with 2010 as the base year)
*(ep value for 2011 = 115.894129124433)

gen ep = (ep2010/115.894129124433)*100
```

Transform the variables into natural logarithms

```
gen lei=ln(ei)
gen lkpub=ln(kpub)
gen lkpriv=ln(kpriv)
gen lypc=ln(ypc)
gen lco2pc=ln(co2pc)
gen lep=ln(ep)
```

Transform the variables into first differences of logarithms

```
gen dlei=d.lei
gen dlkpub=d.lkpub
gen dlkpriv=d.lkpriv
gen dlypc=d.lypc
gen dlco2pc=d.lco2pc
gen dlep=d.lep
```

Transform the variables into natural logarithms lagged once

```
gen l_lei=l.lei
gen l_lkpub=l.lkpub
gen l_lkpriv=l.lkpriv
gen l_lypc=l.lypc
gen l_lco2pc=l.lco2pc
gen l_lep=l.lep
```

Generate trend

```
gen trend = year - 1969
```

Descriptive statistics

```
sum lei lkpub lkpriv lypc lco2pc lep
sum dlei dlkpub dlkpriv dlypc dlco2pc dlep
```

Cross section dependence (CD) test

```
xtcd lei lkpub lkpriv lypc lco2pc
xtcd dlei dlkpub dlkpriv dlypc dlco2pc
```

Panel unit root test (CIPS)

```
multipurt lei lkpub lkpriv lypc lco2pc, lags(3)
multipurt dlei dlkpub dlkpriv dlypc dlco2pc, lags(3)
```

EViews:

****Transform the variable ep into natural logarithms****

Quick > Generate Series > lep=log(ep) > OK

****ADF unit root test****

Quick > Series Statistics.> Unit-Root Test.> Introduce the variable name: lep > OK > On test type choose: Augmented Dickey-Fuller > Choose the lag length criterion > Choose test unit root for level, first or second differences > Choose the inclusion of constant, constant and trend, or none > OK

****KPSS unit root test****

Quick > Series Statistics.> Unit-Root Test.> Introduce the variable name: lep > OK > On test type choose: Kwiatkowski–Phillips–Schmidt–Shin > Choose the lag length criterion > Choose test unit root for level, first or second differences > Choose the inclusion of constant, constant and trend, or none > OK

STATA:

****Westerlund cointegration test**

```
set matsize 800
xtwest lei lkpub lkpriv lypc lco2pc lepi, lags(1)lrwindow(3) bootstrap(800)
xtwest lei lkpub lkpriv lypc lco2pc lepi, lags(1)lrwindow(3) bootstrap(800) constant
xtwest lei lkpub lkpriv lypc lco2pc lepi, lags(1) lrwindow(3) bootstrap(800) constant trend
```

****Correlation matrices****

```
pwcorr lei lkpub lkpriv lypc lco2pc lep
pwcorr dlei dlkpub dlkpriv dlypc dlco2pc dlep
```

****VIF statistics****

```
qui: reg lei lkpub lkpriv lypc lco2pc lep
estat vif
qui: reg dlei dlkpub dlkpriv dlypc dlco2pc dlep
estat vif
```

****Hausman test (FE vs. RE)****

```
qui: xtreg dlei trend dlkpub dlkpriv dlypc dlco2pc dlep l_lei l_lkpub l_lkpriv l_lypc l_lco2pc l_lep,fe
estimates store fixed
qui: xtreg dlei trend dlkpub dlkpriv dlypc dlco2pc dlep l_lei l_lkpub l_lkpriv l_lypc l_lco2pc l_lep,re
estimates store random
hausman fixed random
hausman fixed random, sigmamore
hausman fixed random, sigmaless
```

****Hausman test (MG vs. PMG vs. Pooled)****

```
qui: xtpmg dlei trend dlkpub dlkpriv dlypc dlco2pc dlep, lr(l.lei l.lkpub l.lkpriv l.lypc l.lco2pc l.lep) ec(ecm) replace mg
estimates store mg
qui: xtpmg dlei trend dlkpub dlkpriv dlypc dlco2pc dlep, lr(l.lei l.lkpub l.lkpriv l.lypc l.lco2pc l.lep) ec(ecm) replace pmg
estimates store pmg
qui: xtpmg dlei trend dlkpub dlkpriv dlypc dlco2pc dlep, lr(l.lei l.lkpub l.lkpriv l.lypc l.lco2pc l.lep) ec(ecm) replace dfe
estimates store dfe
hausman mg pmg,sigmamore alleqs constant
hausman pmg dfe,sigmamore alleqs constant
hausman mg dfe,sigmamore alleqs constant
```

****Time-fixed effects test****

```
qui:xtreg dlei trend dlkpub dlkpriv dlypc dlco2pc dlep l_lei l_lkpub l_lkpriv l_lypc l_lco2pc l_lep i.year,fe
testparm i.year
```

Modified Wald test

```
qui: xtreg dlei trend dlkpub dlkpriv dlypc dlco2pc dlep l_lei l_lkpub l_lkpriv l_lypc l_lco2pc l_lep,fe
xttest3
```

Pesaran, Frees and Friedman tests of cross-sectional independence

```
qui: xtreg dlei trend dlkpub dlkpriv dlypc dlco2pc dlep l_lei l_lkpub l_lkpriv l_lypc l_lco2pc l_lep,fe
xtcsd, pesaran abs
xtcsd, frees
xtcsd, friedman
```

Breusch_Pagan LM test

```
qui: xtreg dlei trend dlkpub dlkpriv dlypc dlco2pc dlep l_lei l_lkpub l_lkpriv l_lypc l_lco2pc l_lep,fe
xttest2
```

Wooldridge test

```
xtserial dlei trend dlkpub dlkpriv dlypc dlco2pc dlep l_lei l_lkpub l_lkpriv l_lypc l_lco2pc l_lep
```

PARDL estimation with the Driscoll and Kraay estimator

```
xtscc dlei trend dlkpub dlkpriv dlypc dlco2pc dlep l_lei l_lkpub l_lkpriv l_lypc l_lco2pc l_lep,fe
```

Long-run elasticities

```
qui:xtscc dlei trend dlkpub dlkpriv dlypc dlco2pc dlep l_lei l_lkpub l_lkpriv l_lypc l_lco2pc l_lep,fe
nlcom (ratio1: -_b[l_lkpub]/_b[l_lei])
nlcom (ratio1: -_b[l_lkpriv]/_b[l_lei])
nlcom (ratio1: -_b[l_lypc]/_b[l_lei])
nlcom (ratio1: -_b[l_lco2pc]/_b[l_lei])
nlcom (ratio1: -_b[l_lep]/_b[l_lei])
```

Hausman test (FE vs. RE) (parsimonious)

```
qui: xtreg dlei trend dlypc dlep l_lei l_lkpriv l_lkpub l_lco2pc l_lep,fe
estimates store fixed
qui: xtreg dlei trend dlypc dlep l_lei l_lkpriv l_lkpub l_lco2pc l_lep,re
estimates store random

hausman fixed random
hausman fixed random, sigmamore
hausman fixed random, sigmaless
```

Hausman test (MG vs. PMG vs. Pooled) (parsimonious)

```
qui: xtpmg dlei trend dlypc dlep, lr(l.lei l.lkpub l.lkpriv l.lco2pc l.lep) ec(ecm) replace mg
estimates store mg
qui: xtpmg dlei trend dlypc dlep, lr(l.lei l.lkpub l.lkpriv l.lco2pc l.lep) ec(ecm) replace pmg
estimates store pmg
qui: xtpmg dlei trend dlypc dlep, lr(l.lei l.lkpub l.lkpriv l.lco2pc l.lep) ec(ecm) replace dfe
estimates store dfe

hausman mg pmg,sigmamore alleqs constant
hausman pmg dfe,sigmamore alleqs constant
hausman mg dfe,sigmamore alleqs constant
```

Time-fixed effects test (parsimonious)

```
qui:xtreg dlei trend dlypc dlep l_lei l_lkpriv l_lkpub l_lco2pc l_lep i.year,fe
testparm i.year
```

Modified Wald test (parsimonious)

```
qui: xtreg dlei trend dlypc dlep l_lei l_lkpriv l_lkpub l_lco2pc l_lep,fe
xttest3
```

Pesaran, Frees and Friedman tests of cross-sectional independence (parsimonious)

```
qui: xtreg dlei trend dlypc dlep l_lei l_lkpriv l_lkpub l_lco2pc l_lep,fe
xtcsd, pesaran abs
xtcsd, frees
xtcsd, friedman
```

Breusch_Pagan LM test (parsimonious)

```
qui: xtreg dlei trend dlypc dlep l_lei l_lkpriv l_lkpub l_lco2pc l_lep,fe
xttest2
```

Wooldridge test (parsimonious)

```
xtserial dlei trend dlypc dlep l_lei l_lkpriv l_lkpub l_lco2pc l_lep
```

PARDL estimation with the Driscoll and Kraay estimator (parsimonious)

```
xtscc dlei trend dlypc dlep l_lei l_lkpub l_lkpriv l_lco2pc l_lep,fe
```

Long-run elasticities (parsimonious)

```
qui:xtscc dlei trend dlypc dlep l_lei l_lkpub l_lkpriv l_lco2pc l_lep,fe
nlcom (ratio1: -_b[l_lkpub]/_b[l_lei])
nlcom (ratio1: -_b[l_lkpriv]/_b[l_lei])
nlcom (ratio1: -_b[l_lco2pc]/_b[l_lei])
nlcom (ratio1: -_b[l_lep]/_b[l_lei])
```

Wipe out the cyclical component

```
pfilter lei, method(hp) trend(lei1) smooth(400)
```

Log t regression for the convergence test

```
logtreg lei, kq(0.3) nomata
```

Identifying convergence clubs

```
psecta lei1, name(id) kq(0.3) gen(club) noprt
matrix b=e(bm)
matrix t=e(tm)
matrix result1=(b \ t)
matlist result1, border(rows) rowtitle("log(t)") format(%9.3f) left(4)
```

Perform possible club merging

```
scheckmerge lei1, kq(0.3) club(club) mdiv
matrix b=e(bm)
matrix t=e(tm)
matrix result2=(b \ t)
matlist result2, border(rows) rowtitle("log(t)") format(%9.3f) left(4)
```

Averages of energy intensity for the convergence clubs

```
bysort year: egen club1=mean(lei1)if club == 1
bysort year: egen club2=mean(lei1)if club == 2
bysort year: egen club3=mean(lei1)if club == 3
bysort year: egen club4=mean(lei1)if club == 4
```

Descriptive statistics of the energy intensity of the convergence clubs

```
sum club1 club2 club3 club4
```

Average relative energy intensity transition paths

```
twoway connected club1 club2 club3 club4 year, sort msymbol(s d t o) xlabel(1970(5)2015)
```

Ordered Logit (Specification 1)

ologit dv ei_1 kpub_m kpriv_m, vce (robust)
brant, detail

Ordered Logit (Specification 2)

ologit dv ei_1 kpub_g kpriv_g, vce (robust)
brant, detail

References

Ahmad, M., Hengyi, H., Rahman, Z., Khan, Z., Khan, S., Khan, Z., 2018. Carbon emissions, energy use, gross domestic product and total population in China. Ekon. Środowisko 2 (2), 32–44. Available at: http://bazekon.icm.edu.pl/bazekon/element/bwmeta1.element.ekon-element-000171529260.

Araújo, J.T., Vostroknutova, E., Wacker, K.M., Clavijo, M., 2016. Understanding the Income and Efficiency Gap in Latin America and the Caribbean. Directions in Development-Countries and Regions. World Bank Publications, Washington, DC, https://doi.org/10.1596/978-1-4648-0450-2.

Bai, C., Mao, Y., Gong, Y., Feng, C., 2019. Club convergence and factors of per capita transportation carbon emissions in China. Sustainability. https://doi.org/10.3390/su11020539.

Balatsky, A.V., Balatsky, G.I., Borysov, S.S., 2015. Resource demand growth and sustainability due to increased world consumption. Sustainability 7 (3), 3430–3440. https://doi.org/10.3390/su7033430.

Balza, L.H., Espinasa, R., Serebrisky, T., 2016. Lights on?: Energy Needs in Latin America and the Caribbean to 2040. Inter-American Development Bank, Washington, DC. Available at: https://publications.iadb.org/en/lights-energy-needs-latin-america-and-caribbean-2040.

Belke, A., Dobnik, F., Dreger, C., 2011. Energy consumption and economic growth: New insights into the cointegration relationship. Energy Econ. 33, 782–789. https://doi.org/10.1016/j.eneco.2011.02.005.

Belsley, D.A., Kuh, E., Welsch, R.E., 1980. Regression diagnostics: identifying influential data and sources of collinearity. Wiley, New York. https://doi.org/10.1002/0471725153.

Bertoldi, P., Cayuela, D.B., Monni, S., De Raveschoot, R.P., 2010. Guidebook: how to develop a Sustainable Energy Action Plan (SEAP). Publication Office of the European Union, Luxemburg. https://doi.org/10.2790/20638.

Breusch, T.S., Pagan, A.R., 1980. The Lagrange multiplier test and its applications to model specification in econometrics. Rev. Econ. Stud. 47 (1), 239–253. https://doi.org/10.2307/2297111.

Burns, A., 2009. Technology diffusion in the developing world. In: Braga, C.A.P., Padoan, P.C., Chandra, V., Erocal, D. (Eds.), Innovation and Growth: Chasing a Moving Frontier. OECD Publishing, Paris, https://doi.org/10.1787/9789264073975-10-en.

Dean, T.J., McMullen, J.S., 2007. Toward a theory of sustainable entrepreneurship: reducing environmental degradation through entrepreneurial action. J. Bus. Ventur. 22, 50–76. https://doi.org/10.1016/j.jbusvent.2005.09.003.

Deichmann, U., Reuter, A., Vollmer, S., Zhang, F., 2019. The relationship between energy intensity and economic growth: new evidence from a multi-country multi-sectorial dataset. World Dev. 124, 104664. https://doi.org/10.1016/j.worlddev.2019.104664.

Dickey, D.A., Fuller, W.A., 1981. Likelihood ratio statistics for autoregressive time series with a unit root. Econometrica 49 (4), 1057–1072. https://doi.org/10.2307/1912517.

Driscoll, J.C., Kraay, A.C., 1998. Consistent covariance matrix estimation with spatially dependent panel data. Rev. Econ. Stat. 80 (4), 549–560. https://doi.org/10.1162/003465398557825.

Du, K., 2017. Econometric convergence test and club clustering using stata. Stata J. 17 (4), 882–900. https://doi.org/10.1177/1536867X1801700407.

European Commission, 2011. Roadmap to a Resource Efficient Europe. Communication from the Commission to the European Parliament, the Council, The European Economic and Social Committee and the Committee of the Regions. COM(2011) 571, Brussels. Available at: https://eur-lex.europa.eu/legal-content/EN/TXT/PDF/?uri=CELEX:52011DC0571&from=EN.

Faruqee, H., 2016. Regional Economic Outlook, April 2016, Western Hemisphere Department: Managing Transitions and Risks. International Monetary Fund, Washington, DC, https://doi.org/10.5089/9781498329996.086.

Fay, M., Morrison, M., 2007. Infrastructure in Latin America and the Caribbean: Recent Developments and Key Challenges. Directions in Development-Infrastructure. World Bank, Washington, DC, https://doi.org/10.1596/978-0-8213-6676-9.

Filipović, S., Verbič, M., Radovanović, M., 2015. Determinants of energy intensity in the European Union: a panel data analysis. Energy 92, 547–555. https://doi.org/10.1016/j.energy.2015.07.011.

Frees, E.W., 1995. Assessing cross-sectional correlations in panel data. J. Econ. 64, 393–414. https://doi.org/10.1016/0304-4076(94)01658-M.

Frees, E.W., 2004. Longitudinal and Panel Data: Analysis and Applications in the Social Sciences. Cambridge University Press, https://doi.org/10.1017/CBO9780511790928.

Friedman, M., 1937. The use of ranks to avoid the assumption of normality implicit in the analysis of variance. J. Am. Stat. Assoc. 32, 675–701. https://doi.org/10.1080/01621459.1937.10503522.

Fuinhas, J.A., Marques, A.C., Couto, A.P., 2015. Oil-growth nexus in oil producing countries: macro panel evidence. Int. J. Energy Econ. Policy 5 (1), 148–163. https://www.econjournals.com/index.php/ijeep/article/view/990/575.

Fuinhas, J.A., Marques, A.C., Koengkan, M., 2017. Are renewable energy policies upsetting carbon dioxide emissions? The case of Latin America countries. Environ. Sci. Pollut. Res. 24, 15044–15054. https://doi.org/10.1007/s11356-017-9109-z.

Galor, O., 1996. Convergence? Inferences from theoretical models. Econ. J. 106, 1056–1069. https://doi.org/10.2307/2235378.

Goñi, E., Maloney, W.F., 2017. Why don't poor countries do R&D? Varying rates of factor returns across the development process. Eur. Econ. Rev. 94, 126–147. https://doi.org/10.1016/j.euroecorev.2017.01.008.

Greene, W.H., 2002. Econometric Analysis, fifth ed. Prentice Hall, Upper Saddle River, ISBN: 978-0130661890.

Hatzigeorgiou, E., Polatidis, H., Haralambopoulos, D., 2011. CO_2 emissions, GDP and energy intensity: a multivariate cointegration and causality analysis for Greece, 1977–2007. Appl. Energy 88 (4), 1377–1385. https://doi.org/10.1016/j.apenergy.2010.10.008.

Hausman, J.A., 1978. Specification tests in econometrics. Econometrica 46 (6), 1251–1271. https://doi.org/10.2307/1913827.

Hodrick, R.J., Prescott, E.C., 1997. Postwar U.S. business cycles: an empirical investigation. J. Money Credit Bank 29 (1), 1–16. https://doi.org/10.2307/2953682.

IEA, 2018. The Role of Energy Efficiency. International Energy Agency, Paris. Available at: https://www.iea.org/reports/the-role-of-energy-efficiency.

IMF, 2017. Estimating the Stock of Public Capital in 170 Countries. Fiscal Affairs Department, International Monetary Fund, Washington, DC. Available at: https://www.imf.org/external/np/fad/publicinvestment/pdf/csupdate_aug19.pdf.

Jimenez, R., Mercado, J., 2014. Energy intensity: a decomposition and counterfactual exercise for Latin American countries. Energy Econ. 42, 161–171. https://doi.org/10.1016/j.eneco.2013.12.015.

Kennedy, C., 2020. Energy and capital. J. Ind. Ecol. 24, 1047–1058. https://doi.org/10.1111/jiec.13014.

Khan, Z., Ali, S., Umar, M., Kirikkaleli, D., Jiao, Z., 2020. Consumption-based carbon emissions and International trade in G7 countries: the role of Environmental innovation and Renewable energy. Sci. Total Environ. 730, 138945. https://doi.org/10.1016/j.scitotenv.2020.138945.

Khan, Z., Sisi, Z., Siqun, Y., 2019. Environmental regulations an option: asymmetry effect of environmental regulations on carbon emissions using non-linear ARDL. Energy Sourc. A: Recov. Util. Environ. Effects 41 (2), 137–155. https://doi.org/10.1080/15567036.2018.1504145.

Koengkan, M., 2018. The positive impact of trade openness on consumption of energy: fresh evidence from Andean community countries. Energy 158, 936–943. https://doi.org/10.1016/j.energy.2018.06.091.

Koengkan, M., Santiago, R., Fuinhas, J.A., 2019. The impact of public capital stock on energy consumption: empirical evidence from Latin America and the Caribbean region. Int. Econ. 160, 43–55. https://doi.org/10.1016/j.inteco.2019.09.001.

Kwiatkowski, D., Phillips, P.C.B., Schmidt, P., Shin, Y., 1992. Testing the null hypothesis of stationarity against the alternative of a unit root: How sure are we that economic time series have a unit root? J. Econ. 54 (1), 159–178. https://doi.org/10.1016/0304-4076(92)90104-Y.

Lardé, J., Sánchez, R., 2014. The Economic Infrastructure Gap and Investment in Latin America. FAL Bulletin, No. 332, Economic Commission for Latin America and the Caribbean (ECLAC), Santiago. Available at: http://hdl.handle.net/11362/37381.

Malaczewski, M., 2018. Complementarity between energy and physical capital in a simple model of economic growth. Econ. Res. 31 (1), 1169–1184. https://doi.org/10.1080/1331677X.2018.1456353.

Martínez, D.M., Ebenhack, B.W., Wagner, T.P., 2019. Chapter 1—Introductory concepts. In: Martínez, D.M., Ebenhack, B.W., Wagner, T.P. (Eds.), Energy Efficiency. Elsevier, pp. 1–33, https://doi.org/10.1016/B978-0-12-812111-5.00001-9.

Matysiak, G., Olszewski, K., 2019. A panel analysis of polish regional cities residential price convergence in the primary market. SSRN Electron. J., 1–30. https://doi.org/10.2139/ssrn.3408797.

Moynihan, G.P., Barringer, F.L., 2017. Energy efficiency in manufacturing facilities: assessment, analysis and implementation. In: Yap, E.H. (Ed.), Energy Efficient Buildings. IntechOpen, https://doi.org/10.5772/64902.

Murad, M.W., Alam, M.M., Noman, A.H.M., Ozturk, I., 2019. Dynamics of technological innovation, energy consumption, energy price and economic growth in Denmark. Environ. Prog. Sustain. Energy 38, 22–29. https://doi.org/10.1002/ep.12905.

O'Rielly, K., Jeswieta, J., 2014. Strategies to improve industrial energy efficiency. Procedia CIRP 15, 325–330. https://doi.org/10.1016/j.procir.2014.06.074.

OECD, 2015. Policy Guidance for Investment in Clean Energy Infrastructure: Expanding Access to Clean Energy for Green Growth and Development. OECD Publishing, Paris, https://doi.org/10.1787/9789264212664-en.

Ozturk, I., 2013. Energy dependency and energy security: the role of energy efficiency and renewable energy sources. Pak. Dev. Rev. 52 (4), 309–330. https://doi.org/10.30541/v52i4Ipp.309-330.

Pablo-Romero, M.P., De Jesús, J., 2016. Economic growth and energy consumption: The Energy-Environmental Kuznets Curve for Latin America and the Caribbean. Renew. Sust. Energ. Rev. 60, 1343–1350. https://doi.org/10.1016/j.rser.2016.03.029.

Patterson, M.G., 1996. What is energy efficiency? Concepts, indicators and methodological issues. Energy Policy 24 (5), 377–390. https://doi.org/10.1016/0301-4215(96)00017-1.

Perrotti, D., 2011. The Economic Infrastructure Gap in Latin America and the Caribbean. FAL Bulletin, No. 293, Economic Commission for Latin America and the Caribbean (ECLAC), Santiago. Available at: http://hdl.handle.net/11362/36339.

Pesaran, M.H., 2004. General diagnostic tests for Cross Section Dependence in Panels. Cambridge Working Papers in Economics, No. 435, Faculty of Economics, University of Cambridge, https://doi.org/10.17863/CAM.5113.

Pesaran, M.H., 2007. A simple panel unit root test in the presence of cross-section dependence. J. Appl. Econ. 22 (2), 265–312. https://doi.org/10.1002/jae.951.

Pesaran, M.H., Smith, R., 1995. Estimating long-run relationships from dynamic heterogeneous panels. J. Econ. 68 (1), 79–113. https://doi.org/10.1016/0304-4076(94)01644-F.

Pesaran, M.H., Shin, Y., Smith, R., 1999. Pooled mean group estimation of dynamic heterogeneous panels. J. Am. Stat. Assoc. 94 (446), 621–634. https://doi.org/10.1080/01621459.1999.10474156.

Phillips, P.C.B., Sul, D., 2007. Transition modeling and econometric convergence tests. Econometrica 75 (6), 1771–1855. https://doi.org/10.1111/j.1468-0262.2007.00811.x.

Ravillard, P., Carvajal, F., Lopez, D., Chueca, J.E., Hallack, M., 2019. Towards Greater Energy Efficiency in Latin America and the Caribbean: Progress and Policies. Inter-American Development Bank, Washington, DC, https://doi.org/10.18235/0002070.

Samargandi, N., 2019. Energy intensity and its determinants in OPEC countries. Energy 186, 115803. https://doi.org/10.1016/j.energy.2019.07.133.

Santiago, R., Fuinhas, J.A., Marques, A.C., 2020. The impact of globalization and economic freedom on economic growth: the case of the Latin America and Caribbean countries. Econ. Chang. Restruct. 53 (1), 61–85. https://doi.org/10.1007/s10644-018-9239-4.

Schnurbus, J., Haupt, H., Meier, V., 2017. Economic transition and growth: a replication. J. Appl. Econ. 32 (5), 1039–1042. https://doi.org/10.1002/jae.2544.

Sineviciene, L., Sotnyk, I., Kubatko, O., 2017. Determinants of energy efficiency and energy consumption of Eastern Europe post-communist economies. Energy Environ. 28 (8), 870–884. https://doi.org/10.1177/0958305X17734386.

Singh, J., Limaye, D.R., Henderson, B., Shi, X., 2009. Public Procurement of Energy Efficiency Services: Lessons From International Experience. Directions in Development-Energy and Mining. World Bank, Washington, DC, https://doi.org/10.1596/978-0-8213-8062-8.
UN, 2018. Overcoming Barriers to Investing in Energy Efficiency. ECE Energy Series, No. 56, UN, New York, https://doi.org/10.18356/fcabb6f3-en.
UN General Assembly, 2015. Resolution Adopted by the General Assembly on 25 September 2015 (A/RES/70/1). UN, New York. Available at: https://digitallibrary.un.org/record/808134.
UNEP, 2011. Decoupling Natural Resource Use and Environmental Impacts From Economic Growth. A Report of the Working Group on Decoupling to the International Resource Panel. Available at: http://hdl.handle.net/20.500.11822/9816.
Voigt, S., De Cian, E., Schymura, M., Verdolini, E., 2014. Energy intensity developments in 40 major economies: structural change or technology improvement? Energy Econ. 41, 47–62. https://doi.org/10.1016/j.eneco.2013.10.015.
Westerlund, J., 2007. Testing for error correction in panel data. Oxf. Bull. Econ. Stat. 69, 709–748. https://doi.org/10.1111/j.1468-0084.2007.00477.x.
Wooldridge, J.M., 2002. Econometric Analysis of Cross Section and Panel Data. The MIT Press Cambridge, MA, ISBN: 978-0262232586.
Yu, Y., Zhang, Y., Song, F., 2015. World energy intensity revisited: a cluster analysis. Appl. Econ. Lett. 22 (14), 1158–1169. https://doi.org/10.1080/13504851.2015.1013603.
Yuan, X., Mu, R., Zuo, J., Wang, Q., 2015. Economic development, energy consumption, and air pollution: a critical assessment in China. Hum. Ecol. Risk Assess. Int. J. 21 (3), 781–798. https://doi.org/10.1080/10807039.2014.932204.

CHAPTER

5

Energy transition in the Latin America region: Initiatives and challenges

JEL codes F62, Q43, Q5

5.1 Introduction

Energy transition has been accelerating and gaining space in the political arena, where it has become an area of great concern for many governments and policymakers. The process of energy transition does not arise in a vacuum. This process was shaped and influenced over time by a broader and deeper shift. Indeed, energy transition does not regard energy, a change in energy sources or merely a process of replacement of technology with other more efficient means. It is a paradigm shift that has been changing the energy world profoundly, in which there exists a change in the values of security, robustness, and reliability (Koengkan and Fuinhas, 2020). However, the existing energy systems that were based on security, robustness, and reliability values have been replaced by new energy system based on new values (see Fig. 5.1).

All these will allow new means of producing and consuming energy. The literature does not offer a precise definition of the term 'energy transition'. According to Smil (2010), there is no precise or widely accepted meaning for this term. However, according to the same author, this term can be often used to describe changes in the energy matrix from fossil fuels energy sources to renewable ones. This change, according to Koengkan and Fuinhas (2020), has been taking place progressively from the established energy system (fossil fuels energy sources) to the new energy system (renewable energy sources). This process can be analysed from a global to a local perspective.

Moreover, it can also be used to indicate a structural transformation in the energy sector. There is a growing trend of sharing renewable energy sources combined with the promotion of energy efficiency to reduce the consumption of nonrenewable energy (Hauff et al., 2014). Certainly, this structural transformation is an indication of a clear objective of reducing environmental degradation (Koengkan and Fuinhas, 2020).

Nevertheless, this term is misinterpreted much of the time, reducing its conceptual scope and consequently emphasising one type of change, namely that of nonrenewable energy sources to renewable. In fact, there is no single energy transition but rather various local experiences (Koengkan and Fuinhas, 2020). As we already know, the accelerated process of energy transition is related to the uncontrolled increase in the level of carbon dioxide emissions (CO_2) that has consequently set off a worldwide alarm signal. Indeed, CO_2 emissions contribute most to greenhouse gas emissions (GHGs), of which it contributes 77%. Other gases such as methane (CH_4), nitrous oxide (N_2O), and ozone (O_3) contribute 14%, 8%, and 1%, respectively (see Fig. 5.2).

Indeed, these gases contribute to global warming and climate change. As a result, several initiatives have emerged, such as the United Nations Framework Convention on Climate Change (UNFCCC), the Earth Summit in 1992 and the Kyoto Protocol in 1997, as well as the more recent Paris Agreement that was approved by 195 countries during the United Nations Framework Convention on Climate Change (UNFCCC) at the 21st Conference of the Parties (COP 21) in December 2015 to mitigate global warming and climate change. All these initiatives have the main goal of limiting the increase in temperature levels substantially to lower than 2°C during this century, as well as efforts to limit that increase to 1.5°C. That is, it aims to take the temperature to preindustrial levels. All countries that ratify this agreement will move towards a low-carbon economy.

https://doi.org/10.1016/B978-0-12-824429-6.00001-2

FIG. 5.1 New values for energy systems. This figure was created by the authors and based on explanations by Koengkan and Fuinhas (2020).

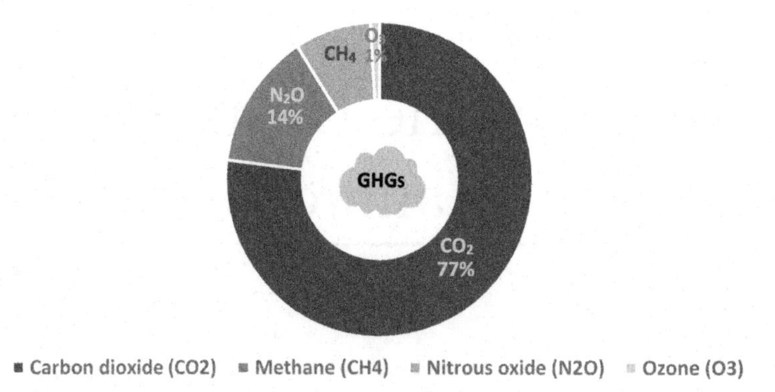

FIG. 5.2 Composition of GHGs. This figure was created by the authors and was based on explanations by Koengkan and Fuinhas (2020) and Khan et al. (2014).

Each country that accepts these agreements will have to make different contributions that will take their historical and current emission standards into account. Thus, each country will adopt the best means to reduce its emissions by setting policies to decarbonise the various segments of the economy (e.g. energy, industry, transportation, and land use). Global GHGs—mainly the CO_2 emissions—have been increasing since the 1970s (see Fig. 5.3).

However, between 1990 and 2014, these emissions grew rapidly, and in 1990, CO_2 emissions were at 3.0991 metric tons per capita and reached a value of 4.9807 metric tons per capita in 2014. That is, during the period these emissions grew from 33 megatons of CO_2 equivalent ($MtCO_2eq$) in 1990 to 48 $MtCO_2eq$ in 2014, an increase of 1.5% during this period (Koengkan and Fuinhas, 2020; Bárcena et al., 2019). All sectors (except agriculture, forestry, and land use (AFOLU—agriculture, forestry, and other land use)) have been showing increased emissions since the 2000s. In 2010, the energy sector contributed 25%, AFOLU 24%, industry 21%, transport 14%, other energy sources 10%, and the buildings sector 6% of this growth, respectively (see Fig. 5.4).

Indeed, most of these emissions caused by electricity and heat production emanate from residential and industrial sectors. That is, GHGs come about through direct emissions from the combustion of fossil fuels for cooking, heating, cooling, and providing power (Khan et al., 2014).

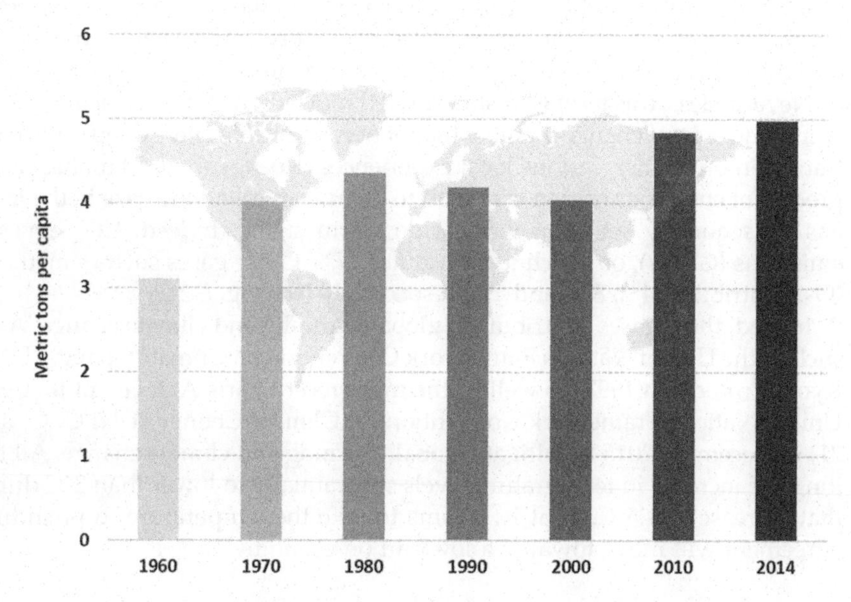

FIG. 5.3 World CO_2 emissions (metric tons per capita), between 1960 and 2014. This figure was created by the authors and was based on the World Bank Open Data (2020).

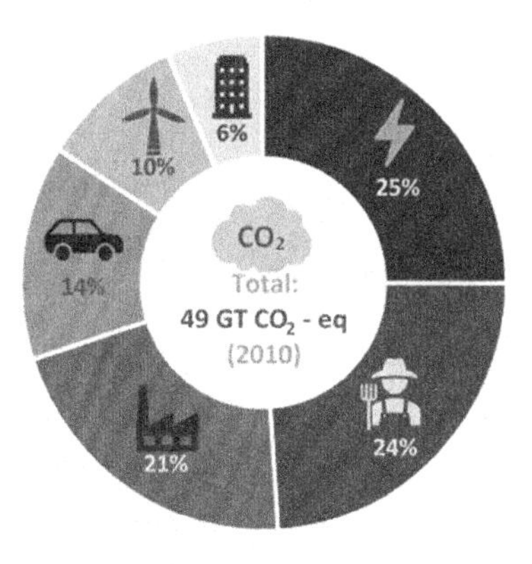

FIG. 5.4 Direct world GHGs per sector in 2010. Total GHGs in $GTCO_2$—EQ/Year. This figure was created by the authors and based on Koengkan et al. (2020).

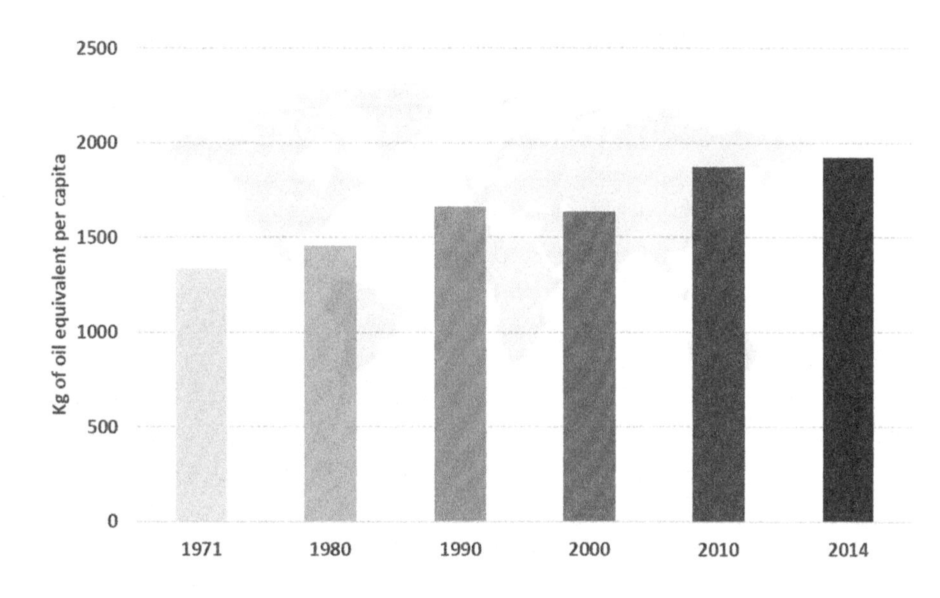

FIG. 5.5 World energy use (kg of oil equivalent per capita), between 1971 and 2014. This figure was created by the authors and was based on the World Bank Open Data (2020).

As previously mentioned, energy consumption accounted for 25% of total world GHG emissions in 2010. That is, the increase in CO_2 emissions is related to the growth in world energy use. Energy use has been increasing since the 1970s: in 1971 energy use (in kg of oil equivalent per capita) was 1337.00, and reached a value of 1922.48 in 2014 (see Fig. 5.5).

However, 94% of this world energy use in 1970 was from fossil fuel energy sources (e.g. oil, coal, and gas), and only 6.45% was from renewable energy sources (e.g. hydropower, nuclear, geothermal, biomass, and waste). In 2014 this value underwent a small decrease, reaching 86% of total energy use. Indeed, this reduction is related to the increase of the share of renewable energy sources (e.g. hydropower, nuclear, geothermal, biomass, waste, solar, and wind) in energy use, reaching a value of 14.35% (see Fig. 5.6).

In the region of Latin America and the Caribbean (LAC), the situation is no different from that of the rest of the world, where GHG emissions in 1970 were 1.7831 metric tons per capita and reached a value of 3.1016 metric tons per capita in 2014 (see Fig. 5.7).

Moreover, Koengkan and Fuinhas (2020) add that during the period between 1990 and 2014, these emissions increased by 0.7%. In 1990 a value of 3.414 $MtCO_2eq$ was registered, and emissions reached a value of 4.020 $MtCO_2eq$ in 2014. Indeed, 70% of this increase is related to the consumption of energy—35%, and AFOLU—35%. In this consumption of energy in the LAC region, liquid fuels account for 60.8%, while coal is only a modest contributor, with 7.6% in 2013 (Koengkan et al., 2019b). Regarding the structure of GHGs in the LAC region, in 1990 electricity and heat production had a 29% share, AFOLU 66%, industry 2%, and waste 3%. However, in 2014 the share of electricity and heat production was 48%, AFOLU 23%, industry 4%, and waste 6% (see Fig. 5.8).

Despite the 0.7% growth in emissions between 1990 and 2014, the region is a small contributor per capita to the world's GHG, accounting for about 11% of total global emissions (Fuinhas et al., 2017). Indeed, the increase in CO_2 emissions in the LAC region is directly related to the increase in consumption of energy, caused by an increase

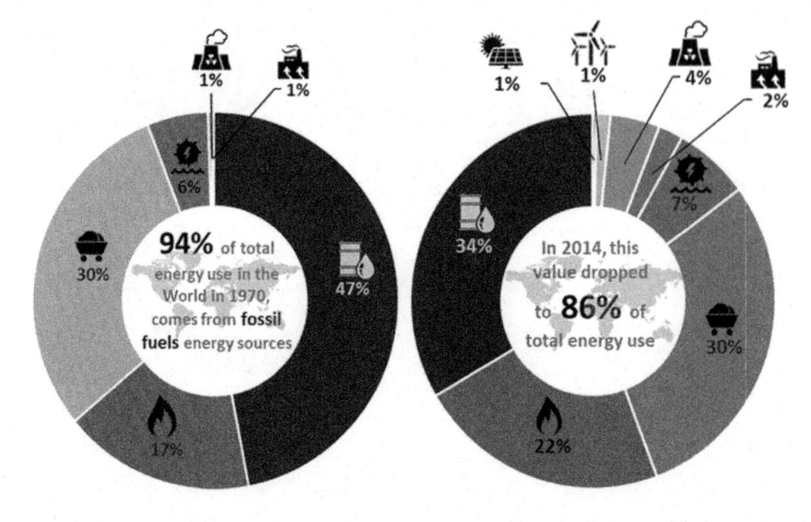

FIG. 5.6 World energy consumption by source in 1970 and 2014. Energy consumption is measured in terawatt-hours (TWh); other renewables include geothermal, biomass, and waste energy; this figure was created by the authors and was based on the Our World in Data (2020).

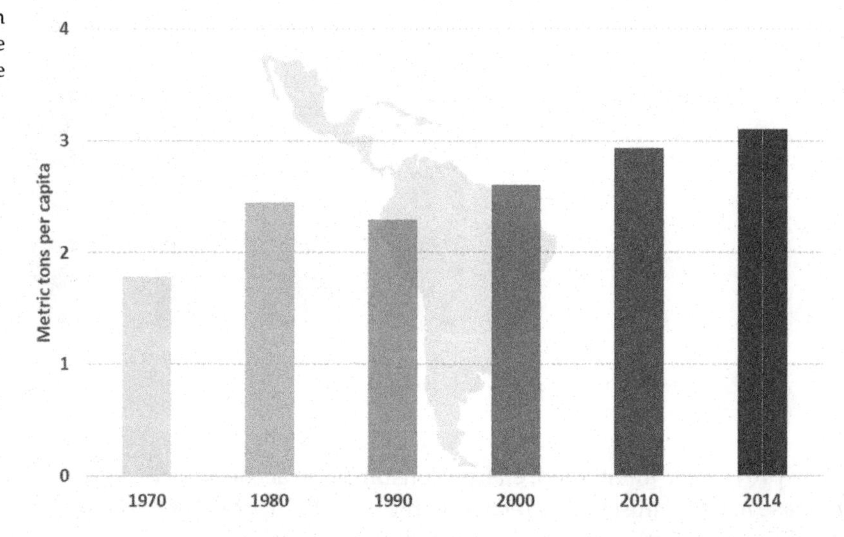

FIG. 5.7 CO_2 emissions (metric tons per capita) in the LAC region, between 1970 and 2014. This figure was created by the authors and was based on the World Bank Open Data (2020).

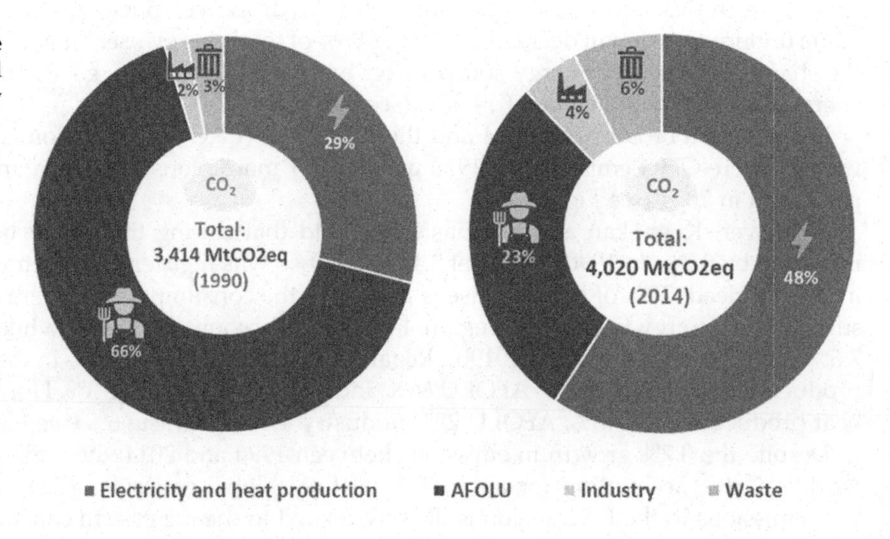

FIG. 5.8 Direct GHGs in Latin America and the Caribbean region per sector in 1990 and 2014.Total GHGs in $MtCO_2eq$/year. This figure was created by the authors and based on Bárcena et al. (2019).

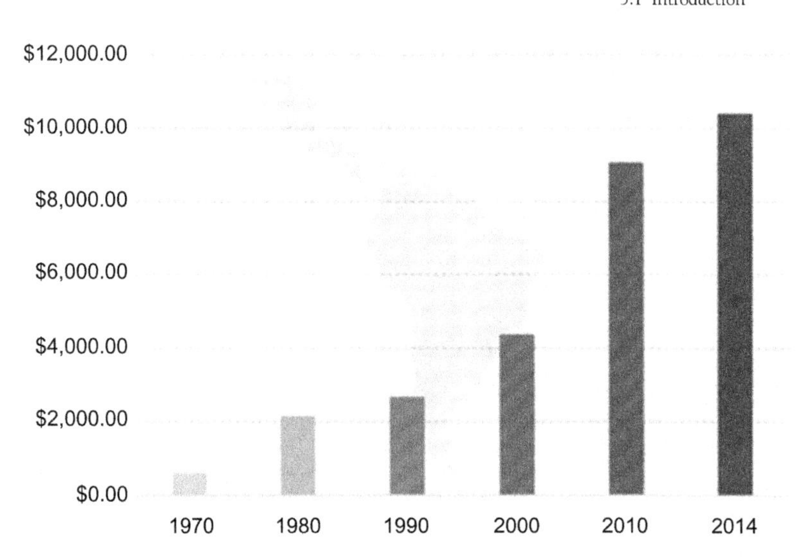

FIG. 5.9 GDP per capita (current US$) in Latin America and the Caribbean, between 1970 and 2014. This figure was created by the authors and was based on the World Bank Open Data (2020).

in economic growth. As can be seen in Fig. 5.9, the gross domestic product (GDP) per capita (current US$) in 1970 was US$612.40 and reached a value of US$10,407.80 in 2014.

In this period, the LAC's GDP per capita (annual %) had an average annual growth rate of approximately 3.75%. Certainly, this increase is related to the structural and stabilisation programmes imposed on Latin American countries by the International Monetary Fund (IMF). These adjustment programmes are neoliberal policies that consisted mainly of the complete opening of their economies to international trade and capital, deregulation of the economy, privatisation, reduction of public expenditure, creation of appropriate conditions for foreign investment and reduction of the role of the state in the economy. Moreover, the 'commodities boom' that occurred between the beginning of the 2000s and end of 2014 also accelerated the process of openness as well as economic growth in the region (Koengkan, 2020; Koengkan and Fuinhas, 2020).

Therefore, this rapid economic growth registered between 1970 and 2014 impacted energy use: in 1971 energy use (in kg of oil equivalent per capita) was 822.6 and reached a value of 1358.20 in 2014, as can be seen in Fig. 5.10.

The increase in energy use in the LAC region, according to Balza et al. (2016), is 220% higher than in the early 1970s and represented an average annual growth rate of 2.8%, where the total energy used increased from 190 million tonnes of oil equivalent (MTOE) in 1971 to 610 MTOE in 2013. Of this increase, the industrial and transport sectors contributed

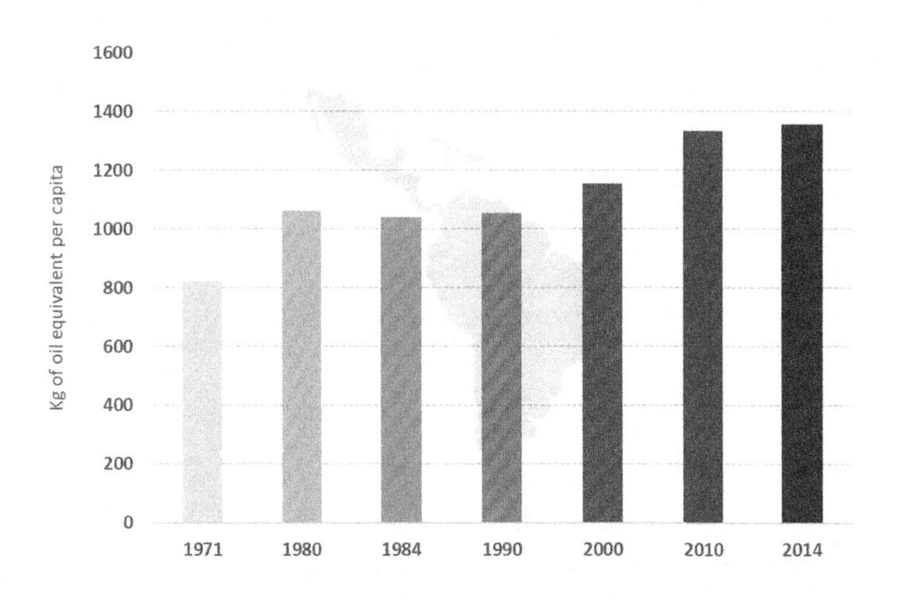

FIG. 5.10 Energy use (kg of oil equivalent per capita) in Latin America and the Caribbean, between 1971 and 2014. This figure was created by the authors and was based on the World Bank Open Data (2020).

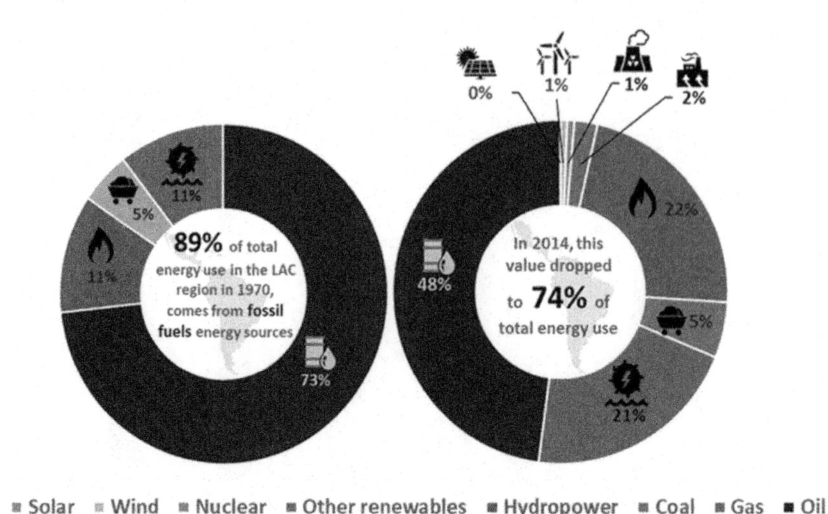

FIG. 5.11 Energy consumption by source in Central and South America in 1970 and 2014. Energy consumption is measured in terawatt-hours (TWh); other renewables include geothermal, biomass, and waste energy; this figure was created by the authors and was based on the Our World in Data (2020).

more than 302 MTOE. To get a sense of this, according to the explanations by Balza et al. (2016), the transport sector had a yearly increase of around 3.5% in energy use from the 1970s, while the industrial sector had an increase of 3% in the same period.

Indeed, 89% of this energy use in the LAC region in 1970 emanates from fossil fuel energy sources (e.g. oil, coal, and gas), and 11% comes from renewable energy sources (e.g. hydroelectric). In 2014 this value dropped to 74% of total energy use in the region. This reduction is related to increase the share of the renewable energy sources (e.g. hydropower, nuclear, geothermal, biomass, waste, solar, and wind) that had a 25.04% share of total energy use (see Fig. 5.11).

The high share of fossil fuel energy sources in total energy use in the LAC region makes it clear how important initiatives are in the energy sector to reduce the consumption of nonrenewable energy and emissions. Therefore, energy planning must consider a scenario of climate change, where additional efforts directed to limiting the emissions from the energy sector are necessary, especially in developing countries such as the LAC countries, where energy demand is expected to increase in subsequent decades. That is, it is estimated that energy use in the LAC region will continue to grow steadily over the coming decades. This growth will accompany economic growth and the rise of the middles classes in the countries in the region. Therefore, total energy use is projected to expand by more than 81.2% through 2040 at an average annual rate of 2.2.%, reaching over 1538 MTOE by the end of the outlook period (see Table 5.1).

The energy transition is certainly a part of this solution and will play an essential role in mitigating the consumption of fossil fuels that are responsible for GHG emissions, environmental degradation, and global climate change. This is one of several reasons for making renewable energies and energy efficiency the main international action priorities to increase clean domestic energy supplies and energy efficiency to mitigate the effects of global warming and climate change (see Fig. 5.12).

In 2020, the energy leaders from 104 countries indicated that renewable energies are an important focal point for boosting electricity production, with the expansion of this kind of energy sources, especially wind and solar, included in energy

TABLE 5.1 Energy use forecast to 2040.

Country	2013	2040	Growth	CAGR
Argentina	81	123	52.6%	1.6%
Brazil	294	577	96.6%	2.5%
Chile	39	99	154.7%	3.5%
Colombia	32	67	110.3%	2.8%
Mexico	191	400	109.2%	2.8%
Venezuela (RB)	69	104	50.7%	1.5%
Other countries	144	169	17.3%	0.6%
LAC region	849	1538	81.2%	2.2%

Notes: Energy use (MTOE) forecast to 2040 in the LAC region. CAGR denotes 'compound annual growth rate'; this figure was created by the authors and was based on data from Balza et al. (2016).

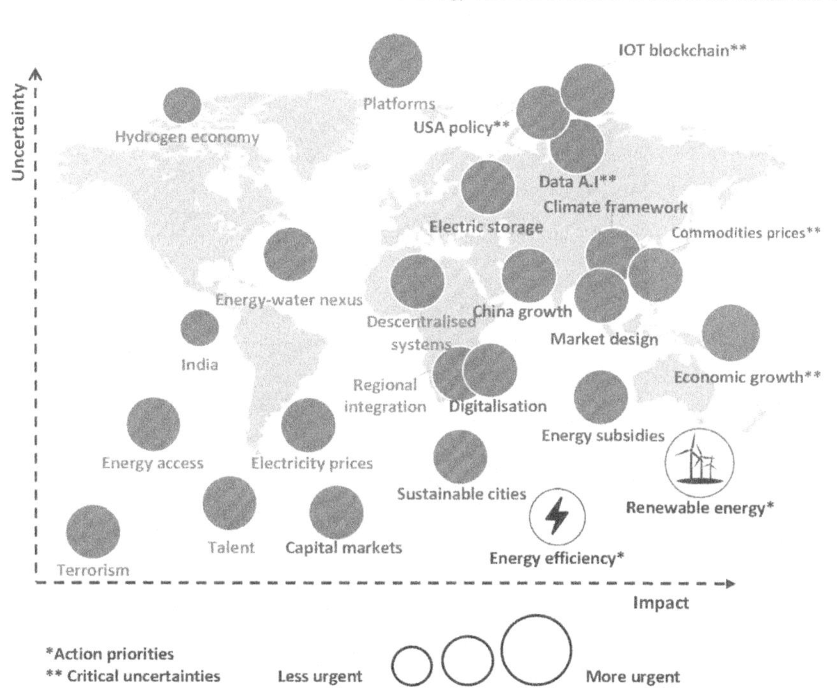

FIG. 5.12 Global Energy Perspectives Monitor. This figure was created by authors and was based on data from the World Energy Council (2020).

strategies of many countries to increase clean domestic energy supplies (World Energy Council, 2020). Energy efficiency also is included in these action priorities and appears as a key theme with clear measures being universally adopted. Indeed, the low costs of energy efficiency and the high potential impacts are recognised in nearly all countries and have led to the adoption of various plans to improve the performance of buildings, gas, and electricity distribution grids, and appliances (World Energy Council, 2020). This global energy perspectives monitor indicates the shifting patterns of this monitor, shaping the process of the energy transition. In the LAC region it is no different (see Fig. 5.13), with renewable energies, energy efficiency, hydroelectricity, biofuels, and energy affordability also being the main action priorities.

As can be seen in Fig. 5.13 in the LAC region decarbonisation and renewable technologies, as well as energy efficiency, appear as a priority focus for action and solutions for improving the sustainability and affordability of the energy sector in the region, as indicated by Koengkan and Fuinhas (2020). Indeed, in the LAC region, renewable energy sources have been an action priority since the 1970s, with several countries in the region using green energies as their main energy sources. An increase in the consumption of this kind of energy and its share in the energy matrix is expected in the coming years. Moreover, the region has significant wind and solar potential that can help to reduce the dependence on large hydropower dams and gas-fired power plants during the period of droughts or other weather events that make the production of energy from the hydropower dams impossible. Energy efficiency in the LAC region is now associated with the decarbonisation efforts mentioned before, with building efficiency standards and appliance labelling programmes implemented across all countries in the region. These action priorities also are part of an energy security strategy to reduce the need for imports and increase energy sector resiliency.

5.2 Energy transition initiatives in Latin America and the Caribbean region

As mentioned before in the introduction, there is no single energy transition but rather various local experiences. Given this, what are the energy transition initiatives that have arisen in the LAC over the last 40 years? Several initiatives have emerged (see Table 5.A1 in Appendix). However, in this section, we comment on the most important energy transition initiatives that arose between 1973 and 2018 in the LAC. Therefore, the process of energy transition in the region began in the 1970s or precisely in 1973 in Brazil and Paraguay with development of the Itaipu Treaty that resulted in the construction of the Itaipu dam from 1974 to 1984, which is a large source of hydropower. Indeed, this large hydropower dam is situated on the Paraná River, approximately 14 km from the international bridge that connects Ciudad del Este in Paraguay and Foz do Iguaçu in Brazil. In the construction, the project proposal included 14 units of 765 megawatts (MW) generators with a total of 10.7 gigawatts (GW)—now expanded to 14 GW. Moreover, the cost of the project was estimated at US$100 million, which will be owned equally by Centrais Elétricas Brasileiras (ELECTROBRAS) and Administración Nacional de Electricidad (ANDE) (IEA, 2020).

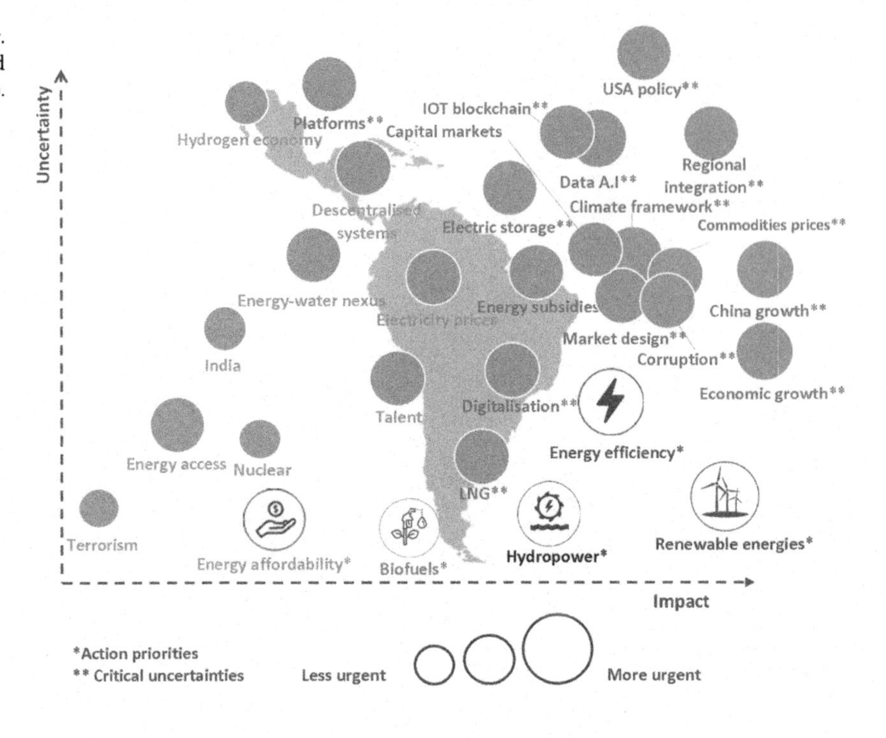

FIG. 5.13 LAC Energy Perspectives Monitor. This figure was created by authors and was based on the data from the World Energy Council (2020).

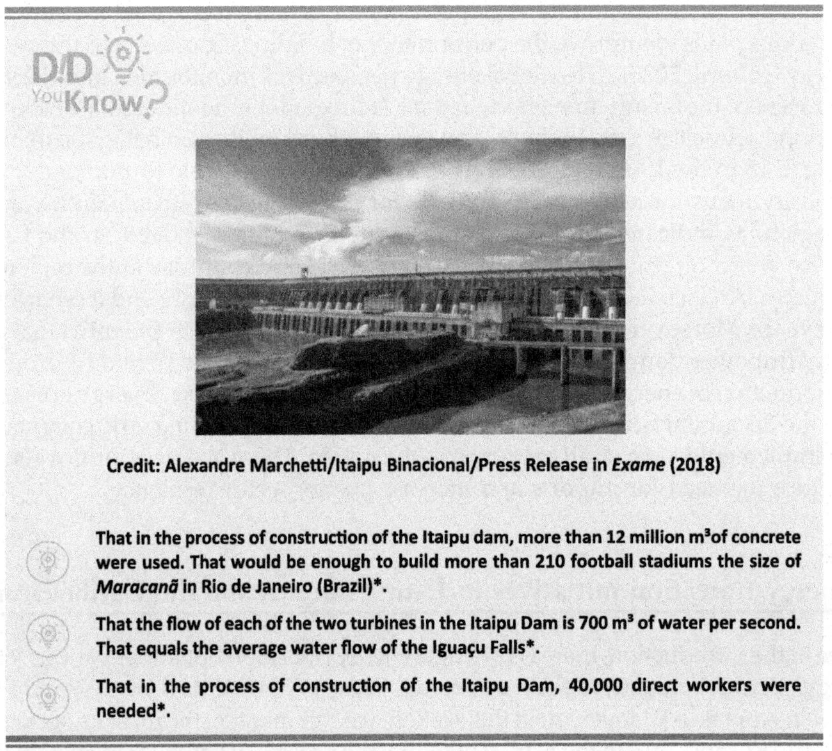

Notes:*This information was retrieved from Itaípu Turismo (2020).

The construction was made to attend to the great energy demand caused by the Brazilian Miracle (Brazilian Portuguese: *milagre econômico brasileiro*), a period of extraordinary economic growth and development in Brazil from 1968 to 1973. Indeed, during this time, the average annual GDP growth was close to 10% during the rule of the Brazilian military government (Veloso et al., 2008).

Other energy transition initiatives arose in the region in the same decade, such as the *Proalcool* programme that started in 1975 in Brazil after the first oil shock that occurred in 1973. Indeed, this programme is a combination of policy instruments that evolve and are mainly used to substitute imported petroleum and to address the needs of both supply and demand sides (Koengkan and Fuinhas, 2020; Gielen et al., 2019; Solomon and Krishna, 2011). Additionally, this programme was driven by biomass-based ethanol demand, but the sector's long-term success continues to be impacted by economic cycles and changing government priorities (Gielen et al., 2019). Moreover, in the same year in Mexico, the Public Electricity Service Law (*Ley del Servicio Público de Energía Eléctrica*) came into force. This law deals with all aspects of renewable energy for public service, including generation, transmission, distribution, transformation and supply of wind, solar, solar photovoltaic, solar thermal, and marine energy sources. Moreover, this legislation establishes that only the public electricity service is of the exclusive competence of the Mexican State, with only the national electricity companies providing electricity on a least-cost basis (IEA, 2020).

In the 1980s these initiatives spread across the LAC region and in 1985, Guatemala established Decree-Law 17 of 1985, regulated by the Governmental Accord 420-1985. This law has the objective of increasing the use of biofuels in the transport sector while reducing the imports of fossil fuels. Moreover, this legislation establishes a blending mandate for bioethanol of 5% (IEA, 2020).

However, in the 1990s energy transition initiatives turned to self-production of renewable energy, with the first country establishing legislation authorising self-produced renewable energy being Costa Rica in 1990. The law authorising self-production of renewable energy (*Ley que Autoriza la Generación eléctrica autónoma o paralela*) regulates the utility-scale private sector projects in Costa Rica (IEA, 2020). In the 2000s, more precisely in 2002, Uruguay established the law of national interest in bioenergy, where bioenergy was declared as national interest for Uruguay. Indeed, this declaration is aligned with the national energy policy (IEA, 2020).

In 2018, Argentina established the RenovAR 3 auction round called 'MiniRen' for renewable power capacity from wind, solar, and hydropower energy sources procurement. Indeed, around 400 MW of renewable energy capacity was opened for competition (IEA, 2020).

All these initiatives were made in the LAC, the only region in the world with a greater share of renewable sources in total energy consumption. In 2014, the share of renewable energy (e.g. hydropower, nuclear, geothermal, biomass, waste, solar, and wind) in energy consumption in the region was 25%, a sizable proportion when compared with the world average of 14.35% (see Figs 5.6 and 5.11). Indeed, of this 25% of renewable energy sources, 21% comes from hydroelectricity (e.g. small and large hydrodams) (see Fig. 5.11). However, the share of energy consumption from hydro has declined since the end of the 1990s due to the development of other energy sources from natural gas and new renewable energy sources (e.g. geothermal, biomass, waste, solar, and wind) (Flavin et al., 2014).

Energy consumption from new renewable energy sources has undergone rapid growth since the end of the 1990s. In 2018 the consumption of this kind of energy comprised only 5.03% of total energy consumption in the LAC, with other renewables including geothermal, biomass, and waste had a share of 2.51%, wind 2.12%, and solar 0.40% (see Fig. 5.14).

This increase reflects investments in the installed capacity of energy sources which occurred in the region. Indeed, investments in this kind of energy sources more than doubled between 2006 and 2012. In 2006, the installed capacity was 11.3 GW and reached a value of 26.6 GW in 2012. Biomass, waste, and wind energy sources make up most of this growth (Koengkan and Fuinhas, 2020). Indeed, this increase is a result of large investments that were made in new renewable energy sources and in Brazil, these investments added up to a value of US$2.4 billion in 2005, and reached a value of US$3.5 billion in 2018 (Fig. 5.15).

Indeed, the increased investment in renewable energy in some LAC made their share of GDP increase (see Fig. 5.16).

As can be seen in the figure above, in some LAC countries, renewable energy investment as a percentage of GDP reached a value of 1.4%, as in the case of Chile. This value is related to several investments in solar photovoltaic generation. However, for other countries in the region, this value is very modest as in the cases of Brazil 0.4%, and Mexico 0.3%.

Given the evidence, what factor has been driving the process of the energy transition in the LAC region? The rapid expansion of energy transition in the LAC region is associated with the fast process of globalisation that is occurring in the region, and that exerts a positive effect on economic growth and consequently increases the energy demand. To meet this demand, new investments in renewable energy technologies are necessary (Koengkan and Fuinhas, 2020; Koengkan et al., 2019c). That is, the process of globalisation has facilitated access to technological advances via

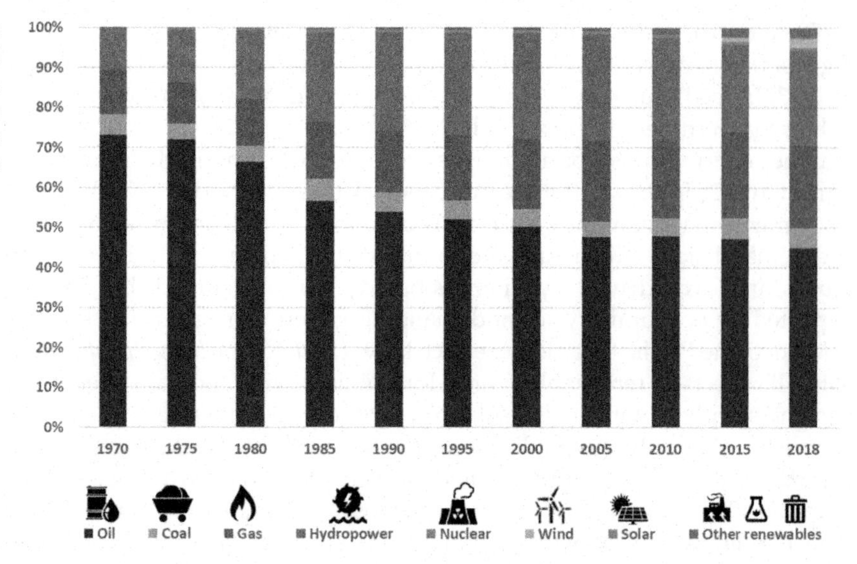

FIG. 5.14 Energy consumption by source in Central and South America between 1970 and 2018. Energy consumption is measured in terawatt-hours (TWh); other renewables includes geothermal, biomass, and waste energy; this figure was created by the authors and was based on the Our World in Data (2020).

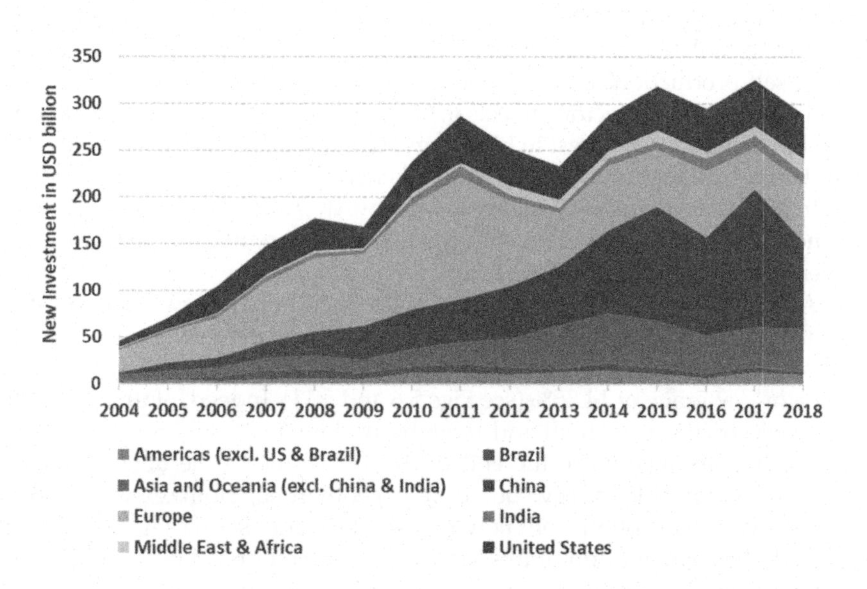

FIG. 5.15 Global trends in renewable energy investment between 2004 and 2018. New investment in billions of United States Dollars (US$). This figure was created by the authors and was based on the IEA (2020).

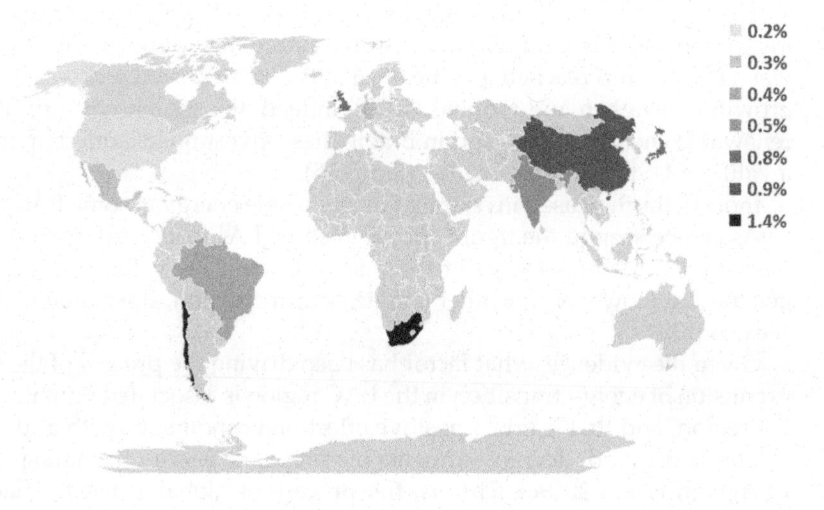

FIG. 5.16 Renewable energy investment (% of GDP) in 2015. This figure was created by the authors and was based on the Our World in Data (2020).

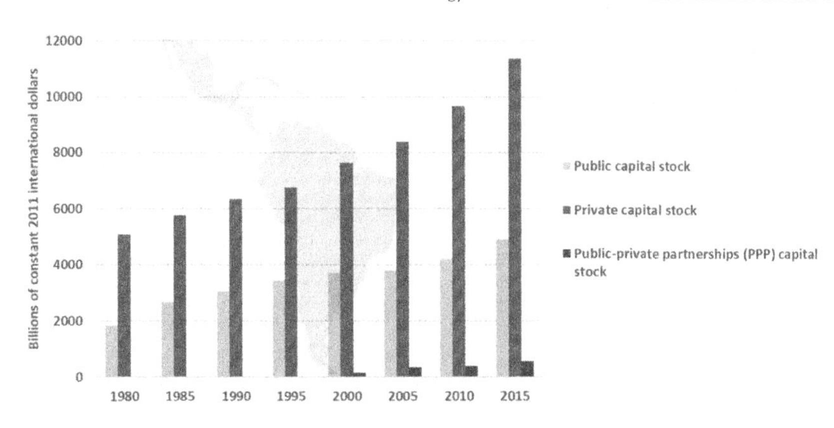

FIG. 5.17 Public, private, and PPP capital stocks in billions of constant 2011 international dollars in the LAC region between 1980 and 2015. This figure was created by the authors and was based on the IMF (2017).

financial and trade liberalisation that consequently contribute to the increase in the installed capacity of renewable energy in the region as referenced by Koengkan et al. (2019b) in their research.

Financial liberalisation increases the public, private, and public-private partnerships (PPP) capital stock (see Fig. 5.17) and consequently reduces the cost of financing, facilitating the investment in renewable energy technologies as pointed out in the studies of Mazzucato and Semieniuk (2018), Kim and Park (2016), and Sbia et al. (2014). This reduction in costs is one of the driving forces that has encouraged investment in renewable energy technologies in recent years and consequently has been reducing their costs significantly.

On the other hand, trade liberalisation also allows LAC countries to import energy-saving technologies, products, and/or processes from developed countries that consume less energy. As indicated by Ghani (2012), evidence of this is visible in developing countries, where the reduction of energy consumption is more evident than in developed countries because developing countries have more capacity to absorb transferred technologies than developed ones.

The economic integration caused by financial and trade liberalisation in the LAC region allows them to access new energy technologies and consequently adopt them in their industries. However, this technology transfer to renewable energy in the LAC region depends on the production and absorption capacity and the path dependency of significant investment in new technologies.

The process of financial and trade liberalisation in the LAC region began in the 1970s in Chile with the profound shift towards free-market economies during the dictatorship of Pinochet (Ahumada and Andrews, 1998). Many other countries from the region, according to Koengkan et al. (2019a), implemented the same neoliberal economic policies during the process of the 'Washington Consensus' and the ''Brady Plan', which were a combination of policies to promote 'macroeconomic adjustment' and restructuring of external debt.

The adoption of these policies occurred between 1980 and 1992, with several countries for example Mexico and Costa Rica in 1989, Venezuela (RB) in 1990, Uruguay in 1991, Argentina and Brazil in 1992, passing schemes for an in-depth process of financial and trade openness, liberalisation of foreign investment, privatisation of substantial portions of the public sector, and reduction of import barriers (Koengkan and Fuinhas, 2020; Aizenman, 2005; Vásquez, 1996). Indeed, before this adjustment, the annual growth rate of the region was approximately 0.35% in 1990. After 'macroeconomic adjustment', the LAC's GDP per capita had an annual compound growth rate of 4.58% in 1994 (see Fig. 5.18).

In the 1990s, the LAC grew again and integrated into the world, with most LAC countries adopting unilateral opening policies, reducing their tariffs and eliminating other trade restrictions. This integration process resulted in a period of several regional agreements, such as *Asociación Latinoamericana de Integración* (ALADI), American Free Trade Agreement (NAFTA) which Mexico joined, and the Common Market of the South (Mercosur) created by Argentina, Brazil, Paraguay, and Uruguay. After these regional agreements, LAC imports grew at an average rate of 11% while exports increased at an average rate of 8.1%, between 1990 and 1999, improving its share in world trade (Terra, 2003).

Due to this, trade in the 1990s in the LAC region has been characterised by imports growing at rates greatly above those of exports. This occurred due to the fact that imports had been drastically reduced in the wake of the debt crisis that followed the Mexican financial crisis of 1982 (Ventura-Dias et al., 1999). Indeed, in this period, imports had a vital role to play in the modernisation of the production process, where modern machines and better industrial inputs contributed to the technological upgrading of the industrial base in the region (Ventura-Dias et al., 1999). The importance of trade to the LAC economy is visible in the share of trade in GDP; in 1985 before the macroeconomic reforms the share of trade on GDP was 28.67%, while after the adjustments the share grew to 38.34% in 2000 and reached a value of 43.75% in 2015 (see Fig. 5.19).

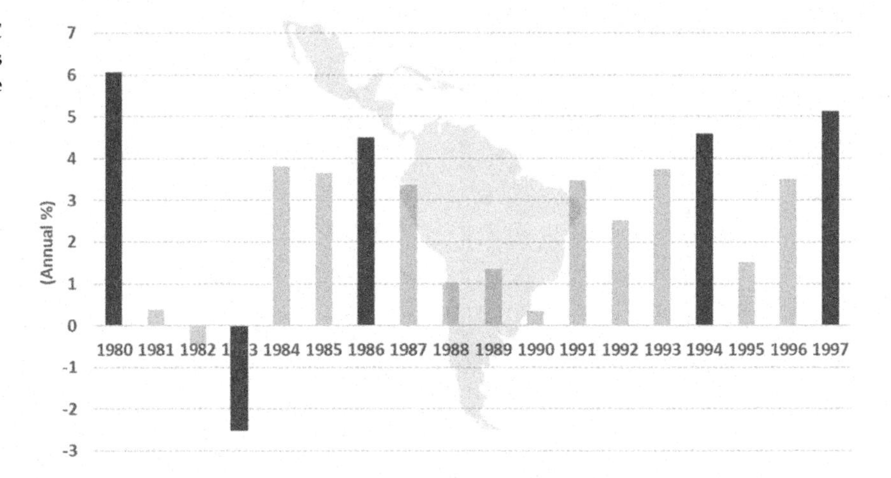

FIG. 5.18 GDP growth (annual %) in the LAC region between 1980 and 1997. This figure was created by the authors and was based on the World Bank Open Data (2020).

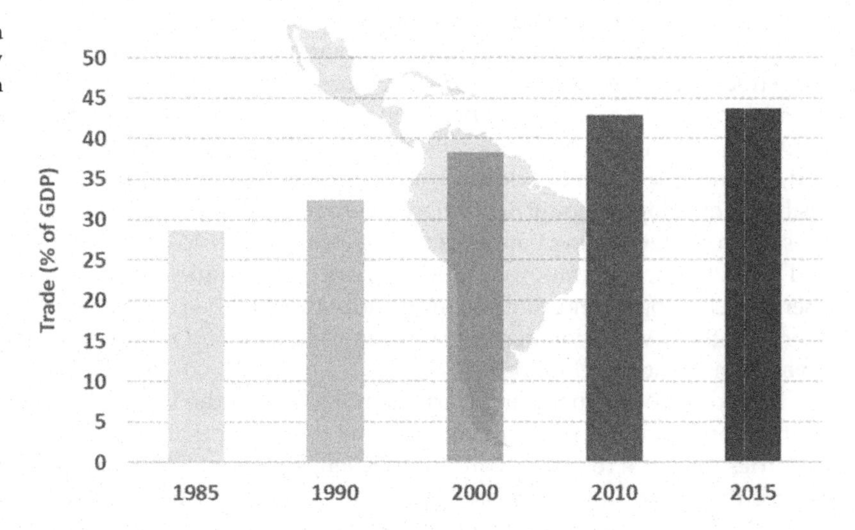

FIG. 5.19 Trade (% of GDP) in the LAC region between 1985 and 2015. This figure was created by the authors and was based on the World Bank Open Data (2020).

Financial liberalisation in the region followed the same way of trade liberalisation, with the inflow of capital in the LAC region resuming after the Brady Plan in the early 1990s. The magnitude of the financial liberalisation in the LAC region can be grasped with the index of capital mobility; in the 1980s the index capital mobility had a value of 40 and in 1990s arose to about 75, with normalising, completely free capital mobility being 100 (Aizenman, 2005). Indeed, the financial liberalisation caused by macroeconomic adjustments promoted the entrance of foreign direct investments (FDI) flows. FDI flows worldwide grew dramatically between 1990 and 1997. Indeed, it was the developing countries that received the most of these flows, with their share of these flows being 15% in 1990 and reaching a value of 38% in 1997. Indeed, the increase in FDI flows into the LAC region is visible in the FDI inflows; in 1985 before the macroeconomic reforms these inflows were US$6.44 billion and after the adjustment, these inflows grew to US$9.73 billion in 2000 and reached a value of US$2.91 trillion in 2015 (see Fig. 5.20).

Indeed, these FDI inflows to the LAC region in the 1990s evolved in three phases (Birch and Halton, 2008). Between 1990 and 1993, investors seemed to favour acquiring already existing assets. However, between 1994 and 1996, most investments were directed to large-scale projects via restructuring of existing foreign firms or modernising recently privatised firms. In 1997, the acquisition of existing assets, this time to consolidate the market power became the most common form of foreign investment in the region. In this period, more money was spent on the purchase of already existing private assets than on privatisation.

Moreover, during the 1990s, the LAC region registered an increase in FDI in the industries related to natural resources and energy sector (Birch and Halton, 2008). For example, in Argentina between 1990 and 1996, the energy sector and the gas and water industries were the leading FDI recipients, where the country received 26%; the

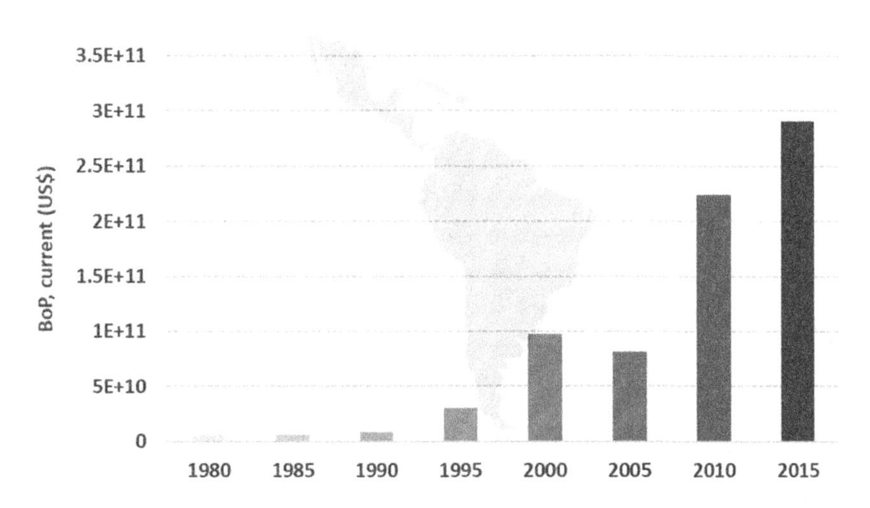

FIG. 5.20 Foreign direct investment, net inflows (balance of payments (BoP), current US$) in the LAC region between 1980 and 2015. This figure was created by the authors and was based on the World Bank Open Data (2020).

petroleum and natural gas industry received 15%, the chemical products industry sector 11%; and food, beverages and tobacco and financial services, each had 10%. In Brazil in 1990, the chemical industry accounted for 14% falling to 11% in 1995. Between 1996 and 1997, investments in electricity, water, and gas soared to 23% of total FDI inflows, mainly due to privatisation over the last 3 years.

The financial sector accounted for 10%, reflecting the restructuring of the Brazilian banking system; Chile accounted for 47% of total FDI inflows between 1974 and 1996 in the mining and quarrying sector. Other services received 25% while manufacturing received 16% of all inflows between 1990 and 1996. The energy, gas, and water sector soared from a 3% share between 1990 and 1996 to 27% in 1997. This change was due in large part to the acquisition of part of the Chilean electric company *Enersis* by the Spanish company Endesa-España; Mexico received 49% of inflows between 1981 and 1993, evenly divided between the manufacturing and services sectors.

Indeed, between 1994 and 1996, the machinery and equipment sector received 24% of these inflows, reflecting the substantial investment in the automotive, electronics, and electrical equipment industries. The food, beverages, and tobacco sector received 12%; the finance and insurance sector 11%, and other services received 10% of these inflows. In 1997, food, beverages, and tobacco received 36% of the total of these inflows. Indeed, the entry of these FDI inflows during the 1980s–1990s for the energy sector in Latin America reduced the high investment and maintenance costs of renewable energy projects via public-private partnerships (PPPs), thus narrowing the gap in financing (Coviello et al., 2012). The reduction in these costs increased the generation of electricity from renewable energy plants; the electricity generation from small and large hydro dams in the LAC region was 55% in the 1970s which reached 67% in the 1990s (Flavin et al., 2014).

However, this process of financial and trade liberalisation that began in the 1970s intensified with the 'commodities boom' that occurred between 2004 and 2014, with the regional average growth rate of 7.40% (Santos, 2015). The cycle of commodity prices in Latin American economies impacted the degree of economic openness or more precisely, the degree of dependence on external demand *vis-à-vis* domestic demand or markets. Indeed, between 1990 and 1993 the degree of economic openness was 28.6, between 1998 and 2001 it was 38.5 on a scale of 0–100, and between 2006 and 2009 it was 44.7, where 100 represents an open economy. Thus, in this period between 1990 and 2009, the degree of economic openness had 50.71% growth (Carneiro, 2012).

Furthermore, this fact seems to have allowed the region to surpass the problems generated by the 2008–2009 financial crisis. Indeed, the growth in the degree of economic openness was caused by an increase in exports and imports in the region; in 2004 the exports of goods and services (BoP current US$) was US$6 billion and reached a value of US$1.39 trillion in 2018, while imports in 2004 were US$5.40 billion and reached a value of US$1.41 trillion in 2018 (see Fig. 5.21).

Indeed, the growth in the manufacturing sector caused by the commodities boom led to FDI inflows into the largest economies in the region, with 61% of total FDI inflows in Mexico and 38% in Brazil. The renewable energy sector received 5% of these FDI inflows from 2005 to 2007. However, it was from 2015 to 2017 that this sector was the main recipient of new FDI inflows, receiving 26% of them. Investments in new renewable energy, domestic and foreign, reached US$6.2 billion each in Brazil and Mexico in 2017, US$1.8 billion in Argentina, and US$1.5 billion in Chile. Most projects in these countries involved foreign companies. Moreover, most of these investments were made in solar (35%) and wind (32%) technology between 2005 and 2017 (ECLAC, 2018).

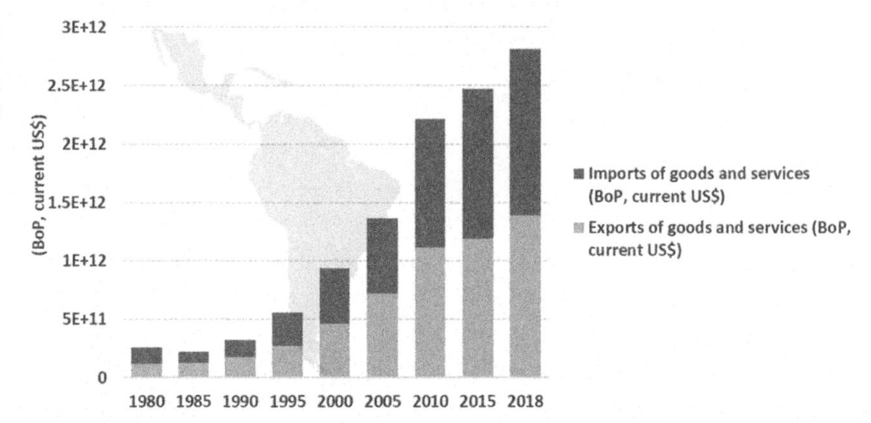

FIG. 5.21 Imports and exports of goods and services (balance of payments (BoP), current US$) in the LAC region between 1980 and 2018. This figure was created by the authors and was based on the World Bank Open Data database (2020).

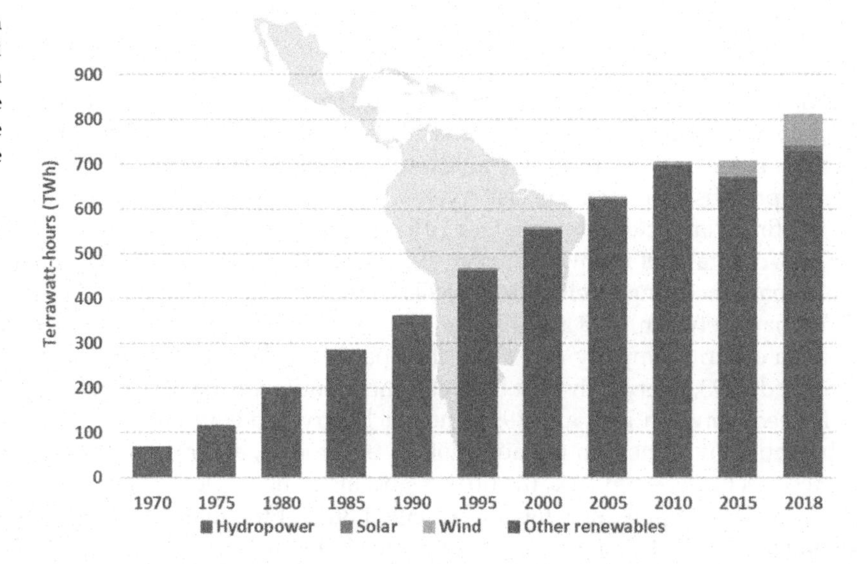

FIG. 5.22 Renewable energy consumption in Central and South America between 1970 and 2018. Energy consumption is measured in terawatt-hours (TWh); other renewables include geothermal, biomass, and waste energy; this figure was created by the authors and was based on the Our World in Data (2020).

These two phases in the process of insertion of LAC economies that occurred from 1989 to 1992 and 2004 to 2014 influenced the degree of openness of economies in the region. Indeed, as can be observed, the process of the energy transition to the consumption of renewable energy sources followed the same trend (see Fig. 5.22).

This increase in renewable energy consumption exerts a positive effect on the environment with the reduction of CO_2 emissions from the consumption of fossil fuels that are responsible for environmental degradation, global warming, and climate change.

5.3 Energy transition challenges in Latin America and the Caribbean

Indeed, the countries in the LAC region have common and unique characteristics that may pose challenges to the process of energy transition. Therefore, what are the challenges to the process of energy transition in this region? The challenges of LAC countries will now be briefly discussed.

5.3.1 Brazil

As we know, there are several initiatives that encourage the process of energy transition in the country. However, these initiatives come across structural problems in the country. First, if we observe the energy expansion in Brazil carefully, there are no targets. That is, long-run projections are made for a possible composition of the energy matrix in the country. Therefore, goals to mitigate climate change are not an objective for energy planners in Brazil in the process of energy expansion.

The lack of these targets, as mentioned before, affects the organisation and dialogue between the numerous economic agents involved. As identified by FGV Energia (2016), Brazil has inefficient communication between the main planner and the energy transition agenda. Indeed, when referring to the main planner, we are referring to the Ministry of the Environment (ME) and the Ministry of Mines and Energy (MME)—and consequently between their regulatory agencies, such as *Agência Nacional de Energia Elétrica* (ANEEL), *Instituto Brasileiro do Meio Ambiente e dos Recursos Naturais Renováveis* (IBAMA), and *Agência Nacional do Petróleo, Gás Natural e Biocombustíveis* (ANP). However, as identified in this same report, there is no structure and interaction between the programmes and the central coordination.

Another challenge that is evidenced in Brazil is related to the lack of a parallel between the various mechanisms of planning, development, and financing of cleaner energy sources with the transport sector. This lack of a parallel between renewable energy sources and the transport sector is visible in Fig. 5.23 where the extensive use of fossil fuels in the transport sector is a major source of CO_2 gases generated in the country.

The evidence shows that the transport sector contributes 11% of the total CO_2 emissions caused by the consumption of energy. That is, this contradicts the Brazilian leadership regarding the development of biofuel technology. It should be noted that Brazil is the only country that produces cars with technology flex engines, where the consumer can interchangeably use gasoline and ethanol.

Indeed, the large share of the transport sector in CO_2 emissions is related to the discovery of the presalt layer in the middle of the 2000s. This discovery made the government's resources and private attention to be focused on the potential of oil and the production of its by-products in the country. That is, investment in renewable energy sources agenda was put aside. It is evident that there is a lack of engagement and public planning to promote the use of renewable energy sources in Brazil. Moreover, this lack is correlated with the transport sector deficit, where this sector has a small rail network and uses very little of its coastal shipping potential. Other factors, such as the low economic activity in the last 5 years (see Fig. 5.24), have been causing a marked slowdown in renewable energy projects (see Fig. 5.15).

5.3.2 Argentina

As for Argentina, even with some ambitious proposals for energy transition, its capacity for achieving them is questionable due to the weak economic growth in recent years (see Fig. 5.25).

Indeed, the search for social and economic stability, which has been falling in recent years, has been overshadowing the energy transition agenda in the country. This lack of attention for the renewable energy agenda has reflected in the structured planning for the energy sector. According to the Institute of the Americas (2016), the Argentinian government planned for renewable energy consumption to increase from 6.6% today to 14.6% in 2025, and the consumption of oil to decrease from 32.6% to 23.7%, as well as natural gas from 51.1% to 49.6%. However, these intentions have not yet been formalised due to the low economic growth that consequently impacted investment in renewable energy sources.

The annual investment in renewable energy from 2003 to 2012 was 550 MW on average, well over half of the 1000 MW annual investment target for this kind of energy, though far from reaching the goal of 10 GW of installed

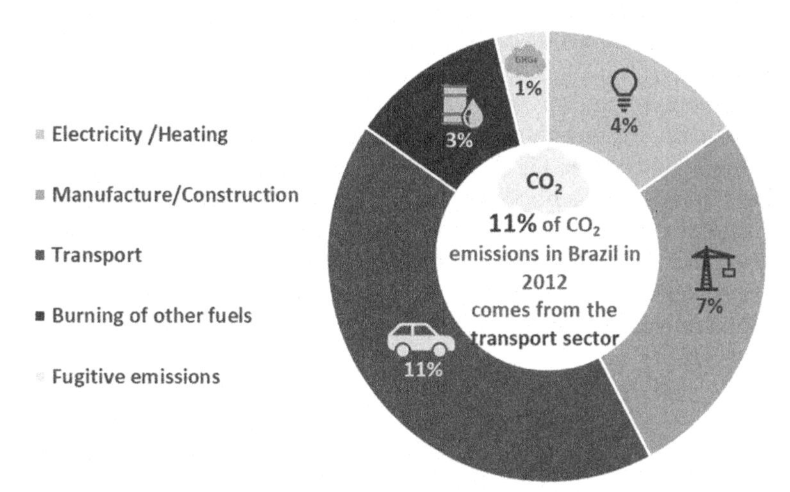

FIG. 5.23 CO_2 emissions from energy consumption per sector in Brazil in 2012. This figure was created by the authors and based on FGV Energia (2016).

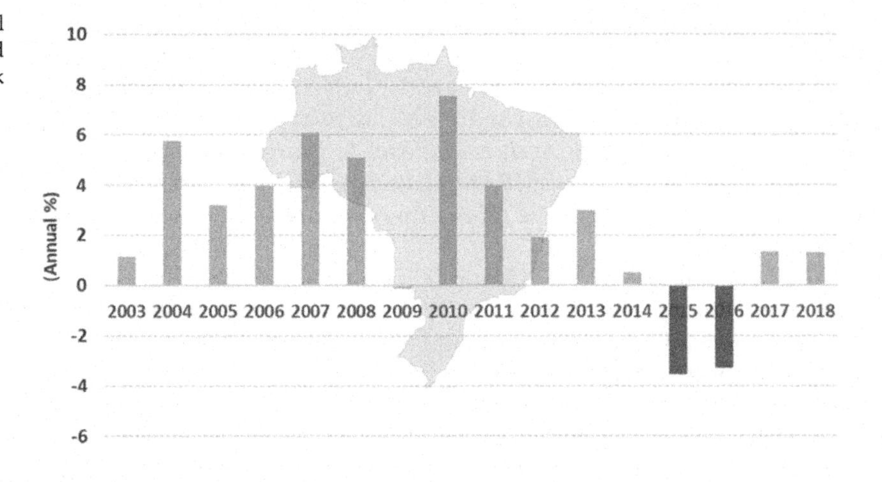

FIG. 5.24 GDP growth (annual %) in Brazil between 2003 and 2018. This figure was created by the authors and was based on the World Bank Open Data (2020).

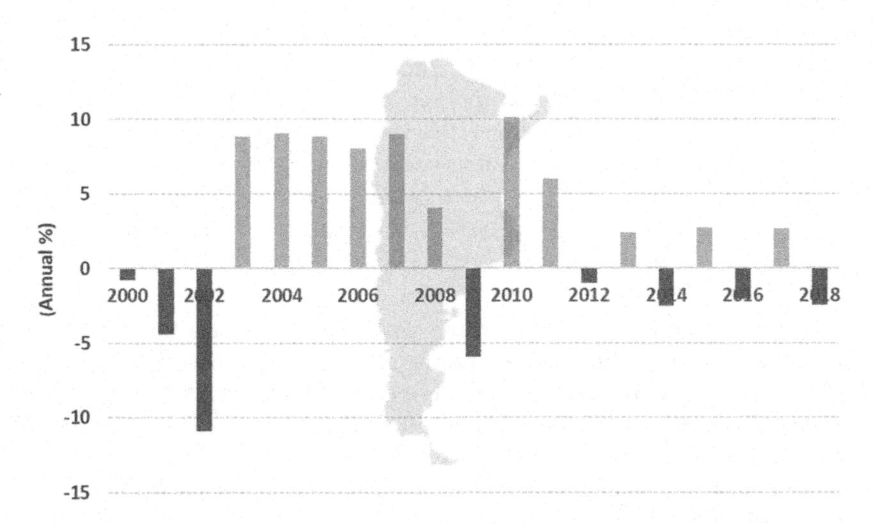

FIG. 5.25 GDP growth (annual %) in Argentina between 2000 and 2018. This figure was created by the authors and was based on the World Bank Open Data (2020).

capacity in 2025 (FGV Energia, 2016). This low investment was reflected in the government's decision to declare a state of emergency in the electricity sector in 2015 due to fuel shortages. This state of emergency in the electricity sector lasted until the end of 2017. This evidence shows the weakness of the Argentine energy plan, as well as one of the possible obstacles/or challenges for the process of the energy transition in the country.

Indeed, another challenge for the process of the energy transition in Argentina is the allocation of natural resources and the energy structure, which are based on fossil fuel energy sources, as can be seen in Fig. 5.26 with 64% of energy production in Argentina coming from fossil fuels in 2013.

Therefore, it is necessary to engender a great effort in planning and investment to change this situation. Moreover, the transport sector is not at the same stage of development as the energy sector. In Argentina, the transport sector is responsible for around 12% of total CO_2 emissions in the country (see Fig. 5.27).

Despite being the third-largest emitter of these gases, the sector has no mitigation targets, plans to increase energy efficiency, or plans for changing the fuels used in the transport sector. This reflects the poor coordination between the environmental planners and the energy industry planners in the country. Additionally, the existence of several barriers such as governance and institutional and macroeconomic factors, according to FGV Energia (2016), makes the establishment of an environmental and energy transition agenda difficult.

5.3.3 Mexico

The environmental agenda in Mexico is regarded as a matter of international relations, where there is no integration with domestic planning bodies as also identified by FGV Energia (2016). This lack of integration between the

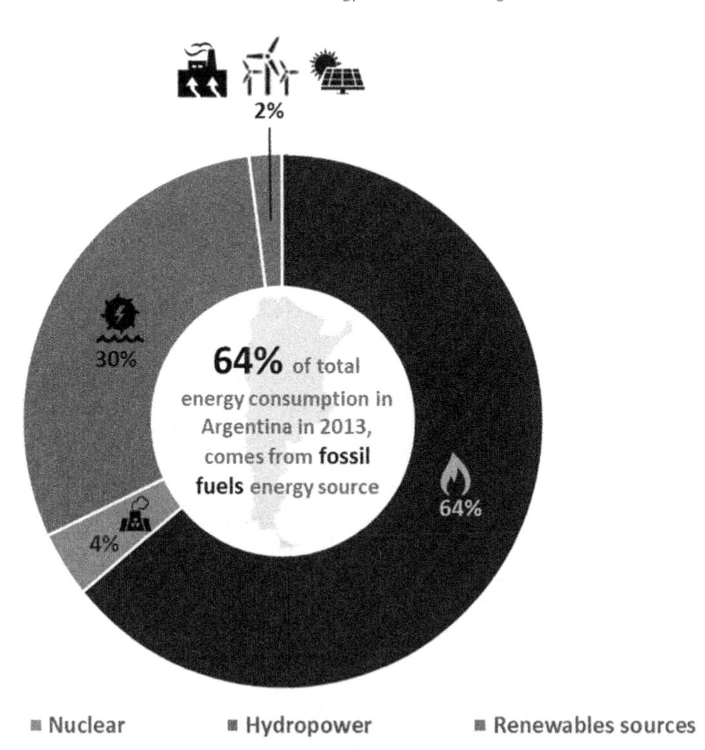

FIG. 5.26 Energy consumption by source in Argentina in 2012. This figure was created by the authors and based on FGV Energia (2016).

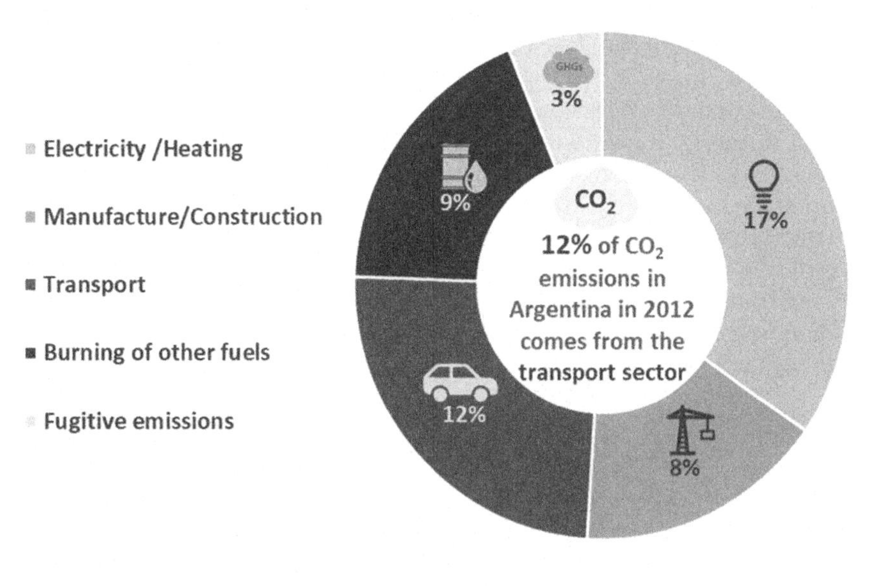

FIG. 5.27 CO_2 emissions from energy consumption per sector in Argentina in 2012. This figure was created by the authors and based on FGV Energia (2016).

environmental agenda and local government is reflected in the lack of consistency in the renewable energy plan in Mexico. Indeed, this inconsistency is also related to the official definition of clean energy by the Mexican government, where it comprises renewable sources, nuclear energy, hydropower, natural gas, and a high-efficiency cogeneration process.

However, considering natural gas as renewable energy by the government of Mexico raises questions from environmentalists. Indeed, the participation of this kind of energy source in energy consumption in the country is about 44.6% (see Fig. 5.28).

Indeed, the substantial share of natural gas in energy use in Mexico is caused by three factors. The first factor is related to the significant reduction of CO_2 emissions in the energy sector caused by the replacement of thermal oil by natural gas. The second factor is related to the international environmental debate, which indicates that natural gas emits fewer pollutants than oil and its derivates. The third factor is related to the low cost of gas and lobbying

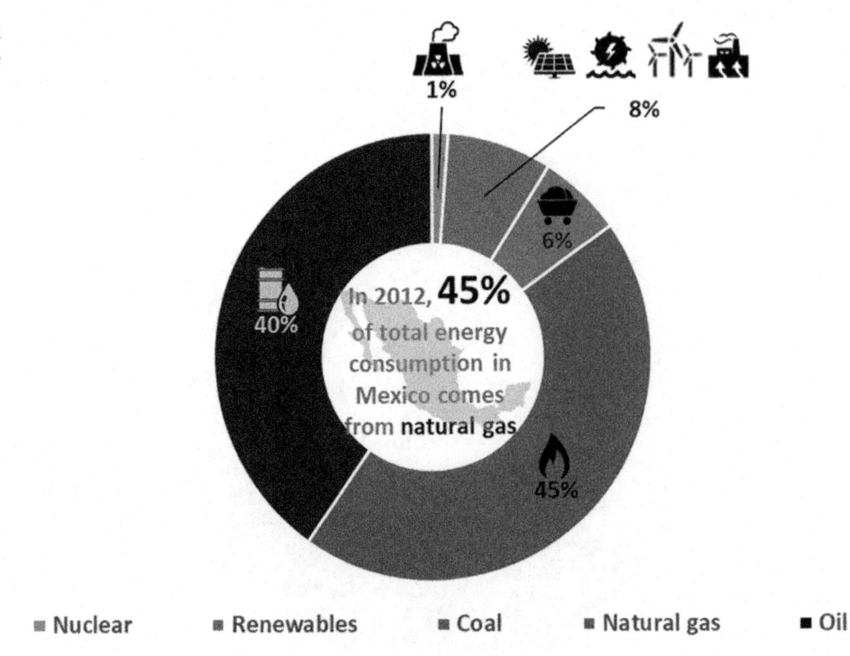

FIG. 5.28 Energy consumption by the source in Mexico in 2012. This figure was created by the authors and based on FGV Energia (2016).

by industries, encouraging the use of this kind of source. All these factors diminish the appeal of the inclusion of renewable energy sources in the energy matrix, such as wind and solar, where the country has immense potential for these two energy sources.

Another area that is not widely covered in the energy transition and environmental planning is the transport sector. Like other LAC countries, such as Argentina and Brazil, the transport sector in Mexico is the second largest CO_2 emitter at about 20% (see Fig. 5.29).

The high share of the transport sector in CO_2 emissions in Mexico contributes to severe air pollution problems, which occur repeatedly in Mexico City. The lack of energy transition to a more efficient one in the transport sector in Mexico, particularly from the energy point of view, could solve this problem in the big cities in Mexico. Nevertheless, the discussion on energy transition for transport is inevitably left aside. At the moment, the oil price is low and in the last years, the Mexican government has been reducing the tax revenues from the process of exploration and production of oil, making this kind of energy more economically competitive than renewable energy sources.

Among initiatives and challenges of the energy transition that has been briefly discussed this chapter, now we can highlight a case of the success of energy transition in the LAC region. Indeed, this success comes from the capacity to make clean energy sources available equally to the entire population, with the existence of well-structured legal and

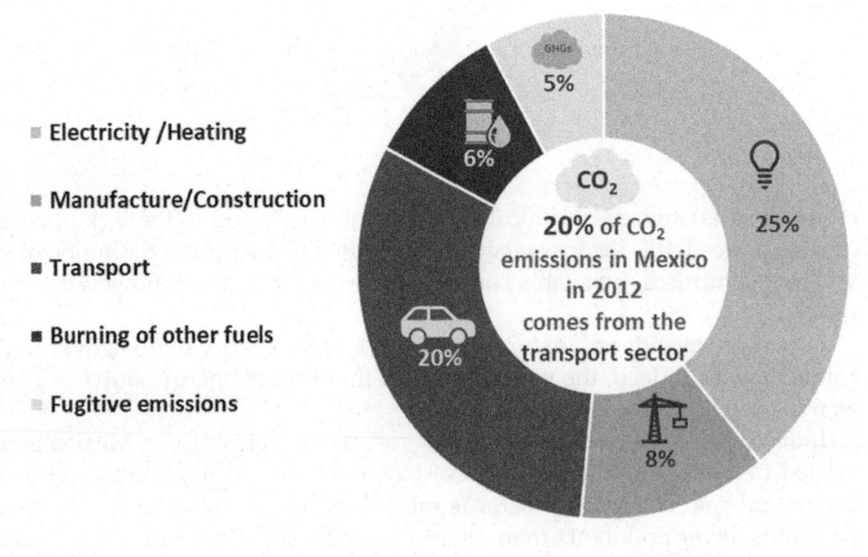

FIG. 5.29 CO_2 emissions from energy consumption per sector in Mexico in 2012. This figure was created by the authors and based on FGV Energia (2016).

regulatory processes that encourage the development of renewable energy technologies, via the robustness of institutions and economic development, and with the capacity to devise strategies between private and public companies in the development of clean energy sources. The country with all these feats accomplished is Uruguay, and we will briefly discuss their case of success in the process of the energy transition.

5.4 Uruguayan energy transition—A success case

As mentioned earlier, Uruguay has the capacity to make clean energy sources available equally. This is due to its development of strategies between private and public companies, as well as the existence of well-structured regulatory and legal processes, and the robustness of institutions and economic growth since 2004 (see Fig. 5.30). This has encouraged the development of renewable energy technologies in the country.

According to FGV Energia (2016), these factors have also attracted private capital investment from abroad that has consequently helped in the process of incorporation of new business models for the energy sector in the country. Moreover, the development and implementation of a clean energy transition plan with long-term energy goals, such as the 'National Energy Policy 2005–2030' plan, which was designed by the Uruguayan government (IRENA, 2015) have also facilitated the attraction of these investments. The objective of this plan is to create diversity in the energy matrix of the country, reduce dependency on nonrenewable energy sources, improve energy efficiency, and increase the use of renewable energy sources. Indeed, this plan sets a target of 50% of primary energy to come from renewable energy sources for electricity generation, in industry, domestic heating and the transport sector by 2015 (IRENA, 2015). Therefore, this energy transition plan is evidence of a well-structured plan that contributes to the realisation of a successful energy transition.

As a result of this well-structured energy transition plan, renewable energy sources accounted for about 93% of Uruguay's energy consumption in 2015, where hydropower electricity accounted for about 58% of energy consumption and biofuels used in the transport sector accounted for about 18% and other renewable energy sources that include solar, wind, and solar photovoltaic accounted for 15.90% (see Fig. 5.31).

Due to this, in 2012, the country ranked first place among the countries that invested most in renewable energy technologies per unit of GDP. Additionally, the country is considered the second green energy leader in the LAC region, due to the existence of a successful renewable energy plan focused on the promotion of renewable energy technologies and energy efficiency (FGV Energia, 2016). Given the existence of a successful energy agenda focused on energy transition, Uruguay was also considered a successful study case at COP 21, having established a reduction target of 25% in CO_2 emissions from consumption of electricity and heat production by 2030, based on 1990 values. Indeed, the success of Uruguay in reducing these emissions can be seen in Fig. 5.32.

Indeed, it is important to make it clear that Uruguay is a small country, with less diversity and political and economic complexity when compared with its neighbours, such as Argentina and Brazil, which have been suffering from severe economic and political problems in recent years. Moreover, it is worth remembering that the electrical integration of the country is less complicated compared with other LAC countries, and therefore, contributed to the development and diffusion of renewable energy technologies.

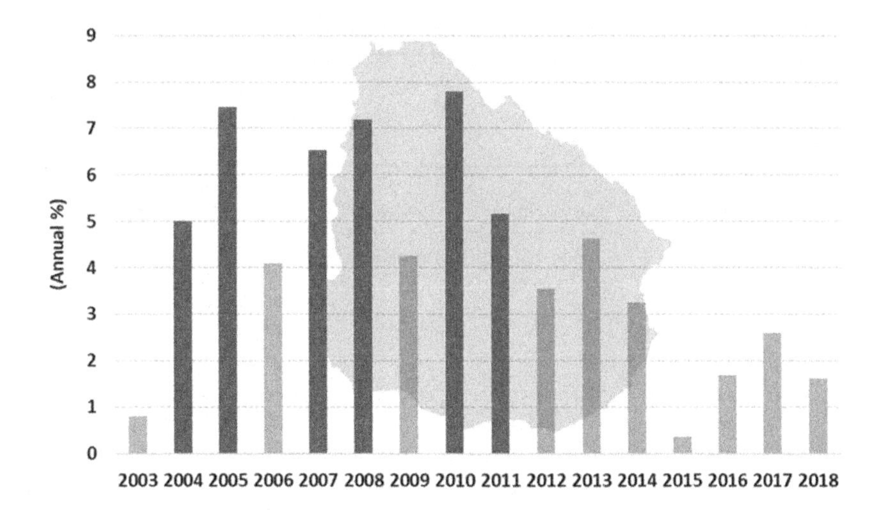

FIG. 5.30 GDP growth (annual %) in Uruguay between 2000 and 2018. This figure was created by the authors and was based on the World Bank Open Data (2020).

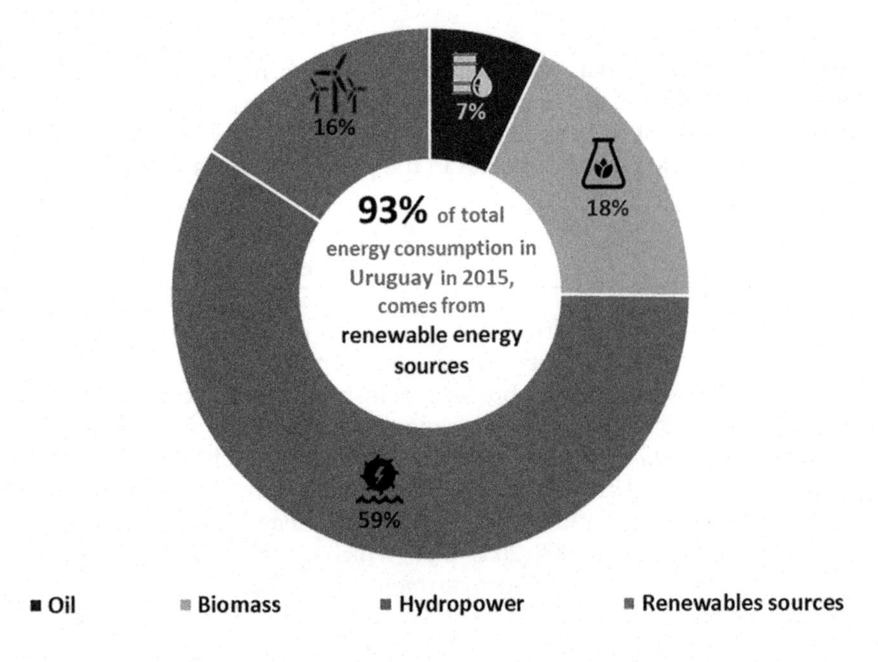

FIG. 5.31 Energy consumption by source in Uruguay in 2015. This figure was created by the authors and based on FGV Energia (2016).

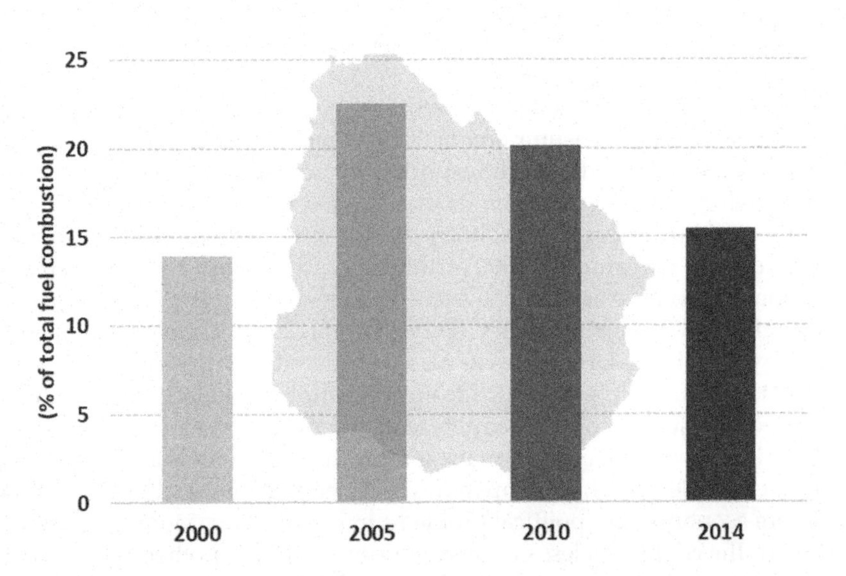

FIG. 5.32 CO_2 emissions from electricity and heat production, total (% of total fuel combustion) in Uruguay between 2000 and 2014. This figure was created by the authors and was based on the World Bank Open Data database (2020).

Appendix

TABLE 5.A1 Energy transition initiatives in the LAC region.

Initiative	Country (ies)	Year	Status	Jurisdiction	Technology
Itaipu Treaty (Paraguay-Brazil)	Paraguay-Brazil	1973	In force	International	Hydropower
PROALCOOL program	Brazil	1975	In force	National	Biofuels
Public Electricity Service Law (Ley del Servicio Público de Energía Eléctrica)	Mexico	1975	In force	National	Wind, solar photovoltaic, solar thermal, and marine energy
Biofuel Law (Ley del Alcohol Carburante)	Guatemala	1985	In force	National	Biofuels

Continued

TABLE 5.A1 Energy transition initiatives in the LAC region—cont'd

Initiative	Country (ies)	Year	Status	Jurisdiction	Technology
Law Authorising Autoproducers (Ley que Autoriza la Generación eléctrica autónoma o paralela)	Costa Rica	1990	In force	National	Multiple renewable energy sources
General Environmental Law (Ley General Ambiental) Law 99	Colombia	1993	In force	National	Multiple renewable energy sources
Ethanol Blending Mandate	Brazil	1993	In force	National	Biofuels
Law on the Energy Sector (Ley de Régimen del Sector Eléctrico)	Ecuador	1996	In force	National	Multiple renewable energy sources
Organic Law on Geothermal Resources (Ley Organica de Recursos Geotermicos)	Peru	1997	In force	National	Geothermal
National Interest in Bioenergy	Uruguay	2002	In force	National	Biofuels
Programme of Incentives for Alternative Electricity Sources—Programa de Incentivo a Fontes Alternativas de Energia Elétrica—PROINFA	Brazil	2002	In force	National	Multiple renewable energy sources
National Energy Policy Green Paper (2006–2020)	Jamaica	2006	In force	National	Multiple renewable energy sources
Costa Rica National Programme on Biofuels (Programa Nacional de Biocombustibles) 2008	Costa Rica	2008	In force	National	Biofuels
Program for Rural and Social Energy (PERYS)	Chile	2009	In force	National	Multiple renewable energy sources
Fund for Energy Transition and Sustainable Energy Use—FOTEASE	Mexico	2009	In force	National	Multiple renewable energy sources
Solar Thermal Energy	Uruguay	2011	In force	National	Solar thermal
Biodiesel blending mandate (Decree 1303)	Ecuador	2012	In force	National	Biofuels
National Biofuels Policy	Brazil	2017	In force	National	Biofuels
Argentina Renewable Energy Auctions—RenovAr Program (Round 3)—MiniRen Round	Argentina	2018	In force	National	Wind, solar, and hydropower

Notes: This figure was created by the authors and was based on the data from the IEA (2020).

References

Ahumada, C., Andrews, C.W., 1998. The impact of globalization on Latin American states: the cases of Brazil and Colombia. Adm. Theory Praxis 20 (4), 452–467. http://www.jstor.org/stable/25611309.

Aizenman, J., 2005. Financial Liberalizations in Latin America in the 1990s: A Reassessment. NBER Working Paper Series, 11145, pp. 1–30. https://www.nber.org/papers/w11145.

Balza, L.H., Espinasa, R., Serebrisky, T., 2016. Lights on? Energy Needs in Latin America and the Caribbean to 2040. Inter-American Development Bank, pp. 1–39. https://publications.iadb.org/en/publication/17053/lights-energy-needs-latin-america-and-caribbean-2040.

Bárcena, A., Samaniego, J., Galindo, L.M., Carbonell, J.F., Alatorre, J.E., Stockins, P., Reyes, O., Sánchez, L., Mostacedo, J., 2019. A Economia da Mudança Climática na América Latina e no Caribe. CEPAL, pp. 1–61. https://repositorio.cepal.org/bitstream/handle/11362/44486/1/S1801217_pt.pdf.

Birch, M.H., Halton, G., 2008. Foreign direct investment in Latin America in the 1990s: old patterns, new trends, and emerging issues. Lat. Am. Bus. Rev. 2 (1–2), 13–31. https://doi.org/10.1300/J140v02n01_03.

Carneiro, R.M., 2012. Commodities, Choques Externos e Crescimento: Reflexões Sobre a América Latina. vol. 117 CEPAL, pp. 1–47. ISSN:1680-8843 http://www.eco.unicamp.br/cecon/images/arquivos/observatorio/Commodities_choques_externos_crescimento.pdf.

Coviello, M.F., Gollán, J., Pérez, M., 2012. Public-Private Partnerships in Renewable Energy in Latin America and the Caribbean. ECLAC, pp. 1–63. https://www.cepal.org/es/publicaciones/4016-alianzas-publico-privadas-energias-renovables-america-latina-caribe.

Economic Commission for Latin America and the Caribbean (ECLAC), 2018. Foreign Direct Investment in Latin America and the Caribbean Moves Away From Natural Resources. ECLAC, pp. 1–4. ISSN:2522-7424 https://repositorio.cepal.org/bitstream/handle/11362/43423/1/S1800258_en.pdf.

Exame, 2018. Itaipu Binacional recupera US$ 6,5 milhões desviados em caso de corrupção. https://exame.abril.com.br/economia/itaipu-binacional-recupera-us-65-milhoes-desviados-em-caso-de-corrupcao/.

FGV Energia, 2016. A Comparative Analysis of Energy Transition in Latin America and Europe. pp. 1–72. http://www.fgv.br/fgvenergia/paper_kas-fgv_ingles/files/assets/common/downloads/Paper_KAS-FGV_Ingl_Web.pdf.

Flavin, C., Gonzalez, M., Majano, A.M., Ochs, A., Rocha, M., Tagwerker, P., 2014. Study on the development of the renewable energy market in Latin America and the Caribbean. IDB 2 (14), 1–79. https://publications.iadb.org/publications/english/document/Study-on-the-Development-of-the-Renewable-Energy-Market-in-Latin-America-and-the-Caribbean.pdf.

Fuinhas, J.A., Marques, A.C., Koengkan, M., 2017. Are renewable energy policies upsetting carbon dioxide emissions? The case of Latin America countries. Environ. Sci. Pollut. Res. 24 (17), 15044–15054. https://doi.org/10.1007/s11356-017-9109-z.

Ghani, G.M., 2012. Does trade liberalization effect energy consumption? Energy Policy 43, 285–290. https://doi.org/10.1016/j.enpol.2012.01.005.

Gielen, D., Boshell, F., Saygin, D., Bazilian, M.D., Wagner, N., Gorini, R., 2019. The role of renewable energy in the global energy transformation. Energy Strat. Rev. 24, 38–50. https://doi.org/10.1016/j.esr.2019.01.006.

Hauff, J., Bode, A., Neumann, D., Haslauer, F., 2014. Global Energy Transitions a Comparative Analysis of Key Countries and Implications for the International Energy Debate. World Energy Council, pp. 1–30. https://www.extractiveshub.org/resource/view/id/13542.

IMF, 2017. Investment and Capital Stock Dataset. http://www.imf.org/external/np/fad/publicinvestment/data/info122216.pdf.

Institute of the Americas, 2016. Argentina's Energy Transition: The Macri Government's Vision.

International Energy Agency (IEA), 2020. Policies Database. https://www.iea.org/policies?region=Central%20%26%20South%20America&page=1.

IRENA, 2015. Renewable Energy Policy Brief: Uruguay. pp. 1–10. www.irena.org.

Itaípu Turismo, 2020. Curiosities. https://www.turismoitaipu.com.br/en/content/curiosities.

Khan, M.A., Khan, M.Z., Zaman, K., Naz, L., 2014. Global estimates of energy consumption and greenhouse gas emissions. Renew. Sust. Energy Rev. 29, 336–344. https://doi.org/10.1016/j.rser.2013.08.091.

Kim, J., Park, K., 2016. Financial development and renewable energy diffusion. Energy Econ. 59, 238–250. https://doi.org/10.1016/j.eneco.2016.08.012.

Koengkan, M., 2020. Capital stock development and their effects on investment expansion in renewable energy in Latin America and the Caribbean region. J. Sustain. Finance Invest., 1–16. https://doi.org/10.1080/20430795.2020.1796100.

Koengkan, M., Fuinhas, J.A., 2020. Exploring the effect of the renewable energy transition on CO_2 emissions of Latin American & Caribbean countries. Int. J. Sustain. Energy, 1–25. https://doi.org/10.1080/14786451.2020.1731511.

Koengkan, M., Santiago, R., Fuinhas, J.A., 2019a. The impact of public capital stock on energy consumption: empirical evidence from Latin America and the Caribbean region. Int. Econ., 1–20. https://doi.org/10.1016/j.inteco.2019.09.001.

Koengkan, M., Fuinhas, J.A., Poveda, Y.E.M., 2019b. Globalisation as a motor of renewable energy development in Latin America countries. GeoJournal, 1–12. https://doi.org/10.1007/s10708-019-10042-0.

Koengkan, M., Fuinhas, J.A., Vieira, I., 2019c. Effects of financial openness on renewable energy investments expansion in Latin American countries. J. Sustain. Finance Invest., 1–19. https://doi.org/10.1080/20430795.2019.1665379.

Koengkan, M., Fuinhas, J.A., Silva, N., 2020. Exploring the capacity of renewable energy consumption to reduce outdoor air pollution death rate in Latin America and the Caribbean region. Environ. Sci. Pollut. Res., 1–15. https://doi.org/10.1007/s11356-020-10503-x.

Mazzucato, M., Semieniuk, G., 2018. Financing renewable energy: who is financing what and why it matters. Technol. Forecast. Soc. Change 127, 8–22. https://doi.org/10.1016/j.techfore.2017.05.021.

Our World in Data, 2020. Energy. https://ourworldindata.org/energy.

Santos, B.G., 2015. O ciclo econômico da América Latina dos últimos 12 anos em uma perspectiva de restrição externa. Revista do BNDES 43:205-251 https://web.bndes.gov.br/bib/jspui/bitstream/1408/6242/2/RB%2043%20O%20ciclo%20econ%C3%B4mico%20da%20Am%C3%A9rica%20Latina_P%20.pdf.

Sbia, R., Shahbaz, M., Hamdi, H., 2014. A contribution of foreign direct investment, clean energy, trade openness, carbon emissions and economic growth to energy demand in UAE. Econ. Model. 36, 191–197. https://doi.org/10.1016/j.econmod.2013.09.047.

Smil, V., 2010. Energy Transitions: History, Requirements, Prospects. Praeger Publishers, Santa Barbara, CA. ISBN-10:0313381771.

Solomon, B.D., Krishna, K., 2011. The coming sustainable energy transition: history, strategies, and outlook. Energy Policy 39 (11), 7422–7431. https://doi.org/10.1016/j.enpol.2011.09.009.

Terra, M.I., 2003. Trade liberalization in Latin American countries and the agreement on textiles and clothing in the WTO. Écon. Int. 94 (95), 137–154. ISSN 1240-8093 https://www.cairn.inforevue-economie-internationale-2003-2-page-137.htm.

Vásquez, I., 1996. The Brady plan and market-based solutions to debt crises. Cato J. 16 (2), 1–11. https://www.cato.org/sites/cato.org/files/serials/files/catojournal/1996/11/cj16n2-4.pdf.

Veloso, F.A., Villela, A., Giambiagi, F., 2008. Determinantes do "milagre" econômico brasileiro (1968–1973): uma análise empírica. Rev. Bras. Econ. 62 (2), 221–246. https://doi.org/10.1590/S0034-71402008000200006.

Ventura-Dias, V., Cabezas, M., Contador, J., 1999. Trade Reforms and Trade Patterns in Latin America. vol. 5 CEPAL, pp. 1–53. ISBN:92-1-121256-1 https://pdfs.semanticscholar.org/56a4/464b039b3e19246894fa2de5fdc8bbf308ef.pdf.

World Bank Open Data, 2020. http://www.worldbank.org/.

World Energy Council, 2020. World Energy: Issues Monitor 2020. pp. 1–179. https://www.worldenergy.org/assets/downloads/World_Energy_Issues_Monitor_2020_-_Full_Report.pdf.

CHAPTER

6

The role of public, private, and public-private partnership capital stock on the expansion of renewable energy investment in Latin America and the Caribbean region

JEL codes E00, N16, H54, Q43

6.1 Introduction

Latin America and the Caribbean (LAC) have received a significative amount of renewable energy investment in recent years. Between 2004 and 2019, investment accumulated in this period exceeded US$18.1 billion (see Fig. 5.5 in Chapter 5). Among the factors contributed strongly to these investment levels in the region are the consolidation of capital stock caused by financial liberalisation and by the 'commodities boom' (Koengkan, 2020; Koengkan et al., 2019a), and renewable energy policies and initiatives in the region, combined with rapid reductions in renewable energy technology costs, as well as the rapid economic growth in the region in the last 30 years (IRENA, 2016).

This chapter starts by providing a brief indication of the recent investment trends in the LAC region. It examines the main components of the capital stock mix such as public, private, and PPP capital for investment in renewable energy. Moreover, this chapter will analyse the effect of the development of the capital stock mix on investment in renewable energy in the LAC region.

6.1.1 Renewable energy investment trend in the LAC region

As mentioned earlier in Chapter 5, the investment in renewable energies began in the 1970s, more precisely in 1974 in Brazil and Paraguay with the construction of the Itaípu dam, a large hydropower dam, in the period 1974–84. Indeed, the participation of large hydropower dam plants in the energy matrix accounted for 10.41% in 1970, and in 2018 the participation of this kind of energy in the energy matrix reached a value of 23.57% (see Fig. 6.1).

Thus, in 1970, the energy generation from large hydropower dam plants was 65.88 terawatt-hours (TWh), and in 2018 reached a value of 731.31 (TWh) an increase of 1.010% (see Fig. 6.2).

The increase in installed capacity of large hydropower dam plants in the region is related to high investment; in 2015, this kind of energy source received US$9 billion in new investment, and in a cumulative period between 2010 and 2015 received US$37 billion (IRENA, 2016). These investments resulted in some important projects in large hydropower dam plants that were developed and/or phases of development/construction in some LAC countries. For example, in Brazil, the Belo Monte hydropower dam complex, located in the northern part of the *Xingu* River in the state of Pará, Brazil, was inaugurated in 2019 with an installed capacity of 11,233 megawatts (MW) (Governo do Brasil, 2019). Construction began in 2011 and was finished in 2019 and had an investment in the plant amounting to US$18.5 billion through public-private partnership (PPP). Indeed, this hydropower dam complex is the second largest hydroelectric dam in Brazil and fourth largest in the world by installed capacity.

https://doi.org/10.1016/B978-0-12-824429-6.00005-X

FIG. 6.1 Share of large hydropower energy generation in the Central and South America energy matrix between 1970 and 2018; this figure was created by the authors and was based on the Our World in Data (2020).

In Colombia, the Ituango hydropower dam plant is currently under construction on the Cauca River near Ituango in the Antioquia Department, Colombia. Its construction began in 2011 and it is expected to start operating in 2021, with a projected installed capacity of 2456 MW when completed. Moreover, until now, investment in the hydroelectric dam plant amounts to US$3.8 billion through a consortium of *Empresas Publicas de Medellin* (EPM) and the Antioquia government (IDB Invest, 2020). Moreover, in Bolivia, the San Jose hydropower dam complex plant that was inaugurated in 2019, located on the eastern slopes of the Andes in the Cochabamba Department, Bolivia, has an installed capacity of 124 MW. The project features two cascade power plants: San José I and San José II, with 55 and 69 MW of installed capacity, respectively. Their construction began in 2011 and was finalised in 2019. Investment in the plant amounts to US$94.8 million, through *Banco de Desarrollo de América Latina* (CAF) and the Cochabamba government (CAF, 2014).

However, investments in large hydropower dam plants have been declining in the LAC in recent decades due to investments in new renewable energy sources (e.g. geothermal, marine, small hydro, solar photovoltaic, solar, waste, and wind) (Koengkan et al., 2019b). Indeed, new renewable energy sources have undergone rapid growth since the end of the 1980s and in 2018 comprised 5.03% of the total energy mix in the region, with wind energy comprising

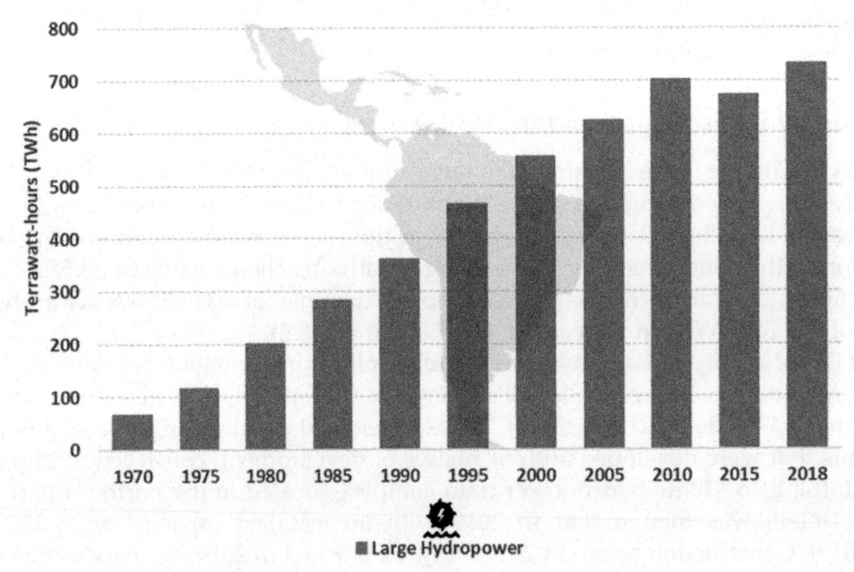

FIG. 6.2 Installed capacity of large hydropower dam plants in the Central and South America energy matrix between 1970 and 2018; the installed capacity is measured in terawatt-hours (TWh); this figure was created by the authors and was based on the Our World in Data (2020).

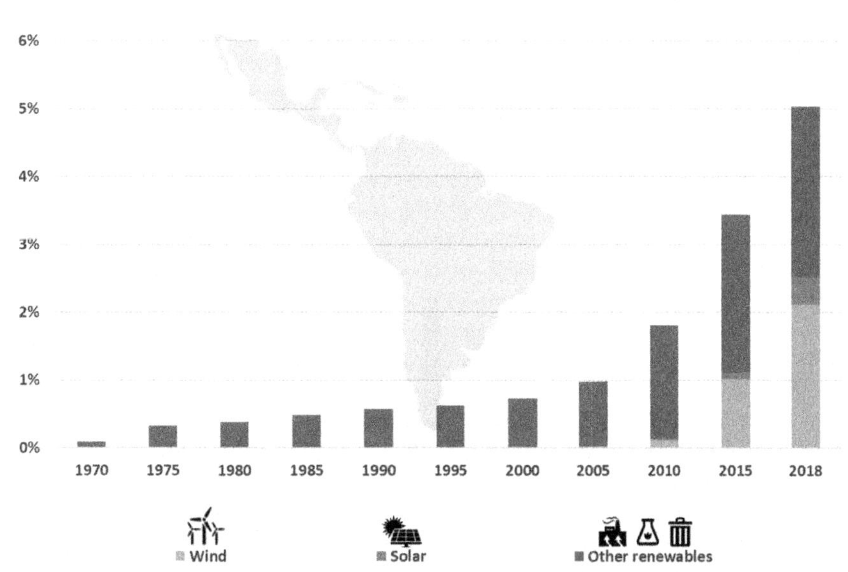

FIG. 6.3 Share of new renewable energy generation in the Central and South America energy matrix between 1970 and 2018. Other renewables include geothermal, biomass, and waste energy. This figure was created by the authors and was based on the Our World in Data (2020).

2.12%, solar energy 0.40%, and other renewable energy that includes geothermal, biomass, and waster comprising 2.51% (see Fig. 6.3).

Moreover, the installed capacity of these energy sources more than doubled between 2005 and 2018. The installed capacity in 2005 was 22.89 TWh and in 2018 it reached a value of 156.31 TWh (see Fig. 6.4). This increase was driven by other renewable energies and by wind energy that had a significant development in the region.

This increase is a result of the high investment that was made in renewable energy sources (excluding large hydropower plants larger than 50 MW); in 2019, the total renewable energy investment in Latin America was US$18.1 billion, or close to 7% of the global total (see Fig. 6.5).

In the LAC region, investment in renewable energy technologies increased by 54% in 2018 and reached a value of US$18.1 billion in 2019. The countries that led these investments were Brazil, Mexico, Chile, and Argentina to finance power capacity additions in the coming years. In 2019, wind energy attracted US$8.9 billion of new investment, an increase of 87% compared with 2018, while solar energy received 31% more investment compared with 2018 and reached a value of US$8.1 billion in 2019 (Bloomberg, 2020).

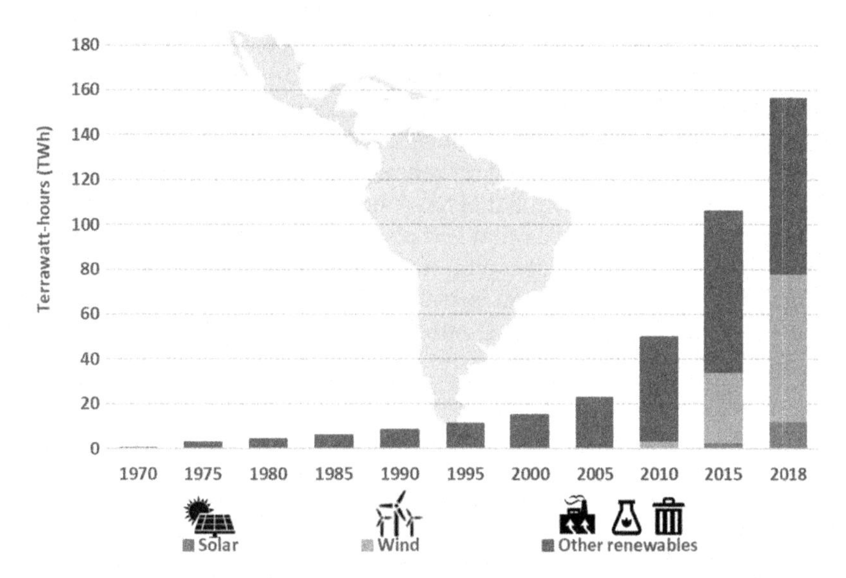

FIG. 6.4 Installed capacity of new renewable energy sources in the Central and South America energy matrix between 1970 and 2018. The installed capacity is measured in terawatt-hours (TWh). Other renewables include geothermal, biomass, and waste energy. This figure was created by the authors and was based on the Our World in Data (2020).

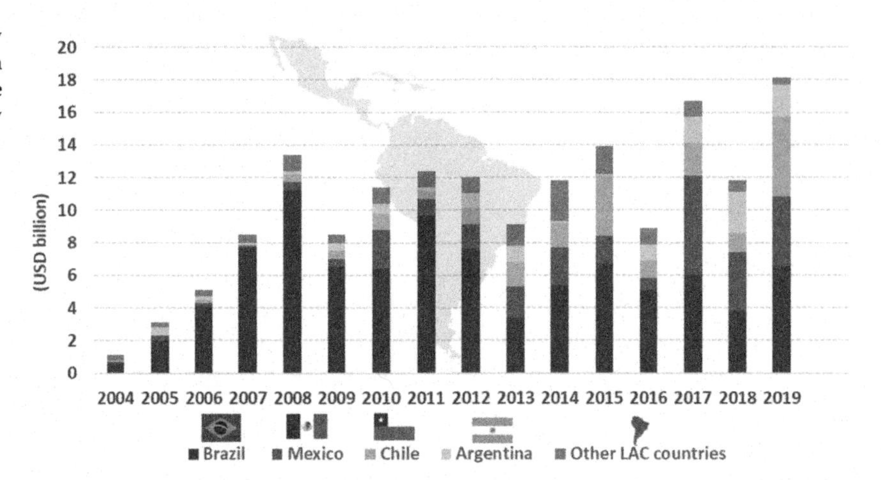

FIG. 6.5 Investment in new renewable energy sources in US$ billion in the LAC region, between 2004 and 2019. This figure was created by the authors and was based on the research by Koengkan (2020).

According to Bloomberg (2020), Brazil invested a total of US$6.5 billion in renewable energy technologies in 2019, 74% more than in 2018. This significant rebound is due to the large wind power auctions that occurred in the country between 2018 and 2019. As can be seen in the figure above, the gap in investment between the rest of the LAC countries and Brazil has been historically notable. Nonetheless, this gap has undergone a reduction in recent years. Another important country in the LAC region that reached a significative amount of investment in renewable energy is Mexico. This country was the second highest destination for investment in 2019, with renewable energy investment doubling from 2014 to reach US$4.3 billion in 2019. Most of these investments were in wind energy technology, which accounts for 79% of total investments. Chile ranked third place in renewable energy investment in the LAC region, with investment in this kind of technology totalling US$4.9 billion, mostly in wind and solar technologies. Moreover, Argentina ranked in fourth place; its investment totalled US$2 billion in 2019, with most of these investments in solar and wind energy technology. Therefore, the rapid growth in renewable energy technologies in Brazil, Mexico, Chile, and Argentina placed these countries on the list of the top 10 largest renewable energy markets globally.

Other countries in the LAC region have received a significant amount of renewable energy investment. For example, Uruguay invested US$1.1 billion in renewable energy technologies, mostly in wind energy technology. Honduras in Central America saw their investments reach a value of US$0.5 billion for the second consecutive year for the development of solar photovoltaic production (IRENA, 2016).

Therefore, the high levels of renewable energy investment in the LAC region illustrate a variety of economic, institutional, and resource conditions that attract these investments. The countries with different degrees of liberalisation in the energy sector, such as Brazil, Chile and Mexico, are considered the top destinations for renewable energy investments in the region. Other countries, such as Costa Rica and Uruguay, are also considered a top destination for new investments in renewable energy sources, due to their vertically integrated utilities and private participation in the energy sector, which occurs via independent power producers (IRENA, 2016).

Although the exact characteristics of the framework for renewable energy investments depend on specific country conditions, there are common and underlying factors that have indicated the success of the development of renewable energy technologies. These factors are the cost of capital and access to financing for renewable energy projects. These factors are strongly influenced by the availability of capital stock caused by financial liberalisation (see Chapter 5) and the capacity to bear the risks of renewable energy projects by public and private sectors. The following section will approach the main components of the capital stock (e.g. public, private, and PPP), which are available for investment in renewable energy technologies in the LAC region.

6.1.2 Development of the capital stock mix in renewable energy investment in the LAC region

In the LAC region, many countries have national public financing institutions; most of them are categorised as development banks and have large financing in renewable energy projects in the region, accounting for over one-third of new renewable energy project finance in the region between 2013 and 2016. The most important national public financing institutions in the LAC region are the Brazilian Development Bank (BNDES) and the Superintendency for the Development of the Northeast (SUDENE), both institutions in Brazil, and Nacional Financiera SNC

BOX 6.1

Participation of BNDES in the renewable energy investments.

The *Banco Nacional de Desenvolvimento Económico e Social* (BNSES) in Brazil has been the most remarkable example of a deployment strategy via public financing in the LAC region. This financing institution offered low-cost, long-term loans for up to 70% of total capital requirements of renewable energy projects in Brazil. In 2014 the bank invested US$3.2 billion in wind, small hydro and biofuel technologies, accounting for approximately 50% of new projects in Brazil (IRENA, 2016). However, the BNDES imposes local content requirement as a condition for awarding the financing, principally regarding the share of locally manufactured equipment. Indeed, this local content policy imposed by BNDES has been one of the main contributions to the development of wind energy technology in all stages, with several international companies developing manufacturing plants in Brazil (IRENA, 2016).

(NAFIN), Banobras, and Bancomext in Mexico. BNDES financed 50% of new renewable energy projects in 2014 in Brazil, while the NAFIN financed close to half of the new renewable energy projects in Mexico in 2015 (Bloomberg, 2015) (Box 6.1).

The role of national public financing institutions in the capital stock mix of renewable energy investment in the LAC region is driven by three main points: support for technological progress, implementing policies, and catalysing financing sources (Fig. 6.6).

However, these objectives are not reciprocally exclusive, and a grant or loan for a single renewable energy project developer usually serves more than one purpose. The high sharing of public financing institutions in the capital stock mix of renewable energy investment is due to public investors accepting a lower return on investment for some renewable energy projects by taking on longer commitments. These longer commitments advance public policy goals, such as the development of a domestic market, and the creation of a local value chain that enhances positive externalities such as social, economic, and environmental benefits for a specific region, unlike the private financial sector that always seeks to maximise their financial returns (IRENA, 2016).

However, countries in the LAC region without national public financing institutions or those whose public institutions have limited financing capability have used foreign public financing institutions to finance their renewable energy projects. Most of them financing these projects are multilateral development banks (MDBs). That is, they are institutions that have the aim of achieving long-term development, and in line with the activities of domestic national public financing institutions (IRENA, 2016). Moreover, these institutions often take on higher risks and offer longer tenors (e.g. term lengths), as well as providing technical assistance for renewable energy projects. The most important foreign public financing institutions in the LAC region are the Central American Bank for Economic Integration (CABEI), the Latin American Development Bank (CAF), and the Inter-American Development Bank (IDB) (Box 6.2).

In addition to national and foreign public financing institutions, the domestic and foreign private financing institutions have shown a significant interest in renewable energy projects in the LAC region. Several private financing institutions have financed renewable energy projects in the LAC region (see Table 6.1).

These private financing institutions tend to have more robust participation in countries with more mature renewable energy technologies and with a steadier flow of projects for example Brazil, Chile, and Mexico. This explains their focus on wind energy technology in these countries. Also, they tend to focus on large projects of wind energy in these countries, due to the lower transaction costs. For this reason, small- and medium-sized renewable energy projects tend

FIG. 6.6 Main points that drive public financing institutions to finance renewable energy projects in the LAC region. This figure was created by the authors and based on explanations from IRENA (2016).

BOX 6.2

The rule of foreign public financiers as catalysers for wind power financing in Mexico.

In Mexico, the financing of wind energy technologies began with the participation of the International Finance Corporation (IFC) and the Inter-American Development Bank (IDB). In total approximately US$525 million was invested in a Mexican wind farm in 2007 (IRENA, 2016). Moreover, development banks, such as NAFIN and Bancomext, co-financed the project, which would provide national private banks with significant experience in financing wind projects. Indeed, the multilateral institutions catalysed financing for wind projects.

to face challenges in obtaining financing for their projects from private financing institutions. Moreover, countries with more stable economies such as Uruguay have been attracting the participation of private banks for the renewable energy projects in recent years. They have been implementing renewable energy policies in the long run, with the existence of well-structured legal and regulatory processes that encourage the development of renewable energy technologies, and have robust institutions (see Chapter 5).

As mentioned earlier, the high investment costs as well as the low economic returns caused by the lack of maturity of some renewable energy technologies in the LAC region create a big challenge in the development of renewable energy projects and the penetration of some renewable energy technologies into the region. These challenges can ward off public and private investors in renewable energy projects. For this reason, it is necessary to create other mechanisms that can help overcome these challenges in the financing phase of these projects. One way to narrow the financing problems in renewable energy projects is to incorporate private capital through public-private partnerships (PPP). Usually, PPPs are defined as a legally binding contract between public and private firms for the provision of goods and services, with the majority of the responsibilities and risks being transferred between the partners. Moreover, in PPPs, the public and private sectors are both involved in all phases of the project, including financial, construction, and

TABLE 6.1 Activities of selected private financing institutions in the LAC region.

Country	Financing institution	Recent activity in the renewable energy sector
Brazil	Bradesco SA	The second-largest Brazilian private bank was the main private financier in the whole LAC region in 2013, with this private financial institution financing over US$220 million in renewable energy projects
	Itaú SA	Itaú financed US$400 million to fund 42 renewable energy projects in Brazil between 2012 and 2015
Chile	CorpBanca SA	CorpBanca and BBVA financed US$130 million for the construction of a 70 MW solar plant in Chile in 2013
Mexico	Banorte SA	The main Mexican private financing institution financed over US$110 million in renewable energy projects in 2013
Spain	Santander SA	Santander borrowed US$150 million from the European Investment Bank (EIB) to finance solar and biomass energy projects in the LAC region in 2014
	BBVA SA	BBVA financed US$150 million in solar and geothermal energy projects in Chile, and a US$250 million for wind energy projects in Mexico in 2014
United Kingdom	HSBC SA	The private institution financed over US$100 million in renewable energy projects in the LAC region in 2013
Netherlands	Rabobank SA	Rabobank financed approximately US$260 million in solar energy projects in Chile, and another US$70 million in renewable energy projects in other LAC countries in 2013
Germany	Deutsche Bank SA	Deutsche Bank financed US$144 million in wind energy projects in Uruguay in 2015

Notes: This table was created by the authors and was based on information from IRENA (2016).

operation of these projects. This makes the arrangement more attractive to the private sector since the risks are shared by the government (Coviello et al., 2012).

In the LAC region, PPPs in the energy sector were carried out between the 1980s and 1990s. However, PPPs in new renewable energy projects in the region began in the 2000s. The first initiatives of PPPs in renewable energy projects in the LAC region occurred in Brazil with the Alternative Energy Sources Incentive Programme (PROINFA) that was launched in 2002 to promote electricity generation based on renewable sources. This programme was created by Ministry of Mines and Energy via Law 10,438/02 and revised via Law 10,762/03 and Law 11,075/04 to promote the installation of 3300 MW of installed capacity of renewable energy sources (excluding large hydroelectric plants). Up to 70% of financing for the project came from this programme and it provides a special line of credit for corresponding investment. The private investor must guarantee at least 30% of the financing their own capital (Coviello et al., 2012). Moreover, other countries from the LAC region joined this type of financing, such as Chile in 2004, with General Law on Electricity Services (LGSE) introduced by Law 19,940 of 2004; Mexico in 2008 with Law for the Use of Renewable Energy and the Financing of the Energy Transition (LAERFTE), and Peru with Legislative Decree 1002 of 2008, which aims to stimulate the use of renewable energy resources through the promotion of investment for renewable electricity generation; and Uruguay with the 'National Energy Policy 2005–2030', where a new version of the legislation was approved in 2008 (Coviello et al., 2012).

The participation of public, private, and PPP capital stock in the renewable energy projects in the LAC region was made possible by the financial liberalisation process and by the 'commodities boom' that increased the capital stock mix (see Fig. 6.7).

This consequently reduced the cost of financing, facilitating investment in renewable energy technologies as pointed out in the studies of Koengkan and Fuinhas (2020), Koengkan et al. (2019b), Mazzucato and Semieniuk (2018), Kim and Park (2016), and Sbia et al. (2014). However, to identify the real effect of the development of capital stock mix on renewable energy investments, the following section will carry out an empirical analysis of the effect of public, private, and PPP capital stock on renewable energy investment in the LAC region between 1990 and 2015.

6.2 Data and methodology

As mentioned in the previous section, we will carry out an empirical analysis of the effect of public, private, and PPP capital on investment in renewable energy technologies in the LAC region.

6.2.1 Data

To this end, this analysis collected the annual data from 1990 to 2015 of 19 countries from the LAC region: **Argentina, Bolivia, Brazil, Chile, Colombia, Costa Rica, Dominican Republic, Ecuador, El Salvador, Guatemala, Haiti, Honduras, Mexico, Nicaragua, Panama, Peru, Trinidad and Tobago, Uruguay**, and **Venezuela (RB)**. The time series began in the 1990s and ended in 2015. The time series began in the 1990s due to the rapid process of development of capital stock that began in this decade and intensified between 2004 and 2014, and also due to the availability of data for all countries that were used in this empirical analysis. The variables which were chosen to perform this analysis are presented in Table 6.2.

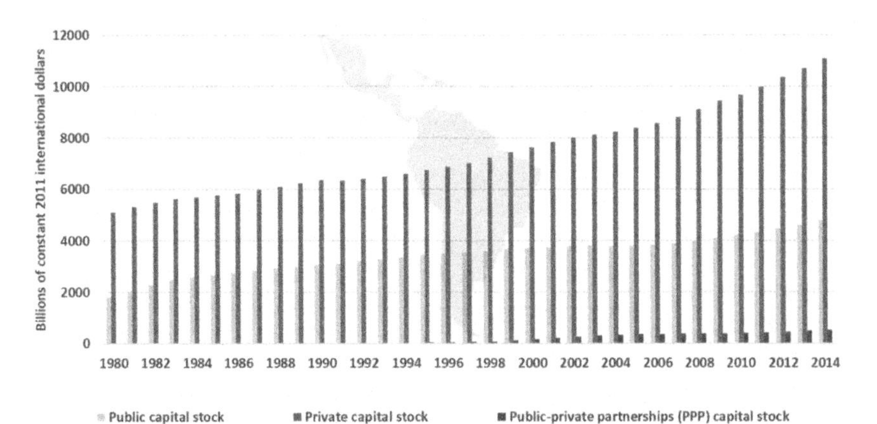

FIG. 6.7 Public, private, and PPP capital stocks in billions of constant 2011 international dollars in the LAC region between 1980 and 2014. This figure was created by the authors and was based on the database from the IMF (2017).

TABLE 6.2 Description of variables and source.

Description of variables		
Variable	Definition	Source
INCRE	The installed capacity of renewable energy in (million kilowatts)	International Energy Agency (IEA) (2019)
GDP_PC	GDP per capita, PPP (constant 2011 international $)	World Bank Open Data (2020)
KGOV	General government capital stock in billions of constant 2011 international dollars	International Monetary Fund (2017)
KPRIV	Private capital stock in billions of constant 2011 international dollars	International Monetary Fund (2017)
KPPP	Public-private partnership (PPP) capital stock in billions of constant 2011 international dollars	International Monetary Fund (2017)
KOFEcGI	Economic globalisation index in the de facto that measure the trade and financial globalisation. Trade globalisation is determined based on trade in goods and services, and financial globalisation includes foreign investment in various categories	KOF Globalisation Index (2020)
TROPEN	Trade (% of GDP) Trade is the sum of exports and imports of goods and services measured as a share of gross domestic product	World Bank Open Data (2020)
RENPOL	Renewable energy policies. This variable includes all policies defined by the International Energy Agency (IEA)	International Energy Agency (IEA) (2019)

In the literature, there are several ways to measure the development of renewable energy investment for example through installed capacity or generation. However, the use of renewable energy generation is influenced by meteorological conditions and equipment performance as well as technical problems, making it biased by external forces that the investor cannot control. Moreover, the installed capacity of renewable energy reflects high levels of investment deployment.

The investment in installed capacity of renewable energy may depend not only on the capital stock to their development but also on other drivers. Therefore, economic growth, economic globalisation, trade openness, and renewable energy policies are likely to be positively correlated with renewable energy investment development. Indeed, higher economic growth increases the consumption of energy, and as the energy demand expands, investment in renewable energy also increases. The process of economic globalisation and trade openness exerts a positive impact on economic growth and consequently increases the consumption of energy and the need for new investments in installed capacity of energy. Moreover, as mentioned in Chapter 5, economic globalisation has facilitated access to technological advances via trade and financial openness that consequently contribute to the increase in renewable energy capacity in the region. In particular, financial development via liberalisation increases public, private, and PPP capital stock, and consequently reduces the cost of financing, which facilitates investment in renewable energy technologies. This reduction in costs is one of the driving forces that have encouraged investment in renewable energy sources in recent years has brought about a significant reduction in their costs. However, only public capital and PPP can promote the rapid renewable energy investment, as the private sector is more risk averse in this context, and public policies have not been able to mobilise the private financing sector.

In addition, trade openness allows developing countries to import energy-saving technologies, products, and processes from developed countries that consume less energy and have high energy efficiency. That is, in developing countries, the reduction of energy consumption is more visible than in developed countries. This is due to the fact that developing countries have greater capacity to absorb transferred technologies than developed ones. Moreover, the economic integration arising from economic globalisation allows access to new energy technologies and consequently their adoption in industries. However, this technology transfer to renewable energy depends on the production and absorption capacity and the path dependency of significant investment in new technologies. Moreover, renewable energy policies have an important role in renewable energy investment development. The existence of well-structured legal and regulatory processes that encourages the development of renewable energy technologies, due to the robustness of institutions and the effectiveness of policies, has encouraged the development of these investments.

All variables used in this empirical analysis were transformed into natural logarithms. Moreover, the variables **KGOV**, **KPRIV**, and **KPPP** were transformed into per capita values using the total population of each cross. This allows for disparities in population growth to be controlled over time and within countries as used by Koengkan et al. (2019c) and Fuinhas et al. (2017). Moreover, the variable economic globalisation measures trade and financial globalisation, where trade globalisation is determined based on trade in goods and services, and financial globalisation includes foreign investment in various categories. Moreover, the variable renewable energy policies include all policies defined by the International Energy Agency (IRENA), namely: **(a) economic instruments; (b) information and education; (c) policy support; (d) regulatory instruments;** and **(e) research, development, and deployment (RD&D)**. This variable was built in accumulated form, where each policy that was created is represented by (1) accumulated over

TABLE 6.3 Summary statistics of variables.

	Summary statistics				
Variables	**Obs.**	**Mean**	**Std.-dev.**	**Min.**	**Max.**
LogINCRE	467	1.3867	2.9658	−2.5181	11.6017
LogGDP_PC	468	9.0456	0.6016	7.3062	10.0209
LogKGOV	468	−12.3696	0.8796	−15.0981	−10.4755
LogKPRIV	468	−11.5651	0.5621	−12.6660	−10.5216
LogKPPP	402	−16.0654	2.0050	−22.8825	−13.2888
LogKOFEcGI	468	3.9081	0.2243	3.2527	4.4145
LogTROPEN	468	3.9939	0.4584	2.6212	5.1161
LogRENPOL	416	1.3725	1.0453	0.0000	3.8712

other policies throughout its useful life or end (e.g. 1, 1, 2, 2, 2, 3,3). Table 6.3 presents the summary statistics of variables that are used in this analysis.

'Log' denotes variables in the natural logarithms, 'Obs' denotes the number of observations in the model, 'Std.-dev' denotes the standard deviation, and 'Min and Max' denote minimum and maximum. These summary statistics were obtained from the command *sum* of **Stata 16.0**. Moreover, the board below shows how to transform the variables into per capita values and into natural logarithms, as well as how to obtain the summary statistics of variables.

How to do:

****Transform the variables kgov, kpriv, and kppp into per capita values****

gen kgov_pc =(kgov/population)

****Transform the variables incre, gdp_pc, kgov, kpriv, kppp, kofecgi, tropen, and renpol into natural logarithms****

gen loggdp_pc =ln(gdp_pc)

****The summary statistics****

sum logincre loggdp_pc logkgov logkpriv logkppp logkofecgi logtropen logrenpol

This section describes the group of countries that were approached and data that were used to carry out this analysis. In the next section, the methodology approach is described.

6.2.2 Methodology

The quantile via moments (**QvM**) model created by Machado and Silva (2019) was used to do this analysis. This model, according to the authors, allows the use of methods that are only valid in the estimation of conditional means. It implies differencing out the individual effect in the panel data models while providing information about how the regressors affect the entire conditional distribution. Machado and Silva (2019) and Koengkan et al. (2021) also added that informational gains are perhaps the most attractive feature of quantile regression. Moreover, this model can be adapted to make estimations in models with endogenous variables and also can be based on the estimation of conditional means, but on moment conditions that identify the conditional means under exogeneity. That is, this model allows us to identify the same structural quantile function as well as can be used in nonlinear models and in models with multiple endogenous variables. Therefore, the **QvM** model is constructed around the following equation:

$$Y_{it} = a_i + y'_{it}\beta + \left(\delta_i + Z'_{it}\gamma\right)U_{it} \tag{6.1}$$

where $\{(Y_{it}, y_{it}')'\}$ from a panel of n individuals $i=1,\ldots, n$ over T time periods with $P\{\delta_i + Z_{it}'\gamma > 0\} = 1$. The parameters (α_1, δ_i), $i=1, \ldots, n$, capture the individual i fixed effects and Z is a k-vector of known differentiable (with probability 1) transformations of the components of y with element l given by $Z_l = Z_l(y)$, $l=1, \ldots, k$. The sequence $\{y_{it}\}$ is *i.i.d.* for any fixed i and independent across t. U_{it} are *i.i.d.* (across i and t), statistically independent of y_{it}, and normalised to satisfy the moment condition $E(U)=0 \wedge E(|U|)=1$.

However, before the realisation of **QvM** model regression, the characteristics of the variables need to be checked. Indeed, undertaking these tests is essential to identify if the **QvM** model is an appropriate methodology to carry out this analysis. Therefore, some **preliminary tests** will be computed for example as summarised in Table 6.4.

Moreover, after the model regression, it is necessary also to apply **postestimations** tests. For this analysis, we used just one test (Table 6.5).

TABLE 6.4 Preliminary tests.

Test	Goal
Variance inflation factor (VIF) test	To check the existence of multicollinearity between the variables in the panel data
Pesaran CD-test	To check the existence of cross-section dependence in the panel data
Panel unit root test (CIPS-test)	To check the presence of unit roots in the variables
Hausman test	To check the presence of heterogeneity i.e. whether the panel has random effects (RE) or fixed effects (FE)
Bias-corrected LM-based test	To check the presence of serial correlation in the fixed-effects panel model

TABLE 6.5 Postestimation test.

Test	Goal
Wald test	To check the global significance of the estimates model

Moreover, all estimations and testing procedures will be accomplished using **Stata 16.0**. This section shows the methodology approach that will be used in this analysis and the preliminary and postestimation tests. In the following section, the empirical results will be shown.

6.3 Empirical results and discussion

This section will show the results from preliminary tests, the **QvM** model regression and the postestimation test, as well as a brief discussion regarding the results obtained. Therefore, to verify the level of multicollinearity between the variables, the VIF test developed by Belsley et al. (1980) was calculated. This test is constructed around the following equation:

$$VIF_i = \frac{1}{1 - R_j^2} \tag{6.2}$$

where R_j^2 is the coefficient of determination of regression of model in step one. Therefore, the results of this test indicate that the presence of multicollinearity is not a concern in the estimation as the VIF and mean VIF values registered are lower than the usually accepted benchmark of 10, in the case of the VIF values, and 6 in the case of the mean VIF values (see Table 6.A1 in Appendix). The results of VIF test were obtained from the command *estat vif* in **Stata 16.0**. The board below shows how to carry out and obtain the results from the VIF test.

How to do:

****The Variance Inflation Factor test****

```
reg logincre loggdp_pc logkgov logkpriv logkppp logkofecgi logtropen logrenpol
estat vif
```

Indeed, in the presence of low multicollinearity between the variables, it is necessary to check the presence of cross-section dependence in the panel data. To this end, the Pesaran CD test developed by Pesaran (2004) was calculated. This test is constructed around the following equation:

$$CD = \sqrt{\frac{2T}{N(N-1)}} \left(\sum_{i=1}^{N-1} \sum_{j=i+1}^{N} \hat{P}ij \right) \tag{6.3}$$

The null hypothesis of this test is the nonpresence of cross-section dependence CD $\sim N(0,1)$ for $N \rightarrow \infty$ and T is sufficiently large. The results of this test point to not rejecting the null hypothesis, thus indicating that all variables used in this study have the presence of cross-section dependence (see Table 6.A2 in Appendix). The results of Pesaran CD test were obtained from the command *xtcd* in **Stata 16.0**. The board below shows how to carry out and obtain the results from the Pesaran CD test.

How to do:

****The Pesaran CD-test****

```
xtcd logincre loggdp_pc logkgov logkpriv

xtcd logkppp logkofecgi logtropen logrenpol
```

The presence of cross-sectional dependence in these variables is an indication that the countries selected in this panel data share the same characteristics. Therefore, in the presence of cross-sectional dependence in the variables, it is necessary to check the order of integration of the variables. To this end, the panel unit root test (CIPS test) developed by Pesaran (2007) was calculated. This test is constructed around the following equation:

$$CIPS(N,T) = t - bar = N^{-1} \sum_{i=1}^{N} ti(N,T) \tag{6.4}$$

where $t_i(N,T)$ is the cross-sectionally augmented Dickey-Fuller statistic for the i, the cross-section unit given by the $t-$ratio of the coefficient of $y_{i,\ t-1}$ in the CADF regression. Therefore, the null hypothesis of this test is that all series have a unit root. The results of this test indicate that some of the variables are on the borderline between the I(0) and I(1) orders of integration such as **LogINCRE**, **LogGDP_PC**, **LogKPRIV**, **LogKPPP**, **LogKOFEcGI**, and **LogRENPOL**, while the variables **LogKGOV** and **LogTROPEN** are I(1) (see Table 6.A3 in Appendix). The results of the CIPS test were obtained from the command *multipurt* in **Stata 16.0**. The board below shows how to carry out and obtain the results from the CIPS test.

How to do:

```
**The CIPS-test**

multipurt logincre loggdp_pc logkgov logkpriv logkppp logkofecgi logtropen logrenpol,lags(1)
```

After identifying the presence of a unit roots in some variables, it is necessary to verify the presence of individual effects in the models. To this end, the Hausman test was calculated to identify these individual effects. This test is constructed around the following equation:

$$H = (\beta_{RE} - \beta_{FE})' \sum^{\wedge} - 1(\beta_{RE} - \beta_{FE}) \sim X^2(k) \tag{6.5}$$

where β_{RE} and β_{FE} are estimators of the parameter β. The null hypothesis of this test is that the difference in coefficients is not systematic, where the random effects are the most sustainable estimator (Koengkan and Fuinhas, 2020; Santiago et al., 2019). The results of the Hausman test indicate that the null hypothesis of this test should be rejected in **Model I** (public capital stock), **Model II** (private capital stock), and **Model III** (PPP capital stock), as the results are statistically significant at 1% and 5% levels (see Table 6.A4 in Appendix). That is, the Hausman test indicates that the fixed effect estimator is the most appropriate for this analysis. Moreover, the **QvM** model approach requires the models to have fixed effects, as mentioned in the above section. The results of the Hausman test were obtained from the command *hausman* with option *sigmaless* in **Stata 16.0**. The board below shows how to carry out and obtain the results from the Hausman test.

How to do:

```
**The Hausman test**

*Model I*

xtreg logincre trend loggdp_pc logkogv logtropen logkofecgi logrenpol,fe
estimates store fixed

xtreg logincre trend loggdp_pc logkogv logtropen logkofecgi logrenpol,re
estimates store random

hausman fixed random, sigmaless
```

Before the realisation of **QvM** regression, it is necessary to carry out the last preliminary test, the bias-corrected LM-based test developed by Born and Breitung (2015) and Wursten (2018), to identify the presence of serial correlation in the fixed effects panel model. This test is constructed around the following equation:

$$e_{it} - \bar{e}_i = \partial(e_{i,t-k} - \bar{e}_i) + \varepsilon_{it} \tag{6.6}$$

In the LM (k), the statistic is fairly straightforward and comes down to a heteroskedasticity and autocorrelation robust, (t) test of $(\partial) = -1/(T-1)$, with (∂) as the coefficient on the kth-order demeaned residuals in (1). Moreover, the e_{it}

residual includes the fixed effects. This leads to the asymptotically equivalent test statistic $\tilde{LM}(k)$ defined by the following equations:

$$z_{k,i} = \sum_{t=k+1}^{T} \left\{ (e_{it} - \bar{e}_i)(e_{i,t-k} - \bar{e}_i) + \frac{1}{T-1}(e_{i,t-k} - \bar{e}_i)^2 \right\} \tag{6.7}$$

$$\tilde{LM}_k = \frac{\sum_{i=1}^{N} z_{k,i}}{\sqrt{\sum_{i=1}^{N} z_{k,i}^2 - \frac{1}{N}\left(\sum_{i=1}^{N} z_{k,i}\right)^2}} \tag{6.8}$$

Therefore, the null hypothesis of this test is nonpresence of serial correlation of order (k). The results of this test indicate that the null hypothesis of this test can be rejected. That is, there is the presence of serial correlation up to the second order (see Table 6.A5 in Appendix). The results of bias-corrected LM-based test were obtained from the command *xtqptest* in **Stata 16.0**. The board below shows how to carry out and obtain the results from the bias-corrected LM-based test.

How to do:

****Bias-corrected LM-based test****

```
xtqptest logincre loggdp_pc logkgov logkpriv logkppp logkofecgi logtropen logrenpol, order(1)
```

After the preliminary tests, the **QvM** regression can be made. The **25th**, **50th**, and **75th** quantiles were, respectively, calculated to assess the impact of public, private, and PPP capital stock on the installed capacity of renewable energy. This model does not allow causality between the variables to be performed; it only allows us to observe the effect at the quantiles. In addition, this investigation will perform the regression separately because the variables **LogKGOV**, **LogKPRIV**, and **LogKPPP** cannot be in the same equation. The inclusion of these three variables together can destroy the model. Table 6.6 presents the results of the **QvM** regression of **Model I** that will identify the effect of public capital stock (**LogKGOV**) on the installed capacity of renewable energy (**LogINCRE**).

The results from **QvM** regression indicate that in the **25th**, **50th**, and **75th** quantiles, public capital stock increases the installed capacity of renewable energy. Furthermore, the quantiles also indicate that economic globalisation encourages investment in this kind of energy, while economic growth, trade openness, and renewable energy policies decrease them. Most of the results of this estimation are statistically significant at the 1% level. The Wald test (Agresti, 1990) verifies the global significance of the estimates model. The null hypothesis of this test is that all the coefficients are equal to zero. The result from the postestimation test indicates that the null hypothesis cannot be rejected in all quantiles, indicating that the estimation of this study is adequate. The results of **QvM** regression of **Model I** and the Wald test (**postestimation test**) were obtained from the commands *xtqreg* and *testparm* in **Stata 16.0**. The board below shows how to carry out and obtain the results from the **QvM** regression and the Wald test.

TABLE 6.6 Estimations for the **Model I.**

	Dependent variable LogINCRE					
	Quantiles					
Independent variables	**25th**		**50th**		**75th**	
TREND	0.0532	***	0.0587	***	0.0657	***
LogGDP_PC	−2.9470	***	−3.3435	***	−3.8443	***
LogKGOV	1.3355	***	1.4867	***	1.6777	***
LogKOFEcGI	0.8902	***	0.7979	***	0.6813	**
LogTROPEN	−0.9533	***	−0.8223	***	−0.6569	**
LogRENPOL	−0.1362		−0.1787	**	−0.2324	**
Obs	415		415		415	
F / Wald test	Chi2(5) = 38.00	***	Chi2(5) = 57.91	***	Chi2(5) = 31.64	***

Notes: *** and ** denotes statistically significant at the 1% and 5% levels.

How to do:

The Quantile via Moments model regression

```
xtqreg logincre trend loggdp_pc logkogv logtropen logkofecgi logrenpol, id(country) quantile(.25)
```

The Wald test

```
testparm logincre loggdp_pc logkogv logtropen logkofecgi logrenpol
```

The Quantile via Moments model regression

```
xtqreg logincre trend loggdp_pc logkogv logtropen logkofecgi logrenpol, id(country) quantile(.50)
```

The Wald test

```
testparm logincre loggdp_pc logkogv logtropen logkofecgi logrenpol
```

The Quantile via Moments model regression

```
xtqreg logincre trend loggdp_pc logkogv logtropen logkofecgi logrenpol, id(country) quantile(.75)
```

The Wald test

```
testparm logincre loggdp_pc logkogv logtropen logkofecgi logrenpol
```

Subsequently to the regression that identifies the effect of public capital stock on investment in installed capacity of renewable energy, it is necessary to identify the effect of private capital stock. Therefore, Table 6.7 presents the results of the **QvM** regression of **Model II** that will identify the effect of private capital stock (**LogKPRIV**) on the installed capacity of renewable energy **(LogINCRE**).

The results from **QvM** regression indicate that in the **25th**, **50th**, and **75th** quantiles, private capital stock does not have any impact on the installed capacity of renewable energy, due to the nonstatistically significant of values. Moreover, the results of the **25th** and **50th** quantiles indicate that economic globalisation increases the investment in installed capacity, while economic growth, trade openness, and renewable energy policies decrease these investments. In the **75th** quantile, economic growth also decreases the installed capacity of renewable energy. Most of the results of this estimation are statistically significant at the 1% and 5% levels. The result from the Wald test indicates that the null hypothesis cannot be rejected in the **25th**, **50th**, and **75th** quantiles, indicating that the estimation of this study is adequate. The results of **QvM** regression of **Model II** and the Wald test (postestimation test) were obtained from the commands *xtqreg* and *testparm* in **Stata 16.0**. The board below shows how to carry out and obtain the results from the **QvM** regression and the Wald test.

TABLE 6.7 Estimations for the **Model II.**

	Dependent variable LogINCRE					
	Quantiles					
Independent variables	**25th**		**50th**		**75th**	
TREND	0.0555	***	0.0594	***	0.0648	*
LogGDP_PC	−1.4920	***	−2.0781	***	−2.8898	**
LogKPRIV	−0.3386	***	0.1199		0.7552	
LogKOFEcGI	1.0376	***	0.9206	**	0.7584	
LogTROPEN	−1.1464	***	−1.0055	***	−0.8103	
LogRENPOL	−0.2129	*	−0.2509	**	−0.3034	
Obs	415		415		415	
F/Wald test	Chi2(5)=24.24	***	Chi2(5)=24.73	***	Chi2(5)=9.99	*

Notes: ***,**, and * denote statistically significant at the 1%, 5%, and 10% levels.

How to do:

```
**The Quantile via Moments model regression**
xtqreg logincre trend loggdp_pc logkpriv logtropen logkofecgi logrenpol, id(country) quantile(.25)
**The Wald test**
testparm logincre loggdp_pc logkpriv logtropen logkofecgi logrenpol
**The Quantile via Moments model regression**
xtqreg logincre trend loggdp_pc logkpriv logtropen logkofecgi logrenpol, id(country) quantile(.50)
**The Wald test**
testparm logincre loggdp_pc logkpriv logtropen logkofecgi logrenpol
**The Quantile via Moments model regression**
xtqreg logincre trend loggdp_pc logkpriv logtropen logkofecgi logrenpol, id(country) quantile(.75)
**The Wald test**
testparm logincre loggdp_pc logkpriv logtropen logkofecgi logrenpol
```

After the regression that identify the effect of public and private capital stock on investment in installed capacity of renewable energy, it is necessary to identify the effect of PPP capital stock. Table 6.8 presents the results of the **QvM** regression of **Model III** that will identify the effect of public-private partnership capital stock (**LogKPPP**) on the installed capacity of renewable energy **(LogINCRE**).

The results from the **QvM** regression indicate that in the **50th** and **75th** quantiles, public-private partnership capital stock increases the installed capacity of renewable energy. Moreover, the **25th** quantile indicates that economic globalisation increases the installed capacity of renewable energy; in contrast, trade openness decreases it. The **50th** quantile shows that economic globalisation increases the installed capacity of renewable energy, while economic growth, trade openness, and renewable energy policies decrease it. The **75th** quantile indicates that economic growth and renewable energy policies decrease the installed capacity of renewable energy. The results of this estimation are statistically significant at 1% and 5% levels. The result from the Wald test indicates that the null hypothesis cannot be rejected in all quantiles. The results of **QvM** regression of **Model III** and the Wald test (postestimation test) were obtained from the commands *xtqreg* and *testparm* in **Stata 16.0**. The board below shows how to carry out and obtain the results from the **QvM** regression and the Wald test.

TABLE 6.8 Estimations for the **Model III.**

	Dependent variable LogINCRE					
	Quantiles					
Independent variables	**25th**		**50th**		**75th**	
LogGDP_PC	−0.5151		−0.7311	**	−0.9627	**
LogKPPP	0.0471		0.0676	**	0.0897	**
LogKOFEcGI	0.7165	***	0.4788	**	0.2239	
LogTROPEN	−0.4445	**	−0.3823	**	−0.3156	
LogRENPOL	−0.1503		−0.1767	**	−0.2050	**
Obs	361		361		361	
F/Wald test	Chi2(5) = 30.92	***	Chi2(5) = 50.76	***	Chi2(5) = 27.83	***

Notes: *** and ** denote statistically significant at the 1% and 5% levels.

How to do:

```
**The Quantile via Moments model regression**
Xtqreg logincre loggdp_pc logkppp logtropen logkofecgi logrenpol, id(country) quantile(.25)
**The Wald test**
testparm logincre loggdp_pc logkppp logtropen logkofecgi logrenpol
**The Quantile via Moments model regression**
xtqreg logincre loggdp_pc logkppp logtropen logkofecgi logrenpol, id(country) quantile(.50)
**The Wald test**
testparm logincre loggdp_pc logkppp logtropen logkofecgi logrenpol
**The Quantile via Moments model regression**
Xtqreg logincre loggdp_pc logkppp logtropen logkofecgi logrenpol, id(country) quantile(.75)
**The Wald test**
testparm logincre loggdp_pc logkppp logtropen logkofecgi logrenpol
```

Fig. 6.8 summarises the impacts of independent variables on the dependent variable. This figure was based on the results from **Models I, II,** and **III**.

After finding that public and PPPs capital stock can encourage investment in renewable energy sources and that private capital stock does not cause any impact, we examined the following questions: What are the explanations for the positive impact of public and public-private partnership capital stock on the installed capacity of renewable energy? Why does private capital stock not cause any impact?

In answer to the first question, the positive impact of public capital stock on the installed capacity of renewable energy is related to the development of public capital supply, which consequently reduces the financing costs and encourages development and investment in renewable energy technologies as mentioned by Koengkan et al. (2019c). Indeed, this explanation also is confirmed by Apergis and Payne (2010) and Narayan and Smyth (2008), where

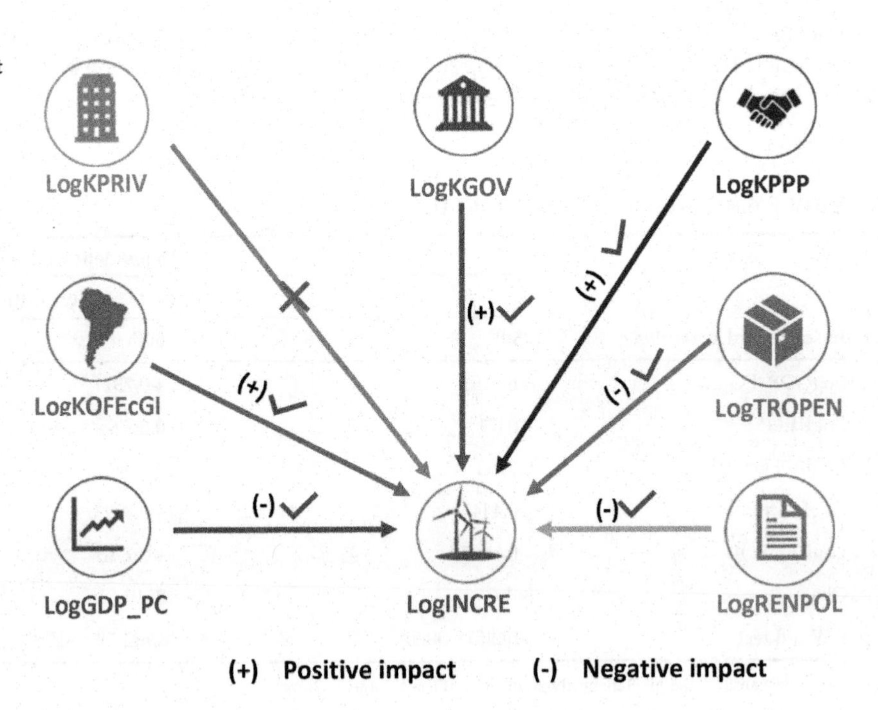

FIG. 6.8 Summary of the impacts of independent variables on the dependent variable.

according to these authors, the development of public capital stock encourages investment in renewable energy because the supply of cheaper credit makes alternative energy sources more feasible.

Additionally, as we already know, financing for renewable energy technologies generally goes through a parallel cycle of maturation from high-risk/high-cost/low-liquidity investment to a low-risk/low-cost/high-liquidity investment. That is, during this cycle, the initial access to capital can be very difficult, and investment in this kind of energy is often supported by public capital for example direct grants and government loan guarantees, which are cheaper than in private institutions, as pointed out by Mendelsohn and Feldman (2013). This idea also is shared by Mazzucato and Semieniuk (2018), where according to these authors, public capital helps to solve the asymmetric information problem that requires taking technologies from proof of concept to commercial scale. That is, public finance can solve these 'market failures'. Therefore, public capital helps renewable energy technologies to mature into early commercialisation phases, and then private capital and project finance are frequently deployed to capitalise deployment (Mendelsohn and Feldman, 2013).

Other possible explanations have been pointed out. The development of public capital stock has an indirect effect on the consumption of energy and consequently new investment in energy is necessary to supply the energy demand, as indicated by Koengkan et al. (2019c) and Lee (2005). This vision is argued by Lee and Chen (2010) and Lee et al. (2008), where according to these scholars, public capital positively impacts investment and industrial production, which consequently increases the energy demand and investment in installed capacity. Moreover, investment in installed capacity of renewable energy can accelerate economic growth and energy consumption, and consequently, more investment in installed capacity is required (Koengkan et al., 2019c).

The positive impact of PPP capital stock on the installed capacity of renewable energy is related to the high investment and maintenance costs, complex construction issues, and economic returns that are not always high in renewable energy projects (Mendelsohn and Feldman, 2013; Coviello et al., 2012). All these factors are a great challenge in the development of renewable energy projects and the penetration of green technologies into the LAC region. However, to work around this challenge, it is necessary to adopt other mechanisms to help overcome the challenge, particularly in the initial financing phase (Coviello et al., 2012). The use of PPPs is one way to narrow the gap in financing in renewable energy projects. The utilisation of this mechanism in the LAC region began during the process of privatisation at the turn of the 1980s to the 1990s with the adoption of restructuring plans such as the 'Washington Consensus' and the 'Brady Plan'. According to Coviello et al. (2012), PPPs were used to increase investment in infrastructure and improve the quality of services that had traditionally been in the hands of the state. In this case, the state then took on the role of regulating the largest private electricity generation, transmission, and distribution market.

Now, in answer to the second question, the nonimpact of private capital stock on the installed capacity of renewable energy is related to the high investment and maintenance costs, complex construction issues, and economic returns that are not always high in the renewable energy projects, as mentioned before by Mendelsohn and Feldman (2013) and Coviello et al. (2012). That is, this discourages private participation in these projects in the LAC region. Another possible explanation is the low private capital supply that consequently increases the financing costs and discourages development and investment in renewable energy technologies.

Regarding the negative impact of economic growth on the installed capacity of renewable energy, it could be related to low economic growth that discourages investment in this kind of source. The negative impact of trade openness is related to low trade openness that makes it difficult to acquire equipment and technologies that produce green energy. Another possible explanation is that trade openness brings more investment to the nonrenewable energy sector, due to the abundance of nonrenewable energy sources in some countries from the LAC region (Sebri and Ben-Salha, 2014). Moreover, the negative impact of renewable energy policies could be related to the low capacity of these policies to encourage renewable energy development (Fuinhas et al., 2017). Another possible explanation is that these policies need time to take effect or the methodology approach is not adequate to capture the positive effect of these policies. Finally, the positive effect of economic globalisation could be related to the financial liberalisation that composes this variable and that consequently encourages capital formation and that subsequently also encourages investment in this kind of energy. Let us recall that the main objective of this analysis is to identify the effect of capital stock development composed by public, private, and PPP capital stock on the installed capacity of renewable energy that is a proxy for renewable energy investment. For this reason, this research does not reveal in-depth explanations about the effects of other variables in the models studied.

6.4 Conclusions

This chapter analysed the effect of public, private, and PPP capital stock on the installed capacity of renewable energy that is a proxy for renewable energy investment in a panel of 18 countries from LAC region in the period between 1990 and 2015. To carry out this research, the quantile via moments methodology was used.

The results of preliminary tests indicate the presence of low multicollinearity in the variables, CSD, stationarity in some variables, the fixed effect in all models that were estimated, and the presence of serial correlation up to the second order. Therefore, the results from the quantile via moments methodology indicate that public and public-private partnership capital stock has a positive effect on the installed capacity of renewable energy. In contrast, private capital stock does not cause any effect on the dependent variable. The positive effect of public and public-private partnership capital stock on the installed capacity of renewable energy is due to the high investment and maintenance costs, complex construction issues, and economic returns that are not always high in renewable energy projects. Indeed, during the initial process of development of these projects, the initial access to capital can be challenging, and investment in this kind of energy is often supported by public and public-private partnership capital, which are cheaper compared with support of private capital only. Moreover, the nonimpact of private capital stock is related to the high investment and maintenance costs of renewable energy projects as well the low private capital supply that consequently increases the financing costs, discouraging private participation in investment in renewable energy technologies.

It is necessary to reduce the cost of capital and the barriers to renewable energy investment in the LAC region. The financial sector has an important role to play in scaling up the level of finance available for investment in renewable energy technologies, and in particular, reducing the risks between public and private investors and increasing the participation of the private sector. To this end, private capital finance needs to be catalysed, whereby it is necessary to create more solid renewable energy policies, instruments to reduce financial risk, and well-structured finance mechanisms; technological progress needs to be enabled, and public financing institutions need to support the initial stage of development of renewable energy technologies in the LAC region. The participation of public financing institutions at this stage is due to the high risks in the deployment phase of renewable energy technologies, given that the private sector may be more reticent in committing resources in the development of renewable energy technologies. Moreover, energy innovation and entrepreneurship needs to be empowered for applications of relevance to the LAC region, where several utilities in the region are spending substantial financial resources for energy efficiency and renewable research and development programmes. Many of these programmes can be turned into commercial projects, as well as creating new markets for renewable energy technologies; finally, the learning curve should be accelerated and it is necessary to disseminate information in the LAC region regarding the information on effective financing mechanisms for renewable energy technologies. This can facilitate access to these sources of finance by local project developers, as well as increase the access of specific credit lines for renewable energy technologies. Dissemination of this information includes knowledge of policymakers, whereby the costs of acquiring information will be reduced and local renewable energy markets can consequently contribute to attracting foreign investment in the sector.

Appendix

TABLE 6.A1 VIF-test.

Dependent variable (LogINCRE)		
Variables	**VIF**	**1/VIF**
LogGDP_PC	5.73	0.1746
LogKGOV	1.69	0.5926
LogKPRIV	5.24	0.1909
LogKPPP	1.67	0.5991
LogKOFEcGI	2.08	0.4814
LogTROPEN	2.11	0.4749
LogRENPOL	1.38	0.7233
Mean VIF	2.84	

TABLE 6.A2 Pesaran CD-test.

Variables	CD-test	*P*-value		Corr	Abs (corr)
LogINCRE	5.97	0.000	***	0.094	0.382
LogGDP_PC	44.38	0.000	***	0.705	0.853
LogKGOV	12.55	0.000	***	0.200	0.689
LogKPRIV	19.19	0.000	***	0.303	0.606
LogKPPP	44.47	0.000	***	0.883	0.883
LogKOFEcGI	14.06	0.000	***	0.280	0.552
LogTROPEN	14.26	0.000	***	0.278	0.487
LogRENPOL	42.44	0.000	***	0.845	0.845

Notes: *** denotes statistical significance at the 1% level.

TABLE 6.A3 Panel Unit Root test (CIPS-test).

	Panel Unit Root test (CIPS) (Zt-bar)				
	Without trend			With trend	
Variables	**Lags**	**Zt-bar**		**Zt-bar**	
LogINCRE	1	−0.642		−1.790	**
LogGDP_PC	1	−4.629	***	−2.951	***
LogKGOV	1	1.281		−3.222	***
LogKPRIV	1	−2.991	***	−2.961	***
LogKPPP	1	−3.691	***	−1.739	**
LogKOFEcGI	1	−0.829		−1.587	**
LogTROPEN	1	−1.668	**	0.542	
LogRENPOL	1	−2.013	**	−1.625	**

Notes: *** and ** denote statistically significant at the 1%, and 5% levels.

TABLE 6.A4 Hausman test.

Variables	(b) Fixed	(B) Random	(b-B) Difference	Sqrt(diag(V_b-V-B)) S.E.
Model I (Dependent variable LogINCRE)				
TREND	0.0592	0.0542	0.0050	0.0018
LogGDP_PC	−3.3793	−2.8584	−0.5208	0.1343
LogKGOV	1.5003	1.1949	0.3054	0.0927
LogKOFEcGI	0.7896	0.8057	−0.0161	0.0283
LogTROPEN	−0.8105	−0.8043	−0.0062	0.0304
LogRENPOL	−0.1825	−0.2030	0.0204	0.0099
Chi2 (5)	21.36***			

Continued

TABLE 6.A4 Hausman test—cont'd

Variables	(b) Fixed	(B) Random	(b-B) Difference	Sqrt(diag(V_b-V-B)) S.E.
Model II (Dependent variable LogINCRE)				
TREND	0.0599	0.0549	0.0049	0.0018
LogGDP_PC	−2.1560	−1.8943	−0.2617	0.0973
LogKPRIV	0.1809	0.1534	0.0275	0.0394
LogKOFEcGI	0.9050	0.8905	0.0145	0.0317
LogTROPEN	−0.9868	−0.9467	−0.0401	0.0298
LogRENPOL	−0.2559	−0.2614	0.0055	0.0093
Chi2 (6)	12.94**			
Model III (Dependent variable LogINCRE)				
LogGDP_PC	−1.1866	−0.6626	−0.5239	0.2587
LogKPPP	0.0453	0.0621	−0.0167	0.0124
LogKOFEcGI	0.5690	0.4948	0.0742	0.0524
LogTROPEN	−0.4373	−0.3602	−0.0770	0.0391
LogRENPOL	−0.2633	−0.1828	−0.0804	0.0487
Chi2 (5)	11.61**			

Notes: *** and ** denote statistically significant at the 1% and 5% levels.

TABLE 6.A5 Bias-corrected LM-based test.

Variables	LM(k)-stat	*P*-value	
LogINCRE	2.71	0.007	***
LogGDP_PC	5.52	0.000	***
LogKGOV	2.49	0.013	**
LogKPRIV	3.66	0.000	***
LogKPPP	4.63	0.000	***
LogKOFEcGI	6.22	0.000	***
LogTROPEN	5.10	0.000	***
LogRENPOL	6.46	0.000	***

Notes: *** and ** denote statistical significance at the 1% and 5% levels.

References

Agresti, A., 1990. Categorical Data Analysis. John Wiley and Sons, New York. ISBN 0-471-36093-7.

Apergis, N., Payne, J.E., 2010. Renewable energy consumption and economic growth: evidence from a panel of OECD countries. Energy Policy 38 (1), 656–660. https://doi.org/10.1016/j.enpol.2009.09.002.

Banco de Desarrollo de América Latina (CAF), 2014. USD 94,8 milhões para a hidroelétrica San José. https://www.caf.com/pt/presente/noticias/2014/01/usd-94-8-milhoes-para-a-hidroeletrica-san-jose/.

Belsley, D.A., Kuh, E., Welsch, E.R., 1980. Regression Diagnostics: Identifying Influential Data and Sources of Collinearity. Wiley, New York, pp. 1–286, https://doi.org/10.1002/0471725153.

Bloomberg, N.E.F., 2015. ClimateScope 2015: The Clean Energy Country Competitiveness Index. Washington, DC https://energydata.info/dataset/d84c8a33-1d3f-4eb6-81b8-844017d79345/resource/29d06e3b-ebb5-4e1f-8b37-fc056630267c/download/climatescope-2017-report-en.pdf.

Bloomberg, N.E.F., 2020. Latin America Hit New Clean Energy Investment Record, 2019. https://about.bnef.com/blog/latin-america-hit-new-clean-energy-investment-record-2019/.

Born, B., Breitung, J., 2015. Testing for serial correlation in fixed-effects panel data models. Econ. Rev. 35 (7), 1290–1316. https://doi.org/10.1080/07474938.2014.976524.

Coviello, M.F., Gollán, J., Pérez, M., 2012. Public-Private Partnerships in Renewable Energy in Latin America and the Caribbean. ECLAC, pp. 1–63. https://www.cepal.org/es/publicaciones/4016-alianzas-publico-privadas-energias-renovables-america-latina-caribe.

Fuinhas, J.A., Marques, A.C., Koengkan, M., 2017. Are renewable energy policies upsetting carbon dioxide emissions? The case of Latin America countries. Environ. Sci. Pollut. Res. 24 (17), 15044–15054. https://doi.org/10.1007/s11356-017-9109-z.

Governo do Brasil, 2019. Energia: Governo Inaugura Belo Monte e dá a Largada Para a Construção de Novas Hidrelétricas. https://www.gov.br/pt-br/noticias/energia-minerais-e-combustiveis/2019/11/governo-inaugura-belo-monte-e-da-a-largada-para-a-construcao-de-novas-hidreletricas.

IDB Invest, 2020. Ituango Hydroelectric Project. https://www.idbinvest.org/en/projects/ituango-hydroelectric-project.

International Energy Agency (IEA), 2019. https://www.iea.org/energyaccess/database/.

International Monetary Fund (IMF), 2017. Investment and Capital Stock Dataset. http://www.imf.org/external/np/fad/publicinvestment/data/info122216.pdf.

IRENA, 2016. Renewable Energy Market Analysis: Latin America. pp. 1–160. https://www.irena.org/-/media/Files/IRENA/Agency/Publication/2016/IRENA_Market_Analysis_Latin_America_2016.pdf?la=en&hash=6D59BCB8265FBECCE7FC2992C38458E1FF6796C6.

Kim, J., Park, K., 2016. Financial development and renewable energy diffusion. Energy Econ. 59, 238–250. https://doi.org/10.1016/j.eneco.2016.08.012.

Koengkan, M., 2020. Capital stock development and their effects on investment expansion in renewable energy in Latin America and the Caribbean region. J. Sustain. Finance Invest., 1–16. https://doi.org/10.1080/20430795.2020.1796100.

Koengkan, M., Fuinhas, J.A., 2020. Exploring the effect of the renewable energy transition on CO_2 emissions of Latin American & Caribbean countries. Int. J. Sustain. Energy, 1–25. https://doi.org/10.1080/14786451.2020.1731511.

Koengkan, M., Fuinhas, J.A., Marques, A.C., 2019a. The role of financial openness and China's income on fossil fuels consumption: fresh evidence from Latin American countries. GeoJournal, 1–15. https://doi.org/10.1007/s10708-019-09969-1.

Koengkan, M., Fuinhas, J.A., Poveda, Y.E.M., 2019b. Globalisation as a motor of renewable energy development in Latin America countries. GeoJournal, 1–12. https://doi.org/10.1007/s10708-019-10042-0.

Koengkan, M., Fuinhas, J.A., Vieira, I., 2019c. Effects of financial openness on renewable energy investments expansion in Latin American countries. J. Sustain. Finance Invest., 1–19. https://doi.org/10.1080/20430795.2019.1665379.

Koengkan, M., Fuinhas, J.A., Silva, N., 2021. Exploring the capacity of renewable energy consumption to reduce outdoor air pollution death rate in Latin America and the Caribbean region. Environ. Sci. Pollut. Res. 28, 1656–1674.

KOF Globalisation Index, 2020. https://kof.ethz.ch/en/forecasts-and-indicators/indicators/kof-globalisation-index.html.

Lee, C.-C., 2005. Energy consumption and GDP in developing countries: a cointegrated panel analysis. Energy Econ. 27 (3), 415–427. https://doi.org/10.1016/j.eneco.2005.03.003.

Lee, C.-C., Chen, P.-F., 2010. Dynamic modelling of energy consumption, capital stock, and real income in G-7 countries. Energy Econ. 32 (3), 564–581. https://doi.org/10.1016/j.eneco.2009.08.022.

Lee, C.-C., Chang, C.-P., Chen, P.-F., 2008. Energy-income causality in OECD countries revisited: the key role of capital stock. Energy Econ. 30 (5), 2359–2373. https://doi.org/10.1016/j.eneco.2008.01.005.

Machado, J.A.F., Silva, J.M.C.S., 2019. Quantiles via Moments. J. Econ. https://doi.org/10.1016/j.jeconom.2019.04.009Get. forthcoming.

Mazzucato, M., Semieniuk, G., 2018. Financing renewable energy: who is financing what and why it matters. Technol. Forecast. Soc. Chang. 127, 8–22. https://doi.org/10.1016/j.techfore.2017.05.021.

Mendelsohn, M., Feldman, D., 2013. Financing U.S. Renewable Energy Projects Through Public Capital Vehicles: Qualitative and Quantitative Benefits. NREL is a national laboratory of the U.S. Department of Energy, Office of Energy Efficiency & Renewable Energy, pp. 1–38. Operated by the Alliance for Sustainable Energy, LLC. NREL/TP-6A20-58315 https://www.nrel.gov/docs/fy13osti/58315.pdf.

Narayan, P.K., Smyth, R., 2008. Energy consumption and real GDP in G7 countries: new evidence from panel cointegration with structural breaks. Energy Econ. 30 (5), 2331–2341. https://doi.org/10.1016/j.eneco.2007.10.006.

Our World in Data, 2020. Energy. https://ourworldindata.org/energy.

Pesaran, M.H., 2004. General Diagnostic Tests for Cross-Section Dependence in Panels. The University of Cambridge, Faculty of Economics, https://doi.org/10.17863/CAM.5113. Cambridge Working Papers in Economics, n. 0435.

Pesaran, M.H., 2007. A simple panel unit root test in the presence of cross-section dependence. J. Appl. Econ. 22 (2), 256–312. https://doi.org/10.1002/jae.951.

Santiago, R., Koengkan, M., Fuinhas, J.A., Marques, A.C., 2019. The relationship between public capital stock, private capital stock and economic growth in the Latin American and Caribbean countries. Int. Rev. Econ., 1–25. https://doi.org/10.1007/s12232-019-00340-x.

Sbia, R., Shahbaz, M., Hamdi, H., 2014. A contribution of foreign direct investment, clean energy, trade openness, carbon emissions and economic growth to energy demand in UAE. Econ. Model. 36, 191–197. https://doi.org/10.1016/j.econmod.2013.09.047.

Sebri, C.M., Ben-Salha, O., 2014. On the causal dynamics between economic growth, renewable energy consumption, CO_2 emissions, and trade openness: fresh evidence from BRICS countries. Renew. Sust. Energy Rev. 39, 14–23. https://doi.org/10.1016/j.rser.2014.07.033.

World Bank Open Data, 2020. http://www.worldbank.org.

Wursten, J., 2018. Testing for serial correlation in fixed-effects panel models. Stata J. Promot. Commun. Stat. Stata 18 (1), 76–100. https://doi.org/10.1177/1536867X1801800106.

CHAPTER

7

The effect of energy transition on economic growth and consumption of nonrenewable energy sources in countries of Latin America and the Caribbean

JEL codes E00, N16, H54, Q4

7.1 Introduction

In the Latin America and the Caribbean (LAC) region, the consumption of energy from renewable and nonrenewable energy sources more than tripled between 1970 and 2018 (see Fig. 5.14). As a result, renewable energy sources (e.g. biofuels, biomass, hydropower, wind, and photovoltaic) have reached a substantial weight in the energy matrix of the region, making it the most significant in the world concerning the share of green energy in the energy matrix (Koengkan and Fuinhas, 2020; Koengkan, 2018a).

According to Fuinhas et al. (2017), the renewable energy sector in the LAC region is dynamic and has experienced rapid growth in both investment and consumption of this kind of energy sources. Indeed, this trend, in line with Koengkan and Fuinhas (2020) and Koengkan (2018a) has been enhanced by the abundance of natural resources, the rapid increase in energy demand, the significant dependence on fossil fuel, high energy prices and energy security concerns, inter alia. Moreover, according to the same authors, the increases in energy consumption in the region have been accompanied by the rapid growth of GDP per capita and by the globalisation process, which in turn is enhanced by several episodes of political liberalisation and economic reforms which have occurred in the last 40 years.

Average annual growth rates were approximately 3.0% between 1989 and 2014 (see Fig. 5.18), with GDP per capita (current US$) evolving from US$2319.05 to US$10,405.48 in 2014 (see Fig. 5.9). This annual growth rate was obtained in a period in which the process of economic opening intensified in the region. The first evidence of economic opening in this area emerged in Chile in the 1970s. Nevertheless, in LAC region, that decade was still a period of low economic growth and high inflation (Haggard and Kaufman, 2008), during which most governments were conservative, nationalistic, and unreceptive to the social and economic changes inherent to the emergence of the globalisation process (Rojas, 2007). Moreover, according to Koengkan and Fuinhas (2020), between 1974 and 1979, a large number of tariff and nontariff barriers were eliminated or reduced in Chile. In 1983, Costa Rica initiated a gradual process of economic liberalisation, followed by Bolivia and Mexico in 1985. Other countries, such as Argentina, Brazil, Colombia, Peru, and Venezuela (RB) joined the liberalisation trend in the early 1990s (Pinto and Lahera, 1993).

In 1991, the Mercosur trade bloc established by the Asunción Treaty was created to promote the free trade of goods and services, and the free flow of capital and people across the associate countries—Argentina, Brazil, Paraguay, Uruguay, and Venezuela (RB) (Koengkan and Fuinhas, 2020; Koengkan, 2018a). Venezuela was excluded in 2014 following the still ongoing political and economic crises (Theodore, 2015). In these countries, the consumption of renewable energy began in the 1970s in Brazil, with hydropower in 1973 and biofuels in 1975. Paraguay and Uruguay started in 1973 with hydropower; Argentina in 1998 with hydropower, biomass, biogas, geothermal, wind, wave, and photovoltaic; and Venezuela (RB) in 2001 with hydropower (IRENA, 2016). The consumption of energy from this kind of

Physical Capital Development and Energy Transition in Latin America and the Caribbean
https://doi.org/10.1016/B978-0-12-824429-6.00010-3

sources represented 20% of total energy consumption in 2009 (Santos, 2015). The investment in renewable energy, which grew 13% between 2000 and 2013, is related to the inflow of foreign direct investment (FDI) to the region. In 2016, the Mercosur bloc received 47.4% of FDI flows (Mercosul, 2019).

The relevance of the events described earlier inspired the main research questions of this chapter, namely: What is the effect of consumption of renewable energy on economic growth and consumption of renewable energy in the Mercosur countries? The specific questions resulting from the deepening of the central questions are objective ones: Are there causal links between consumption of renewable energy and economic growth in the Mercosur countries? What is the causality between the consumption of renewable energy and globalisation? To answer these questions, this chapter will use data for the five Mercosur countries from 1981 to 2014, and the PVAR model developed by Holtz-Eakin et al. (1988), performing a Granger causality assessment with a Wald test.

The chapter is organised as follows: Section 7.2 reviews the relevant literature; Section 7.3 presents the data and the method; Section 7.4 describes the empirical analysis; Section 7.5 discusses the obtained results, and Section 7.6 presents the conclusions and policy implications.

7.2 Literature review

The nexus between the consumption of energy and economic growth has received considerable attention in the energy economic literature (Koengkan et al., 2018). Research that approach this relationship have used variables such as primary energy consumption (including renewable and fossil), fossil fuels, total energy consumption, petroleum consumption, nuclear energy, gross domestic product (GDP) growth (annual %), GDP (current US$), and GDP per capita (current US$). Other studies have used total renewable energy consumption, wind energy consumption, hydroelectricity consumption, and photovoltaic energy consumption as variables (Koengkan, 2018a). Still other studies used the variable GDP in constant local currency units (LCU) (e.g. Koengkan et al., 2018; Koengkan, 2017a,b, 2018b).

Even though several authors have used different variables to investigate the nexus between economic growth and consumption of energy, there is a gap in the literature that needs to be filled concerning the relationship between renewable energy consumption and economic growth. Indeed, this gap exists in the literature due to the existence of the low number of researches that approach this relationship. So, as the focus of this literature review is approaching only studies that used the variable consumption of renewable energy—What conclusions have been reached by the literature regarding the relationship between economic growth and consumption of renewable energy?—What innovations do these studies bring to the literature? In recent years, the energy literature has evolved in four ways. The first argues the existence of a neutral relationship between consumption of renewable energy and economic growth (e.g. Menegaki, 2011); the second argues the existence of a unidirectional relationship between economic growth and renewable energy consumption (e.g. Caraiani et al., 2015); the third, the existence of unidirectional dynamic causality from renewable energy consumption to economic growth (e.g. Menyah and Wolde-Rufael, 2010; Ocal and Aslan, 2013; Pao and Fu, 2013; Zeb et al., 2014; Aslan, 2016; Bélaïd and Youssef, 2017; Destek and Aslan, 2017,b); and the fourth, the bidirectional relationship between economic growth and renewable energy consumption (e.g. Apergis and Payne, 2010; Apergis et al., 2010; Apergis and Payne, 2011, 2012; Tugçu et al., 2012; Al-Mulali et al., 2013; Sebri and Ben-Salha, 2014; Lin and Moubarak, 2014; Amri, 2017).

Indeed, one author in this literature review found the existence of a neutral relationship between economic growth and consumption of renewable energy. For instance, Menegaki (2011) investigated the nexus between consumption of renewable energy and economic growth for 27 European countries from 1997 to 2007. The author used a random effect model as a method. The empirical results of this investigation pointed to the existence of a neutral relationship between the variables. However, a second author discovered the presence of a unidirectional relationship between economic growth and renewable energy consumption. For example, Caraiani et al. (2015) studied the causality between economic growth and the consumption of this kind of source in 28 European Union countries in the period between 1980 and 2013. Granger causality tests and cointegration tests were used as methodology. The empirical results of the investigation indicated the presence of unidirectional causality between economic growth and consumption of renewable energy.

A third group of scholars found the existence of unidirectional causality from consumption of renewable energy to economic growth. Menyah and Wolde-Rufael (2010) studied the causal relationship among CO_2 emissions, consumption of renewable and nuclear energy, and GDP in the period from 1960 to 2007. The Granger causality test was used as a method for this investigation. The authors found the presence of a unidirectional relationship between consumption of energy and economic growth. Ocal and Aslan (2013) observed the nexus between renewable energy consumption and economic growth in Turkey. The autoregressive distributed lag (ARDL) bound test and Granger causality were used. The results showed unidirectional causality running from economic growth to renewable energy consumption. Pao and Fu (2013) investigated the causal relationship among GDP, nonhydroelectric renewable energy consumption (NHREC), nonrenewable energy consumption (NREC), total primary energy consumption (TEC), and total renewable energy consumption (TREC) in Brazil in the period from 1980 to 2010. The cointegration test was used as a methodology. The authors found the presence of a unidirectional relationship between NHREC and economic growth, a bidirectional relationship between economic growth and TREC, and unidirectional causality from economic growth to NREC or TEC.

Zeb et al. (2014) analysed the relationship between renewable energy, carbon dioxide emissions, GDP, natural resource depletion, and poverty in Bangladesh, India, Nepal, Pakistan, and Sri Lanka, over the period from 1975 to 2010. The Granger causality test was used as a method. The results indicated the presence of unidirectional causality between the variables. Aslan (2016) examined the nexus between biomass energy consumption, economic growth, employment, and capital in the United States between 1961 and 2011. The ARDL bound test and Granger causality test were used. The results suggested the presence of unidirectional causality from biomass energy to GDP. Bélaïd and Youssef (2017) explored the dynamic relationship between emissions of CO_2, consumption of renewable and nonrenewable electricity, and economic growth in Algeria by using an autoregressive distributed lag cointegration approach over the period from 1980 to 2012. The results revealed the existence of a unidirectional relationship between renewable and nonrenewable electricity consumption and economic growth. Destek and Aslan (2017,b) studied the impact of renewable and nonrenewable energy consumption on economic growth in 17 emerging economies in the period from 1980 to 2012. Bootstrap panel causality was used. The empirical results indicated the positive impact of the consumption of energy on economic growth.

Moreover, another group of researchers discovered the existence of a bidirectional relationship between economic growth and consumption of renewable energy. Apergis and Payne (2010) also investigated the nexus between consumption of renewable energy and economic growth for 13 countries in Eurasia, over the period from 1992 to 2007. The heterogeneous panel cointegration test was used. The authors discovered the existence of bidirectional causality between renewable energy consumption and economic growth in both the short and long run. Using a panel error correction model, Apergis et al. (2010) examined the causal relationship among nuclear energy consumption, renewable energy consumption, economic growth, and CO_2 emissions in 19 developed and developing countries in the period from 1984 to 2007. The results indicated the existence of a bidirectional relationship between all variables. Apergis and Payne (2011) studied the relationship between the consumption of renewable energy and economic growth for a panel of six Central American countries in the period from 1980 to 2006. The panel cointegration test was used as a methodology. The empirical results of the panel error correction model indicated the bidirectional relationship between the consumption of renewable energy and economic growth in the short and long run. Apergis and Payne (2012) tested the relationship between consumption of renewable and nonrenewable energy and economic growth of 80 countries, over the period from 1990 to 2007. The Pedroni heterogeneous panel cointegration test was used. The test indicated the existence of a bidirectional relationship between the variables. Tugçu et al. (2012) researched the long run and causal relationship between the consumption of renewable and nonrenewable energy and GDP in G7 countries in the period from 1980 to 2009, using the ARDL bounds test. The authors found the existence of a long-run bidirectional relationship in all countries investigated.

Al-Mulali et al. (2013) analysed the bidirectional relationship between the consumption of renewable energy and economic growth in high-income, upper-middle-income, lower-middle-income, and low-income countries. The outcomes indicated that 79% of countries have a bidirectional relationship between the consumption of renewable energy and economic growth. On the other hand, 19% of the countries showed the presence of a unidirectional relationship between consumption of renewable energy and economic growth. Furthermore, 2% pointed to a unidirectional relationship between economic growth and the consumption of renewable energy. Sebri and Ben-Salha (2014) studied the causal nexus between consumption of renewable energy and GDP growth in Brazil, Russia, India, China, and South Africa (BRICS), over the period from 1971 to 2010. The ARDL bound test approach and vector error correction model (VECM) were used. The empirical results indicated the presence of a bidirectional relationship between economic growth and consumption of renewable energy, suggesting the feedback hypothesis. Lin and Moubarak (2014) examined the relationship between renewable energy consumption and economic growth in China for the period from 1977 to 2011. The ARDL model was used as a methodology. The outcomes showed that there is bidirectional causality between consumption of renewable energy and economic growth. Amri (2017) analysed the relationship among economic growth, consumption of renewable energy, and trade for 72 countries for the period from 1990 to 2012. The outcomes demonstrate a feedback linkage between income and renewable energy consumption, between trade and renewable energy consumption, and between trade and income.

Although the literature has used different variables, methods, countries, regions, and time series to explain the relationship between consumption of renewable energy and economic growth, some gaps were identified in the literature review, and this needs to be filled. Among them is the use of GDP in constant local currency units (LCU) as an alternative to constant US dollars. In this literature review, none of the authors used this variable. The noninclusion of the variable globalisation index in the model by other authors is another gap. The inclusion of this variable is essential because globalisation has a positive impact on factor productivity and economic growth, and consequently exerts a positive impact on energy consumption (e.g. renewable and fossil), and also on new investments in renewable technology, whereby the Mercosur countries have to access new green technology. Another gap that was identified was the nonutilisation of the PVAR model as a methodology. All authors just utilised the same methodology, such as the ARDL bounds test and heterogeneous panel cointegration test, complemented with the Granger causality test.

Moreover, there are no investigations that approach the Mercosur countries specifically. Indeed, the investigations that were used in the literature review focused on Africa, Asia, the European Union, Middle East countries, and the global scale leaving aside the Mercosur countries and the Latin America & Caribbean region. In other words, the literature that approaches the relationship between consumption of renewable energy and economic growth has not presented significant innovations, due to the use of methodologies and countries already explored.

To fill these gaps, this investigation will adopt a new approach that includes: (i) the inclusion of GDP in constant local currency units (LCU); (ii) the inclusion of the globalisation index in the model; (iii) the use of PVAR model as a methodology; and (iv) the use of Mercosur countries, given that this group is not addressed in the literature that approaches this topic. Based on the various conclusions and approaches of the literature review—What hypotheses should be raised to answer the central question of this investigation? This chapter puts forward the following four hypotheses so to deal with the central research question (Table 7.1).

These hypotheses have been confirmed by Fuinhas and Marques (2019) and by Ozturk (2010). The summary of the literature presented in this section has discussed the most important research into the relationship between the consumption of renewable energy and economic growth. Moreover, this review specifically focused on evidence and discussed the empirical results found in the literature.

TABLE 7.1 Hypotheses for the nexus between economic growth and energy consumption.

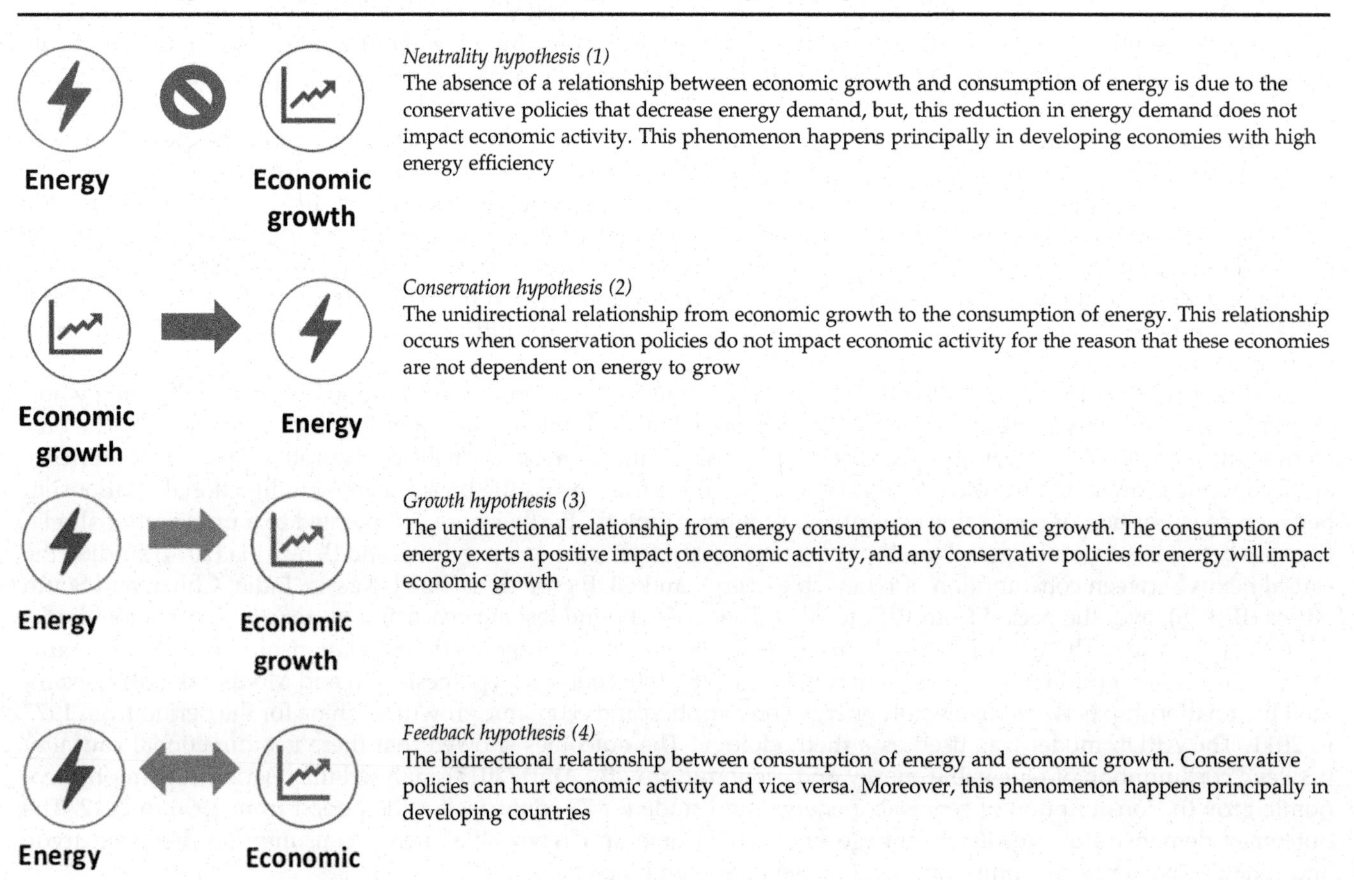

Energy ⊘ Economic growth	*Neutrality hypothesis (1)* The absence of a relationship between economic growth and consumption of energy is due to the conservative policies that decrease energy demand, but, this reduction in energy demand does not impact economic activity. This phenomenon happens principally in developing economies with high energy efficiency
Economic growth → Energy	*Conservation hypothesis (2)* The unidirectional relationship from economic growth to the consumption of energy. This relationship occurs when conservation policies do not impact economic activity for the reason that these economies are not dependent on energy to grow
Energy → Economic growth	*Growth hypothesis (3)* The unidirectional relationship from energy consumption to economic growth. The consumption of energy exerts a positive impact on economic activity, and any conservative policies for energy will impact economic growth
Energy ↔ Economic growth	*Feedback hypothesis (4)* The bidirectional relationship between consumption of energy and economic growth. Conservative policies can hurt economic activity and vice versa. Moreover, this phenomenon happens principally in developing countries

7.3 Data and methodology

This section is organised into two parts. In the first part, the data include variables and database, and the second part describes the methodology that will be used in this study.

7.3.1 Data

Five countries from the Mercosur bloc were selected, namely **Argentina**, **Brazil**, **Paraguay**, **Uruguay**, and **Venezuela (RB)**, to study the nexus between economic growth and consumption of renewable energy. Mercosur is a subregional bloc that was created in 1991, with the purpose of free trade and the fluid movement of goods, people, and currency among the associate countries (Koengkan and Fuinhas, 2020). The period from 1981 to 2014, available for all variables, was used for this chapter. This group of countries was chosen because they have experienced rapid economic growth in the last 34 years as well as a rapid increase in the consumption of renewable energy.

Moreover, another motivation that led us to select the countries from Mercosur was the integration of countries with the rest of the world. Indeed, this integration is related to the globalisation process, where the more a country is integrated with others, the greater the globalisation. The variables which were chosen to perform this investigation are presented in Table 7.2.

'DLog' denotes variables in the first differences of logarithms, 'Obs' denotes the number of observations in the model, 'Std.-dev' denotes the standard deviation, and 'Min. and Max' denote minimum and maximum. The variable **GDP_PC** was transformed into per capita values with the total population of each cross. These summary statistics were obtained from the command *sum* in **Stata 15.0**. Moreover, the board below shows how to transform the variables into per capita values, natural logarithms, and the first differences of logarithms, as well as how to obtain the summary statistics of variables.

TABLE 7.2 Description of variables, source, and summary statistic.

Description of variables			Summary statistic				
Variable	**Definition**	**Source**	**Obs**	**Mean**	**Std.-dev.**	**Min**	**Max**
DLogFOSSIL	Fossil fuel energy consumption (FOSSIL) in billion kilowatt-hours (kWh) from coal, gas and oil	International Energy Agency (IEA, 2018)	170	−16.6500	0.4867	−18.6642	−15.9598
DLogREN	Renewable energy consumption (REN) in billion kilowatt-hours (kWh) from biomass, hydropower, solar, photovoltaic, wind, wave and waste	International Energy Agency (IEA, 2018)	170	−13.1758	0.8537	−15.4224	−11.4340
DLogGDP_PC	Gross Domestic Production (GDP_PC) in constant local currency unity (LCU) and expressed per capita	World Bank Open Data (2018)	169	10.5075	2.6283	7.2285	15.2759
DLogKOFGI	KOF globalisation index De facto (KOFGI) that measures the economic, social and political dimensions of globalisation on a scale from 1 to 100	KOF Index of Globalization (2018)	170	3.9631	0.1891	3.3919	4.2093

How to do:

****Transform the variable gdp_pc into per capita values****

gen gdp_pc =(gdp/population)

****Transform the variables fossil, ren, gdp_pc, and kofgi into natural logarithms****

gen loggdp_pc =log(gdp_pc)

****Transform the variables fossil, ren, gdp_pc, and kofgi into first differences of logarithms****

gen dloggdp_pc =d.log(gdp_pc)

****The summary statistics****

sum dlogfossil dlogren dlgogdp_pc dlogkofgi

7.3.2 Methodology

The best methodology to analyse the nexus between the variables mentioned in Section 7.3.1 is the PVAR model. This methodology was developed by Holtz-Eakin et al. (1988) as an alternative to multivariate simultaneous equation models. The PVAR model is used in several research fields but is most commonly used by macroeconomists working with data for many countries and with a long-time span (Koop and Korobilis, 2016). Canova and Ciccarelli (2009) emphasise that PVARs are an excellent way to model how shocks are transmitted across countries. As Mercosur deepens its integration, the examination of these issues becomes essential for modern applied economists.

According to Abrigo and Love (2016), the PVAR model has the advantage of treating all variables as endogenous. However, the existence of restrictions based on statistical procedures may be imposed on disentangling the impact of exogenous shocks on the system.

Moreover, according to Koengkan and Fuinhas (2020), this methodology is adequate to use in panels with long periods (macro panels), as in our case, the presence of cointegration between the variables and the endogeneity is expected. Indeed, to handle the problem of endogeneity and cointegration, the literature has recommended the use of PVAR models. For this reason, the PVAR model was chosen for undertaking the investigation in this chapter. The PVAR model is represented by the following linear equation:

$$a_{it} = a_{it-1}x_1 + f_{it-p+1}x_{p-1} + g_{it-p+1}x_p + d_{it}l + \varepsilon_{it} \tag{7.1}$$

where a_{it} is the vector of dependent variables that are represented by variables in the first differences of natural logarithms (e.g. **DLogFOSSIL**, **DLogREN**, **DLogGDP_PC**, and **DLogKOFGI**). The variables were used in the first differences of natural logarithms as the PVAR model requires that variables to be of order one. The stationarity of variables can be confirmed by the visual analysis of descriptive statistics and by the second-generation unit root test that is evidenced in Table 7.A3 in Appendix.

g_{it} is the vector of exogenous covariates, and ε_{it} are the vectors of the dependent variable in a panel of fixed effects and idiosyncratic errors, and the matrices x_1, x_{p-1}, x_p and matrix l are parameters to be estimated. Before the realisation of PVAR regression, it is advisable to check the properties of the variables. To this end, some preliminary tests were applied (Table 7.3).

After the PVAR regression, it is necessary to apply the specification tests to verify the characteristics of the model (Table 7.4)

This section shows the data that will be used, the method, and the preliminary and specification tests. In the next section, the results will be shown.

TABLE 7.3 Preliminary tests.

Test	Goal
Variance inflation factor (VIF) test	To check the existence of multicollinearity between the variables in the panel data
Pesaran CD-test	To check the existence of cross-section dependence in the panel data
Panel unit root test (CIPS-test)	To check the presence of unit roots in the variables
Hausman test	To check the presence of heterogeneity i.e. whether the panel has random effects (RE) or fixed effects (FE)
Panel VAR lag-order selection	To report the overall model coefficients of determination

TABLE 7.4 Specification tests.

Test	Goal
The eigenvalue stability condition test	To indicate that the panel-VAR model is stable
The panel Granger causality Wald test	To analyse the causal relationship between the variables of the model
Forecast-error variance decomposition (FEVD) test	To show how a variable responds to shocks in specific variables
Impulse-response function (IRF)	To indicate the impulse-response function of variables of the model

7.4 Results

In line with what was stated earlier, this section shows the outcomes of the preliminary tests, PVAR model, and specification tests. Then, to identify the level of multicollinearity between the variables in the panel data, the VIF test that was developed by Belsley et al. (1980) was calculated. This test is constructed around the following equation:

$$VIF_i = \frac{1}{1 - R_j^2} \tag{7.2}$$

where R_j^2 is the coefficient of determination of regression of model in step one. The results of the VIF test indicate that the values are lower than the usually accepted benchmark of 10, in the case of the VIF values, and six in the case of the mean VIF values (see Table 7.A1 in Appendix). The results of the VIF test were obtained from the command *estat vif* in **Stata 15.0**. The board below shows how to carry out and obtain the results from the VIF test.

How to do:

**** The Variance Inflation Factor test****

```
reg dlogfossil dlogren dlgogdp_pc dlogkofgi
estat vif
```

Moreover, to identify the presence of cross-sectional dependence (CSD) in the panel data, the Pesaran CD test developed by Pesaran (2004) was calculated. This test is constructed around the following equation:

$$CD = \sqrt{\frac{2T}{N(N-1)}}\left(\sum_{i=1}^{N-1}\sum_{j=i+1}^{N}\hat{P}ij\right) \tag{7.3}$$

The null hypothesis of this test is the nonpresence of cross-section dependence $CD \sim N(0,1)$ for $N \rightarrow \infty$ and that T is sufficiently large. The results of CSD test indicate that variables such as **DLogGDP_PC** and **DLogKOFGI** have the presence of cross-section dependence (see Table 7.A2 in Appendix). The results of the Pesaran CD test were obtained from the command *xtcd* in **Stata 15.0**. The board in the following shows how to carry out and obtain the results from the Pesaran CD test.

How to do:

****The Pesaran CD-test****

```
xtcd dlogfossil dlogren dlgogdp_pc dlogkofgi
```

However, the presence of CSD in these variables can be an indication that the selected countries of this study share the same characteristics and shocks (Fuinhas et al., 2017). However, the nonpresence of cross-section dependence in the variable **DLogFOSSIL** and **DLogREN** can be related to each country from the Mercosur trade bloc having its generation of fossil fuels and renewable energy sources characteristics (Fuinhas et al., 2017).

Nevertheless, in the presence of CSD, it is necessary to verify the order of integration of the variables that will be used in the panel-VAR regression. To this end, the panel unit root (CIPS) test developed by Pesaran (2007) was calculated. This test is constructed around the following equation:

$$CIPS(N,T) = t - bar = N^{-1}\sum_{i=1}^{N} ti(N,T) \tag{7.4}$$

where $t_i(N,T)$ is the cross-sectionally augmented Dickey-Fuller statistic for i, the cross-section unit given by the $t-$ ratio of the coefficient of $y_{i,\,t-1}$ in the CADF regression. Therefore, the null hypothesis of this test is that all series have a unit root. The results from the CIPS test obtained indicate that all the variables are of order one, which is a precondition for the use of a panel-VAR estimator as mentioned before (see Table 7.A3 in Appendix). The results of CIPS test were obtained from the command *multipurt* in **Stata 15.0**. The board below shows how to carry out and obtain the results from the CIPS test.

How to do:

****The CIPS-test****

```
multipurt dlogfossil dlogren dlgogdp_pc dlogkofgi,lags(1)
```

The next step of this investigation is to identify the presence of individual effects in the model. To this end, the Hausman test, which compares the random (RE) and fixed effects (FE), was calculated. This test is constructed around the following equation:

$$H = (\beta_{RE} - \beta_{FE})'\sum^{\wedge} - 1(\beta_{RE} - \beta_{FE}) \sim X^2(k) \tag{7.5}$$

where β_{RE} and β_{FE} are estimators of the parameter β. The null hypothesis of this test is that the difference in coefficients is not systematic, where the random effects are the most suitable estimator (Koengkan and Fuinhas, 2020). The results of this test indicate that the null hypothesis should be rejected (**chi2 (3) = 6.81***, statistically significant at 10% level) (see Table 7.A4 in Appendix). The results of the Hausman test were obtained from the command *hausman* with option *sigmaless* in **Stata 15.0**. The board below shows how to carry out and obtain the results from the Hausman test.

How to do:

****The Hausman test****

```
xtreg dlogfossil dlogren dlgogdp_pc dlogkofgi,fe
estimates store fixed

xtreg dlogfossil dlogren dlgogdp_pc dlogkofgi,re
estimates store random

hausman fixed random, sigmaless
```

Next it is necessary to report the overall model coefficients of determination. To this end, the panel VAR lag-order selection test developed by Abrigo and Love (2016) was used. This test is constructed around the following equation:

$$\begin{aligned} MMSC_{BIC,n}(k,p,q) &= J_n(k^2p,k^2q) - (|q| - |p|)k^2 \ln n \\ MMSC_{AIC,n}(k,p,q) &= J_n(k^2p,k^2q) - 2k^2(|q| - |p|) \\ MMSC_{HQIC,n}(p,q) &= J_n(k^2p,k^2q) - Rk^2(|q| - |p|) \ln \ \ln n, R > 2 \end{aligned} \tag{7.6}$$

where $J_n\ (k,p,q)$ is the j statistic of over-identifying restriction for a k-variate PVAR of order p and moment conditions based on q lags of the dependent variables with sample size n. By construction, the above MMSC is available only when $q>p$. As an alternative criterion, the overall coefficient of determination (CD) may be calculated even with just-identified GMM models. Suppose we denote the $(k\ x\ k)$ unconstrained covariance matrix of the dependent variables by Ψ. CD captures the proportion of variation explained by the PVAR model as

$$CD = 1 - \frac{\det\left(\sum\right)}{\det(\Psi)} \tag{7.7}$$

The results of the panel VAR lag-order selection test points to the use of 1 lag in the panel-VAR model (see Table 7.A5 in Appendix). The results of the panel VAR lag-order selection test were obtained from the command *pvarsoc* in **Stata 15.0**. The board below shows how to carry out and obtain the results from the Hausman test.

How to do:

****The PVAR Lag-order selection test****

```
pvarsoc dlogfossil dlogren dlgogdp_pc dlogkofgi, maxlag (3) pvaropts (instl(1/7))
```

After the realisation of preliminary tests, the realisation of the panel-VAR regression is necessary. Table 7.5 provides the results of the first equation from the panel-VAR regression model. The results of this equation will answer the central question of this investigation. The lag length (1) indicated by the panel VAR lag-order selection test was used in this estimation.

The results of PVAR regression points to the existence of endogeneity in the variables. Indeed, the lagged variables in all PVAR equations are at least statistically significant at the 1% level. In addition, only the variables in the first differences were used in the PVAR regression because the respective model requires all variables to be I(1) (see Table 7.A3 in Appendix). Indeed, the results from the PVAR regression were obtained from the command *pvar* with option *lag (1)* of **Stata 15.0**. The board below shows how to carry out and obtain the PVAR regression outcomes.

How to do:

****The PVAR model regression****

```
pvar dlogfossil dlogren dlgogdp_pc dlogkofgi, instl(1/7)
```

After the PVAR estimation, it is advisable to verify the characteristics of the model. To this end, the specification tests developed by Abrigo and Love (2016) were computed. The Granger causality Wald test was used to analyse the causal relationship between the variables in the PVAR model. Table 7.A6 in Appendix presents the results of the panel Granger causality Wald test. Indeed, the results of this test point to the existence of a bidirectional relationship between economic growth and consumption of renewable energy, economic growth and consumption of fossil fuels, economic growth and globalisation, consumption of renewable energy and consumption of fossil fuels, globalisation and consumption of renewable energy, and globalisation and consumption of fossil fuels. The results from the panel Granger causality Wald test were obtained from the command *pvargranger* in **Stata 15.0**.

Fig. 7.1 summarises the causalities between the variables. This figure was based on results from the panel Granger causality Wald test (see Table 7.A6 in Appendix) and the results of PVAR estimation (see Table 7.5).

Indeed, after the Granger causality Wald test, the eigenvalue stability condition was applied. Table 7.A7 in Appendix displays the graph of the eigenvalue stability condition. The eigenvalue test indicates that the PVAR model is stable, because all eigenvalues are inside the unit circle, satisfying the stability condition of the test. The results from the eigenvalue stability condition test were obtained from the command *r* of *pvarstable* with option *graph* in **Stata 15.0**. The FEVD needs to be computed after the eigenvalue test. Table 7.A8 in Appendix shows the outputs of the FEVD test.

TABLE 7.5 PVAR model outcomes.

Response of	Response to							
	DLogFOSSIL$^{(t)}$		DLogREN$^{(t)}$		DLogGDP_PC$^{(t)}$		DLogKOFGI$^{(t)}$	
DLogFOSSIL$_{(t-1)}$	−0.1424	***	0.2060	***	−0.1517	***	0.1803	***
DLogREN $_{(t-1)}$	−0.0654	***	−0.4881	***	0.0231	***	−0.0113	**
DLogGDP_PC$_{(t-1)}$	0.2529	***	0.5878	***	0.3830	***	−0.2414	***
DLogKOFGI$_{(t-1)}$	−0.4305	***	−0.5558	***	0.1069	**	0.6428	***
N. obs	124							
N. panels	5							

Notes: *** and ** denote statistical significance level at the 1% and 5% levels.

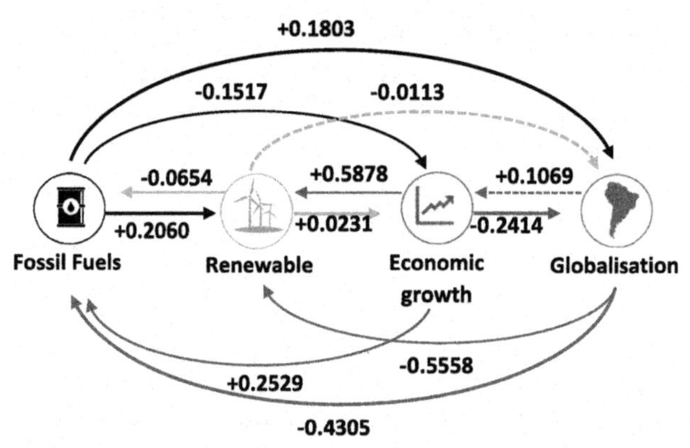

FIG. 7.1 Summary of causality of the variables.

The FEVD test indicates that one period after the shock, the variables themselves explained almost all the forecast error variance.

Then, one period after a shock in the consumption of fossil fuels, the variable explains 68% of forecast error variance, consumption of renewable energy explains 27%, economic growth five periods after a shock explain 5.82%, and globalisation 3.53%. The consumption of renewable energy five periods after a shock explains 72% of forecast error variance, consumption of fossil fuels one period after a shock explains 68%, economic growth one period after a shock explains 2.73%, and globalisation one period after a shock explains 0.91%. Economic growth one period after a shock explains 97% of forecast error variance, consumption of fossil fuels five periods after a shock explains 4.15%, consumption of renewable energy 10 periods after a shock explains 0.45% of forecast error variance, and globalisation 10 periods after a shock explains 2%. Globalisation one period after a shock explains 97% of forecast error variance, consumption of fossil fuels five periods after a shock explains 5%, consumption of renewable energy explains 1.56%, and economic growth 10 periods after a shock explains 18.44% of forecast error variance. The results from the FEVD test were obtained from the command *pvarfevd* with options *mc(1000) st(15)* in **Stata 15.0**.

Moreover, Fig. 7.A1 in Appendix shows the impulse-response functions. In the long run, all variables converge to equilibrium, supporting that the variables of the model are I(0). Then, the impulse-response functions are in concordance with FEDV test. The results from the impulse-response functions test were obtained from the command *pvarirf* with options *mc(1000) oirf byopt(yrescale) st(15)* in **Stata 15.0**. The next section will show the discussion of the empirical results.

7.5 Discussions

The results from the preliminary tests verifying the characteristic of variables that were used in this chapter point to the presence of low-multicollinearity, cross-sectional dependence, stationarity in the first differences of variables, random effects in the model, and the need to use the lag length (1) in the PVAR regression (see Tables 7.A1–7.A5 in Appendix). In the preliminary tests, the variable consumption of renewable energy in the first differences does not have the presence of cross-sectional dependence.

The answer for the nonexistence of cross-sectional dependence in the variable consumption of renewable energy in the first difference is largely country specific and conditional on the intermittence that characterises its generation (Fuinhas et al., 2017). The existence of cross-sectional dependence in the variables of the panel data means that the countries of this study share the same characteristics and shocks (Koengkan, 2018a).

The results of the PVAR model regression indicate that the process of globalisation and consumption of renewable energy increases economic growth. In contrast, the consumption of nonrenewable energy sources (e.g. fossil fuels) reduces it. Consumption of fossil fuels and economic growth encourages the consumption of renewable energy sources, while the globalisation process reduces it. Economic growth increases the consumption of fossil fuels, while the globalisation process and consumption of renewable energy reduce it. Moreover, the consumption of fossil fuels increases the process of globalisation, while economic growth and consumption of renewable energy decrease this process (see Fig. 7.1).

The outcomes of specification indicated the presence of a bidirectional relationship between economic growth and consumption of renewable energy, economic growth and consumption of fossil fuels, economic growth and globalisation, consumption of renewable energy and consumption of fossil fuels, globalisation and consumption of renewable energy, and globalisation and consumption of fossil fuels. The PVAR model is stable. One period after the shock, the variables themselves explained almost all the forecast error variance, and the impulse-response functions of all variables converge to equilibrium, supporting that the variables of the model are I(0) (see Tables 7.A6, 7.A7, 7.A8, and Fig. 7.A1 in Appendix).

The bidirectional relationship between economic growth and the consumption of fossil fuels is in line with several studies that approached this nexus (e.g. Mirza and Kanwal, 2017; Fuinhas et al., 2017; Koengkan, 2017b; Koengkan, 2017c). In Latin American countries, fossil fuels are a vital input. Higher economic growth leads to increases in the consumption of fossil fuels (such as oil, coal, and gas) to supply the demand in these countries. Additionally, Mirza and Kanwal (2017) added that in the Latin American region, fossil fuels are the primary inputs for agriculture and industry.

Koengkan (2017b) affirmed that South American economies are dependent on the consumption of energy, where a 1% increase in the consumption of energy increases economic growth by 0.5%. Fuinhas et al. (2017) found that the high economic dependency on fossil fuels is because Latin American countries are among the major fossil fuel energy producers (e.g. Argentina, Brazil, Venezuela), and others are significant importers, such as Uruguay and Paraguay. Koengkan (2017c) confirmed that the bidirectional relationship between consumption of fossil fuels and economic growth is due to energy use in the LAC countries being very sensitive to changes in economic activity, where rapid economic growth exerts a positive influence on energy demand.

Several authors have confirmed the bidirectional nexus between economic growth and consumption of renewable energy (e.g. Amri, 2017; Destek and Aslan, 2017,b; Kahia et al., 2017; Koengkan, 2017d; Rafindadi and Ozturk, 2017; Lin and Moubarak, 2014; Al-Mulali et al., 2013; Apergis and Payne, 2012; Tugçu et al., 2012; Apergis and Payne, 2011; Apergis and Payne, 2010; Apergis et al., 2010). The vast abundance of renewable sources (e.g. hydropower, solar, photovoltaic, wind, geothermal, and waste) in all countries of the Latin American region stimulate investment in renewable energy and consequently exerts a positive impact on economic activity, and also on the consumption of energy (Apergis and Payne, 2010).

According to Koengkan (2017d), the increase in economic activity exerts a positive impact on renewable energy consumption and on investment in this kind of source to supply the demand in the long run. In addition, the bidirectional relationship between the consumption of renewable energy and fossil fuel is confirmed by Apergis and Payne (2010). The bidirectionality between consumption of renewable energy and fossil fuel is due to both energy sources being substitutes for each other in the energy mix in Latin American countries (Apergis and Payne, 2010).

Finally, the bidirectional relationship between consumption of renewable energy and fossil fuel and globalisation is in line with some authors that studied this nexus (e.g. Koengkan, 2017a; Shahbaz et al., 2015; Leitão, 2014). According to Koengkan (2017a), the globalisation process in Latin American countries has a positive impact on factor productivity and economic growth, and consequently exerts a positive impact on renewable energy consumption and also in new investment in renewable technology. This consequently increases technology efficiency, whereby Latin American countries have to access new green technologies via imports. Leitão (2014) confirms that trade and financial liberalisation, as well as international environmental rules, encourages economies to use renewable energy sources, and consequently reduces the consumption of nonrenewable energy sources.

Moreover, the idea advanced by Koengkan (2017a) and Leitão (2014) is confirmed in Chapter 6, where it was discovered that economic globalisation encourages investment in renewable energy sources. Indeed, economic globalisation increases capital stock and consequently reduces the cost of external financing, encouraging investment in renewable energy technologies. This section showed the possible explanations for bidirectionality in the Mercosur countries, and the next section will show the conclusion and policy implications of this chapter.

7.6 Conclusions and policy implications

The effect of the energy transition on economic growth and consumption of nonrenewable energy was investigated in this chapter. Five Mercosur countries in the period from 1981 to 2014 were analysed. The PVAR model was used as a methodology. Thus, the results of preliminary tests indicated the existence of low-multicollinearity between the variables of the model, cross-sectional dependence, the stationarity of all variables in the first differences of logarithms, and the need to use the lag length (1) in the PVAR regression.

The results of the PVAR model indicated the consumption of renewable energy, which is a proxy for energy transition, increases economic growth, and decreases the consumption of fossil fuels in the Mercosur countries. Moreover, the Granger causality Wald test confirmed hypothesis (4): the existence of a bidirectional relationship between consumption of energy (renewable and also fossil sources) and economic growth. The countries depend on fossil fuels to grow due to the bidirectional relationship between the consumption of fossil fuels and economic growth. The existence of substitutability between consumption of renewable and fossil sources in periods of drought in the reservoirs was found. Indeed, hydropower was substituted by thermoelectric plants that are powered by oil or gas. The process of globalisation in the countries has a negative impact on the consumption of renewable energy and fossil fuels. Thus, the dependency on fossil fuels for growth and the substitutability between renewable and fossil reveals the existence of low energy source diversification in the Mercosur countries. The low energy diversification in these countries is due to low public and private investment in green energy to supply the growing and future demand.

More public policies and incentives should be created to attract more investment in renewable energy and increase the consumption of this kind of source. Policies should be advanced that encourage households and firms to purchase appliances with a high energy-efficiency standard to reduce energy consumption. Policies should be developed that encourage public and private banks to support investment in renewable energy technologies or the purchase of technologies that reduce energy consumption and environmental degradation by firms and households with low-interest rates and credit. The bureaucracy that discourages foreign investment in renewable energy should be reduced, as should the political lobby between governments and large producers of fossil fuels.

These policies need to be implemented to reduce the dependency of Mercosur countries on fossil fuels, as well as to reduce environmental degradation by increasing the consumption of renewable energy. Also, it is advisable to promote economic growth and take advantage of the enormous abundance of renewable energy sources in Mercosur countries. Finally, the empirical findings of this study not only help to advance the existing literature but also warrant attention from governments and policymakers.

Appendix

TABLE 7.A1 VIF-test.

Variables	VIF	1/VIF
DLogREN	1.00	0.9961
DLogGDP_PC	1.00	0.9964
DLogKOFGI	1.01	0.9932
Mean VIF	1.00	

TABLE 7.A2 Pesaran CD-test.

Variables	CD-test	*P*-value		Corr	Abs (corr)
DLogFOSSIL	1.43	0.154		0.079	0.121
DLogREN	0.43	0.668		0.024	0.146
DLogGDP_PC	7.56	0.000	***	0.418	0.418
DLogKOFGI	6.02	0.000	***	0.333	0.333

Notes:*** denotes statistical significance at the 1% level.

TABLE 7.A3 Panel unit root test (CIPS-test).

Variables	Panel unit root test (CIPS) (Zt-bar)				
	Without trend			With trend	
	Lags	Zt-bar		Zt-bar	
DLogFOSSIL	1	−5.990	***	−4.871	***
DLogREN	1	−6.588	***	−5.256	***
DLogGDP_PC	1	−5.759	***	−4.488	***
DLogKOFGI	1	−6.652	***	−6.317	***

Notes: *** denotes statistically significant at the 1% level.

TABLE 7.A4 Hausman test.

Variables	(b) Fixed	(B) Random	(b-B) Difference	Sqrt(diag(V_b-V-B)) S.E.
DLogREN	0.3107	0.3100	0.0006	0.0104
DLogGDP_PC	0.3883	0.3186	0.0697	0.0474
DLogKOFGI	−0.2021	−0.0643	−0.1377	0.0534
Chi2 (3)	6.81*			

Notes: *** denotes statistically significant at the 10% level.

TABLE 7.A5 PVAR Lag-order selection test

Lag	CD	J	J p-value	MBIC	MAIC	MQIC
1	**0.4910**	**100.465**	**0.3574***	**−362.2818**	**−91.5348**	**−201.5186**
2	0.3868	77.0044	0.5741	−308.6181	−82.9956	−174.6488
3	0.0529	69.362	0.3015	−239.136	−58.638	−131.9605

The overall coefficient of determination (CD), Hansen's J statistic (J), p-value (J p-value), MMSC-Bayesian information criterion (MBIC), MMSC-Akaike information criterion (MAIC), and MMSC-Hannan and Quinn information criterion (MQIC) were computed.
Notes: * indicates the best lag-order selection.

TABLE 7.A6 Panel Granger causality Wald test.

Equation\excluded		chi2	Df.	Prob > chi2
DLogFOSSIL	DLogREN	31.314	1	0.000
	DLogGDP_PC	20.393	1	0.000
	DLogKOFGI	22.222	1	0.000
	All	67.347	3	0.000
DLogREN	DLogFOSSIL	11.576	1	0.001
	DLogGDP_PC	36.222	1	0.000
	DLogKOFGI	10.046	1	0.002
	All	77.767	3	0.000
DLogGDP_PC	DLogFOSSIL	51.447	1	0.000
	DLogREN	10.852	1	0.000
	DLogKOFGI	3.908	1	0.048
	All	86.880	3	0.000
DLogKOFGI	DLogFOSSIL	123.472	1	0.000
	DLogREN	6.341	1	0.012
	DLogGDP_PV	57.740	1	0.000
	All	199.475	3	0.000

Notes: *** and ** denote statistical significance at the 1% and 5% levels, respectively.

TABLE 7.A7 Eigenvalue stability condition.

Eigenvalue			Graph
Real	Imaginary	Modulus	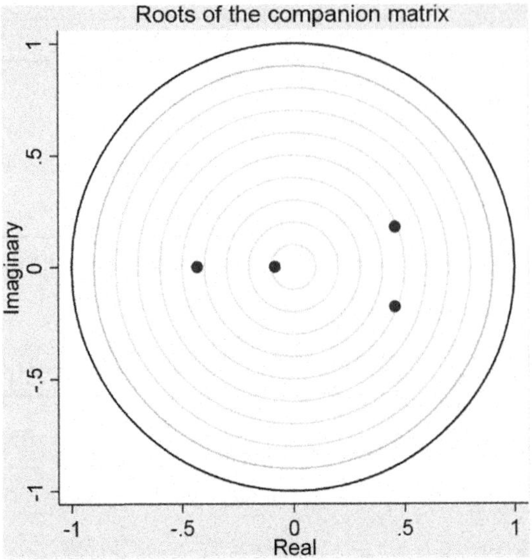
0.4561	0.1778	0.4896	
0.4561	−0.1778	0.4896	
−0.4332	0.0000	0.4332	
−0.0839	0.0000	0.0839	

TABLE 7.A8 Forecast-error variance decomposition.

Response variable and Forecast Impulse Variable Horizon	Impulse variables			
	DLogFOSSIL	DLogREN	DLogGDP_PC	DLogKOFGI
DLogFOSSIL				
0	0	0	0	0
1	0.6880	0.2777	0.0342	0
5	0.6210	0.2854	0.0582	0.0353
10	0.6209	0.2854	0.0583	0.0353
15	0.6209	0.2854	0.0583	0.0353
DLogREN				
0	0	0	0	0
1	0.2968	0.7031	0	0
5	0.2473	0.7161	0.0273	0.0091
10	0.2473	0.7161	0.0273	0.0091
15	0.2473	0.7161	0.0273	0.0091
DLogGDP_PC				
0	0	0	0	0
1	0.0342	0.0002	0.9654	0
5	0.0415	0.0044	0.9336	0.0202
10	0.0416	0.0045	0.9330	0.0207
15	0.0416	0.0045	0.9330	0.0207
DLogKOFGI				
0	0	0	0	0
1	0.0085	0.0052	0.0106	0.9754
5	0.0498	0.0156	0.1831	0.7513
10	0.0498	0.0156	0.1844	0.7500
15	0.0498	0.0156	0.1844	0.7500

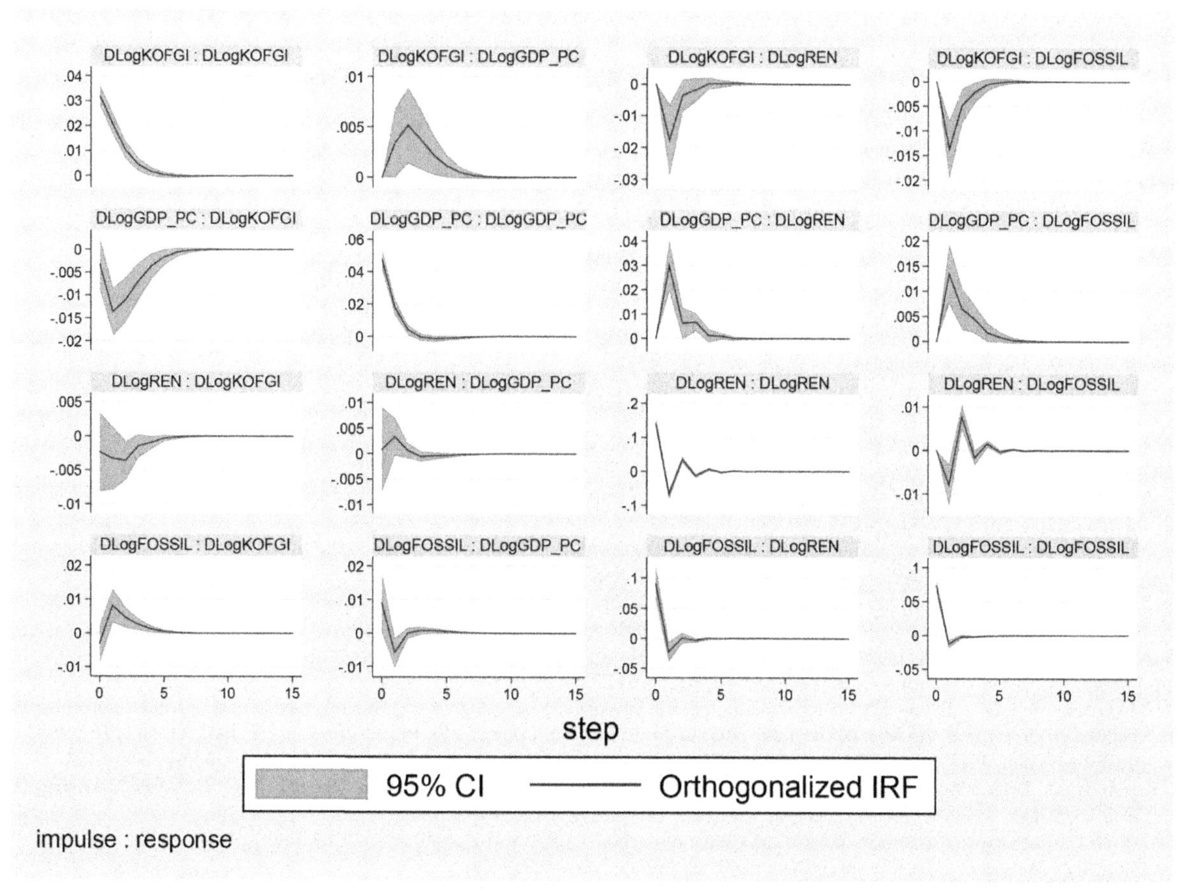

FIG. 7.A1 Impulse-response functions.

References

Abrigo, M.R.M., Love, I., 2016. Estimation of panel vector autoregression in stata: a package of programs. Stata J. 16 (3), 778–804. https://doi.org/10.1177/1536867X1601600314.

Al-Mulali, U., Fereidouni, H.G., Lee, J.Y.M., Sab, C.N.B.C., 2013. Examining the bi-directional long-run relationship between renewable energy consumption and GDP growth. Renew. Sust. Energy Rev. 22, 209–222. https://doi.org/10.1016/j.rser.2013.02.005.

Amri, F., 2017. Intercourse across economic growth, trade and renewable energy consumption in developing and developed countries. Renew. Sust. Energy Rev. 69, 527–534. https://doi.org/10.1016/j.rser.2016.11.230.

Apergis, N., Payne, J.E., 2010. Renewable energy consumption and growth in Eurasia. Energy Econ. 32 (6), 1392–1397. https://doi.org/10.1016/j.eneco.2010.06.001.

Apergis, N., Payne, J.E., 2011. Renewable energy consumption–growth nexus in Central America. Appl. Energy 88 (1), 343–347. https://doi.org/10.1016/j.apenergy.2010.07.013.

Apergis, N., Payne, J.E., 2012. Renewable and non-renewable energy consumption-growth nexus: evidence from a panel error correction model. Energy Econ. 34 (3), 733–738. https://doi.org/10.1016/j.eneco.2011.04.007.

Apergis, N., Payne, J.E., Menyah, K., Wolde-Rufael, Y., 2010. On the causal dynamics between emissions, nuclear energy, renewable energy, and economic growth. Ecol. Econ. 69 (11), 2255–2260. https://doi.org/10.1016/j.ecolecon.2010.06.014.

Aslan, A., 2016. The causal relationship between biomass energy use and economic growth in the United States. Renew. Sust. Energy Rev. 57, 362–366. https://doi.org/10.1016/j.rser.2015.12.109.

Bélaïd, F., Youssef, M., 2017. Environmental degradation, renewable, and non-renewable electricity consumption, and economic growth: assessing the evidence from Algeria. Energy Policy 102, 277–287. https://doi.org/10.1016/j.enpol.2016.12.012.

Belsley, D.A., Kuh, E., Welsch, E.R., 1980. Regression Diagnostics: Identifying Influential Data and Sources of Collinearity. Wiley, New York, pp. 1–286, https://doi.org/10.1002/0471725153.

Canova, F., Ciccarelli, M., 2009. Estimating multi-country VAR models. International. Econ. Rev. 50, 929–959. https://doi.org/10.1111/j.1468-2354.2009.00554.x.

Caraiani, C., Lungu, C.I., Dascalu, C., 2015. Energy consumption and GDP causality: a three-step analysis for emerging European countries. Renew. Sust. Energy Rev. 44, 198–210. https://doi.org/10.1016/j.rser.2014.12.017.

Destek, M.A., Aslan, A., 2017. Renewable and non-renewable energy consumption and economic growth in emerging economies: evidence from bootstrap panel causality. Renew. Energy 111, 757–763. https://doi.org/10.1016/j.renene.2017.05.008.

Fuinhas, J.A., Marques, A.C., 2019. The Extended Energy-Growth Nexus: Theory and Empirical Applications. Academic Press, pp. 1–332. ISBN: 9780128157190.

Fuinhas, J.A., Marques, A.C., Koengkan, M., 2017. Are renewable energy policies upsetting carbon dioxide emissions? The case of Latin America countries. Environ. Sci. Pollut. Res. 24 (17), 15044–15054. https://doi.org/10.1007/s11356-017-9109-z.

Haggard, S., Kaufman, R., 2008. Development, Democracy, and the Welfare State: Latin America, East Asia, and Eastern Europe. Princeton, Princeton University Press, pp. 190–192. ISBN: 9780691135960.

Holtz-Eakin, D., Newey, W., Rosen, H.S., 1988. Estimating vector autoregressions with panel data. Econometrica 56 (6), 1371–1395. https://www.jstor.org/stable/1913103.

International Energy Agency (IEA), 2018. https://www.iea.org/.

IRENA (International Renewable Energy Agency), 2016. Renewable Energy Market Analysis: Latin America. pp. 1–160. ISBN 978-92-95111-49-3.

Kahia, M., Aïssa, M.S.B., Lanouar, C., 2017. Renewable and non-renewable energy use - economic growth nexus: the case of MENA Net Oil Importing Countries. Renew. Sust. Energy Rev. 71, 127–140. https://doi.org/10.1016/j.rser.2017.01.010.

Koengkan, M., 2017a. Is the globalization influencing the primary energy consumption? The case of Latin America and the Caribbean countries. Cadernos UniFOA Volta Redonda 12 (33), 59–69. ISSN: 1809-9475.

Koengkan, M., 2017b. O nexo entre o consumo de energia primária e o crescimento econômico nos países da América do Sul: Uma análise de longo prazo. Cadernos UniFOA Volta Redonda 12 (34), 56–66. ISSN: 1809-9475.

Koengkan, M., 2017c. The nexus between energy consumption, economic growth, and urbanization in Latin American and Caribbean countries: an approach with PVAR model. Revista Valore 6 (2), 202–218. https://doi.org/10.22408/reva22201761%25p.

Koengkan, M., 2017d. The hydroelectricity consumption and economic growth nexus: a long time span analysis. Revista Brasileira de Energias Renováveis* 6 (4), 678–704. https://doi.org/10.5380/rber.v6i4.49181.

Koengkan, M., 2018a. The decline of environmental degradation by renewable energy consumption in the MERCOSUR countries: an approach with ARDL modelling. Environ. Syst. Decis., 1–11. https://doi.org/10.1007/s10669-018-9671-z.

Koengkan, M., 2018b. The positive impact of trade openness on the consumption of energy: fresh evidence from Andean community countries. Energy 158, 936–943. https://doi.org/10.1016/j.energy.2018.06.091.

Koengkan, M., Fuinhas, J.A., 2020. The interactions between renewable energy consumption and economic growth in the Mercosur countries. Int. J. Sustain. Energy 39 (6), 594–614. https://doi.org/10.1080/14786451.2020.1732978.

Koengkan, M., Losekann, L.D., Fuinhas, J.A., 2018. The relationship between economic growth, consumption of energy, and environmental degradation: renewed evidence from Andean community nations. Environ. Syst. Decis., 1–13. https://doi.org/10.1007/s10669-018-9698-1.

KOF Index of Globalization, 2018. https://www.kof.ethz.ch/en/forecasts-and-indicators/indicators/kof-globalisation-index.html.

Koop, G., Korobilis, D., 2016. Model uncertainty in panel vector autoregressive models. Eur. Econ. Rev. 81, 115–131. https://doi.org/10.1016/j.euroecorev.2015.09.006.

Leitão, N.C., 2014. Economic growth, carbon dioxide emissions, renewable energy and globalization. Int. J. Energy Econ. Policy 4 (3), 391–399. ISSN: 2146-4553 https://www.econjournals.com/index.php/ijeep/article/view/830/465.

Lin, B., Moubarak, M., 2014. Renewable energy consumption – economic growth nexus for China. Renew. Sust. Energy Rev. 40, 111–117. https://doi.org/10.1016/j.rser.2014.07.128.

Menegaki, A.N., 2011. Growth and renewable energy in Europe: a random effect model with evidence for neutrality hypothesis. Energy Econ. 33 (2), 257–263. https://doi.org/10.1016/j.eneco.2010.10.004.

Menyah, K., Wolde-Rufael, Y., 2010. CO_2 emissions, nuclear energy, renewable energy and economic growth in the US. Energy Policy 38 (6), 2911–2915. https://doi.org/10.1016/j.enpol.2010.01.024.

Mercosul, 2019. Saiba Mais Sobre o Mercosul. http://www.mercosul.gov.br/saiba-mais-sobre-o-mercosul.

Mirza, F.M., Kanwal, A., 2017. Energy consumption, carbon emissions and economic growth in Pakistan: dynamic causality analysis. Renew. Sust. Energy Rev. 72, 1233–1240. https://doi.org/10.1016/j.rser.2016.10.081.

Ocal, O., Aslan, A., 2013. Renewable energy consumption–economic growth nexus in Turkey. Renew. Sust. Energy Rev. 28, 494–499. https://doi.org/10.1016/j.rser.2013.08.036.

Ozturk, I., 2010. A literature survey on energy-growth nexus. Energy Policy 38 (1), 340–349. https://doi.org/10.1016/j.enpol.2009.09.024.

Pao, H.-T., Fu, H.-C., 2013. Renewable energy, non-renewable energy and economic growth in Brazil. Renew. Sust. Energy Rev. 25, 381–392. https://doi.org/10.1016/j.rser.2013.05.004.

Pesaran, M.H., 2004. General Diagnostic Tests for Cross-section Dependence in Panels. The University of Cambridge, Faculty of Economics, https://doi.org/10.17863/CAM.5113. Cambridge Working Papers in Economics, n. 0435.

Pesaran, M.H., 2007. A simple panel unit root test in the presence of cross-section dependence. J. Appl. Econ. 22 (2), 256–312. https://doi.org/10.1002/jae.951.

Pinto, A., Lahera, E., 1993. Trade Liberalization in Latin America. vol. 50 CEPAL, pp. 41–62. https://www.cepal.org/en/publications/10440-tradeliberalization-latin-america.

Rafindadi, A.A., Ozturk, I., 2017. Impacts of renewable energy consumption on the German economic growth: evidence from combined cointegration test. Renew. Sust. Energy Rev. 75, 1130–1141. https://doi.org/10.1016/j.rser.2016.11.093.

Rojas, E.J.G., 2007. Nación y nacionalismo en America Latina. Universidad Nacional de Colombia, Centro de Estudios Sociales, Bogota, Colombia, pp. 1–276. ISBN: 958-806343-4 http://bdigital.unal.edu.co/1508/.

Santos, H., 2015. Energy security in MERCOSUR+2: challenges and opportunities. OIKOS 14 (2), 05–18. https://www.academia.edu/36080873/Seguran%C3%A7a_Energ%C3%A9tica_no_MERCOSUL_2_desafios_e_oportunidades_Energy_Security_in_MERCOSUR_2_challenges_and_opportunities.

Sebri, M., Ben-Salha, O., 2014. On the causal dynamics between economic growth, renewable energy consumption, CO_2 emissions, and trade openness: fresh evidence from BRICS countries. Renew. Sust. Energy Rev. 39, 14–23. https://doi.org/10.1016/j.rser.2014.07.033.

Shahbaz, M., Khan, S., Ali, A., Bhattacharya, M., 2015. The impact of globalization on CO_2 emissions in China. Munich Personal RePEc Archive 64450, 1–29. https://mpra.ub.uni-muenchen.de/64450/.

Theodore, J.D., 2015. The process of globalization in Latin America. Int. Bus. Econ. Res. J. 14 (1), 193–198. https://doi.org/10.19030/iber.v14i1.9044.

Tugçu, C.T., Ozturk, I., Aslan, A., 2012. Renewable and non-renewable energy consumption and economic growth relationship revisited: evidence from G7 countries. Energy Econ. 34 (6), 1942–1950. https://doi.org/10.1016/j.eneco.2012.08.021.

World Bank Open Data, 2018. http://www.worldbank.org/.

Zeb, R., Salar, L., Awan, U., Zaman, K., Shahbaz, M., 2014. Causal links between renewable energy, environmental degradation and economic growth in selected SAARC countries: Progress towards a green economy. Renew. Energy 71, 123–132. https://doi.org/10.1016/j.renene.2014.05.012.

CHAPTER

8

The consequences of energy transition on environmental degradation of Latin America and the Caribbean

JEL codes F62, Q43, Q5

8.1 Introduction

CO_2 emissions are a significant contributor to global warming. Therefore, these emissions are the most significant potential cause of climate change, along with one of the most significant challenges that human society currently faces. Given this, policymakers and scholars have discussed and developed strategies for reducing these emissions and, consequently, their impacts on global warming.

Indeed, many international initiatives have been made, such as Eco-1992, the Kyoto Protocol in 1997, and COP 21 in 2015. In these commitments, several nations committed themselves to decrease their greenhouse gas emissions. For these countries to be able to accomplish this goal, it then becomes crucial to identify the primary determinants of CO_2 emissions, as well as creating policies that reduce them.

In Latin America and the Caribbean (LAC) region, emissions of CO_2 have grown almost 10-fold since the 1970s (see Fig. 5.3). Indeed, only two countries, Brazil and Mexico, are in the top 20 highest emissions countries of the region. Brazil and Mexico account together for 52.6% of emissions in the region. Other countries from the LAC region emit more than 10 million kilotons of carbon annually, such as Argentina 52.4, Venezuela (RB) 46.2, Chile 19.9, Colombia 18.5, Trinidad and Tobago 13.6, and Peru 11.1, as mentioned by Marland et al. (2011).

The energy sector is the most significant contributor to the increase of CO_2 emissions in the LAC countries, where this sector accounted for approximately 48% of total emissions in 2014 (see Fig. 5.8). A sustainable energy transition to low carbon will play an essential role in reducing environmental degradation and promoting changes in the global climate scenario.

Energy transition implies a radical transformation of the energy sector towards a low-carbon energy system, where renewable energy, energy efficiency technology, and CCS will play an essential role in energy transition in order to reduce the consumption of fossil fuels and consequently environmental degradation, as cited by Tavares (2017). The term 'energy transition' does not have a precise or widely accepted meaning. However, this term is often used to describe changes in the composition of the energy matrix. Indeed, this change takes place gradually from an established energy system to a new one, as mentioned by Smil (2010) in Chapter 5.

The term 'energy transition', according to Hauff et al. (2014) in Chapter 5, refers to a structural change in the energy sector of a country. According to the authors, this term indicates the growing trend of the share of renewable energy combined with the promotion of energy efficiency to reduce the consumption of fossil fuels.

This process of energy transition in the LAC was initiated after the oil shocks of the 1970s when the region began to focus on the development of renewable energy technology. This process started in Brazil and Paraguay in the 1970s or more precisely in 1974 with the construction of Itaípu dam, a large hydropower plant, in 1974–84. Indeed, the participation of large hydropower dam plants in the energy matrix accounted for 10.41% in 1970, and in 2018 the participation of this kind of energy in the energy matrix reached a value of 23.57% (see Fig. 5.1). That is, in 1970 the energy generation from large hydropower dam plants was 65.88 terawatt-hours (TWh), and in 2018 reached a value of 731.31

Physical Capital Development and Energy Transition in Latin America and the Caribbean
https://doi.org/10.1016/B978-0-12-824429-6.00009-7

(TWh) an increase of 1010% (see Fig. 6.2). However, investments in large hydropower dam plants have been declining in the LAC in recent decades due to investments in new renewable energy sources (e.g. geothermal, marine, small hydro, solar photovoltaic, solar, waste, and wind (Koengkan et al., 2019b)). Indeed, new renewable energy sources have had rapid growth since the end of the 1980s and in 2018 comprised 5.03% of the total energy mix in the region, with wind energy comprising 2.12%, solar energy 0.40%, and other renewable energy that includes geothermal, biomass, and waste comprising 2.51% (see Fig. 6.3).

Moreover, the installed capacity of these energy sources more than doubled between 2005 and 2018, where the installed capacity in 2005 was 22.89 TWh and in 2018 reached a value of 156.31 TWh (see Fig. 6.4). In Latin America, the consumption of energy from renewable sources represented 25% of the total energy consumption in 2014 (see Fig. 5.14). The energy mix of the LAC region relies less on fossil fuels than the global average.

According to Hollanda et al. (2016), the LAC region is ahead in energy transition to a low-carbon economy due to the high participation of hydropower and the recent increase in the share of 'new renewables' (e.g. photovoltaic, solar, wind, wave, and biomass). Moreover, the authors add that this process was accomplished due to the region having enormous availability and diverse natural resources, including large hydropower, wind, and solar potential, as well as suitable weather conditions.

Based on this information, the central questions of this chapter are as follows. What is the effect of energy transition on CO_2 emissions in the LAC countries? How does energy transition decrease environmental degradation? The primary objective of this chapter is to study the impact of the energy transition on environmental degradation in 18 LAC countries, over a period ranging from 1990 to 2014. A panel nonlinear autoregressive distributed lag (PNARDL) approach in the form of unrestricted error-correction model (UECM) was used to decompose the positive and negative variations of the independent variables into their short-run impacts and long-run elasticities. Indeed, this process of energy transition is expected to decrease environmental degradation.

In the literature, the impact of energy transition on the environment has not been explored in the literature, which is exceptionally scarce in the LAC countries. Due to the shortage of studies on the impact of the energy transition on environmental degradation, this chapter will opt to use literature-based studies that investigated similar issues as well the use of this structure of the investigation, in this case, chapters that explored the impact of renewable energy consumption on CO_2 emissions (e.g. Koengkan et al., 2018; Fuinhas et al., 2017; Koengkan and Fuinhas, 2017; Bilgili et al., 2016; Shafiei and Salim, 2014; Apergis and Payne, 2014; Sadorsky, 2009; Akella et al., 2009). This literature was reviewed bearing in mind that it is the closest to the topic under discussion.

Although several studies have used different variables, countries, time spans, and methodologies to clarify the impact of the energy transition on environmental degradation, the best approach to achieve it remains without a clear solution. What conclusions have been reached by the literature about the impact of renewable energy consumption on environmental degradation? The literature that approaches this relationship has evolved in two divergent ways. The first argues that renewable energy consumption reduces ecological degradation (e.g. Fuinhas et al., 2017; Koengkan and Fuinhas, 2017; Bilgili et al., 2016; Shafiei and Salim, 2014; Akella et al., 2009). Regarding the negative impact of renewable energy consumption on CO_2 emissions, some authors have stressed some key features.

For instance, Koengkan and Fuinhas (2017) investigated the impact of renewable energy consumption on CO_2 emissions in 10 South American countries in the period from 1980 to 2012. Panel autoregressive distributed lags (PARDL) in the form of UECM was used. The authors found that renewable energy consumption has a negative impact of −0.042 in the short run. Indeed, this negative impact is related to the globalisation process that exerts a positive effect on economic growth and consequently on the consumption of renewable energy and new investment in green technology, and therefore on CO_2 emissions.

Fuinhas et al. (2017) investigated the impact of renewable energy policies on carbon dioxide emissions, for which a panel of 10 Latin American countries was analysed for the period from 1991 to 2012. A PNARDL model was used as the method. The authors have a different opinion about this impact and confirm that this decrease is related to the efficiency of renewable energy policies that encourage the introduction of alternative energy sources in the energy mix. Other authors share this view (e.g. Bilgili et al., 2016; Shafiei and Salim, 2014; Akella et al., 2009).

The second approach in the literature suggests that the consumption of renewable energy causes an increase in emissions (e.g. Koengkan et al., 2018; Apergis and Payne, 2014; Sadorsky, 2009). Therefore, some authors have stressed some key features. For example, Koengkan et al. (2018) studied the impact of hydroelectricity consumption on environmental degradation in seven South American countries from 1966 to 2014. The authors found that the use of this kind of energy increases emissions by 0.0593 in the long run. Indeed, this effect occurs in the first few years after a reservoir is created, when the trees that have died in the process of flooding release CO_2 in their decomposition process, and from turbines and spillways during the process of energy generation. These emissions can be compared with those generated from fossil fuels.

However, Apergis and Payne (2014) examined the determinants of renewable energy consumption for a panel of seven Central American countries from 1980 to 2010. The authors used the fully modified ordinary least squares (FMOLS) model as the methodology. The authors found that the positive impact of renewable energy consumption on environmental degradation is due to several legal and institutional barriers that do not encourage the expansion of renewable energy, as well as the increase in the use of renewable energy sources. Sadorsky (2009) has a different opinion; according to this author, the positive impact of renewable energy on CO_2 emissions is due to the lack of financial incentives that do not encourage the development of renewable energy technologies and consequently the consumption of alternative energy sources.

8.2 Data and methodology

This section is divided into two parts. The first one presents the data and variables that will be used, and the second describes the adopted methodological strategy that will be applied in this chapter.

8.2.1 Data

To explore the asymmetric impact of the ratio of renewable energy that is a proxy for the effect of energy transition on environmental degradation, 18 LAC countries, Argentina, Bolivia, Brazil, Chile, Colombia, Costa Rica, Dominican Republic, Ecuador, El Salvador, Guatemala, Haiti, Mexico, Nicaragua, Panama, Paraguay, Peru, Uruguay, and Venezuela (RB), were carefully chosen in a period from 1990 to 2014. The use of time series between 1990 and 2014 is due to the availability of data until 2014 for the variable renewable energy consumption in kWh per capita that compose the variable ratio of renewable energy in all countries selected.

Therefore, this group of countries was selected for several reasons: (a) they have registered rapid growth in the consumption of renewable and nonrenewable energy; (b) they have experimented with a rapid process of economic growth (see Fig. 5.9); (c) in the last three decades, the CO_2 emissions from LAC countries have more than doubled (see Fig. 5.7); and (d) the existence of a complete database was the main criteria for choosing the countries from the LAC region, as mentioned in Chapters 6 and 7. Additionally, the variables which were chosen to perform the analysis are (Table 8.1):

TABLE 8.1 Description of variables and source.

Description of variables		
Variable	**Definition**	**Source**
CO2	Carbon dioxide emissions (CO_2) in kilotons (kt) per capita from the burning of fossil fuels and the manufacture of cement. These include carbon dioxide produced during the consumption of solid, liquid, and gas fuels and gas flaring	World Bank Open Data (2019)
GDP_PC	Gross domestic product in constant (2010 US$) per capita	World Bank Open Data (2019)
RRE	Ratio of renewable energy (RE), which is the ratio of renewable energy consumption from biomass, hydropower, solar, photovoltaic, wind, wave, and waste in (kWh) per capita divided by the fossil fuel consumption from oil, gas and coal sources in (kWh) per capita	World Bank Open Data (2019)
KGOV	General government capital stock in billions of constant 2011 international dollars	IMF (2019)

The increased effect of greenhouse gas (GHG) concentrations on global temperatures and the earth's climate have consequences for ecosystems, human settlements, agriculture, and other socio-economic activities (UNEP, 2001). GHG emissions have increased because CO_2 emissions are still growing in many countries, despite some progress achieved in decoupling CO_2 emissions from GDP growth (OECD Environment Directorate, 2008). In 1990 a value of 3414 $MtCO_2eq$ was registered, and in 2014 emissions reached a value of 4020 $MtCO_2eq$ (see Fig. 5.7). Therefore, CO_2 emissions are a better indicator of environmental performance because they are the major contributor to the greenhouse effect (OECD Environment Directorate, 2008). The energy sector is the main contributor to CO_2 emissions, whereas, in the LAC countries, the energy sector accounts for approximately 48% of total emissions in 2014 (see Fig. 5.8). This is the justification for our study using CO_2 as the dependent variable.

As mentioned earlier, a relationship exists between CO_2 emissions and GDP growth and this relationship is due to economic growth raising standards of living in most countries; it was also responsible for the increase in CO_2 emissions and reductions in natural resources (Mardani et al., 2019). In Latin American countries, GDP per capita has registered average annual growth rates of approximately 3.0% (see Fig. 5.9). As a consequence of this growth in Latin American countries, electric power consumption (kWh per capita) in the region has followed a similar path (see Fig. 5.10), with the use of energy growing by approximately 5.4%, as stated before by Balza et al. (2016). For this reason, this investigation opted to use **GDP_PC** as an independent variable.

Indeed, economic growth, financial, and trade liberalisation as well as the capital stock accumulation resulting from several economic reforms and political transitions in the last 40 years are responsible for the increased investment in and consumption of energy in the Latin American region. Indeed, the use of renewable energy represented 35% of the total energy consumption in 2013, and investment in renewable energy grew by 13% between 2000 and 2013, as cited before by Koengkan (2018a) and Koengkan et al. (2019a). This increase in the consumption of renewable energy is an indicator of the energy transition process, as mentioned by Hauff et al. (2014). Therefore, to identify the effect of the energy transition on environmental degradation, we opted to use the ratio of renewable energy consumption to fossil fuel consumption. This ratio is obtained by dividing the consumption of renewable energy by the consumption of fossil fuels. This ratio captures the progression of the consumption of renewable energy to the consumption of fossil fuels over time. Moreover, this method for capturing this progression was used before by Fuinhas et al. (2019) to capture the progress of the weight of oil production to consumption of oil consumption over time. For this reason, **RRE** is used as an independent variable in this investigation.

Then, as cited before, the increase of capital stock accumulation is responsible for the rise in investment in energy and its consumption in the Latin American region. The abundance of capital as cited by Lee and Chen (2010) and Lee et al. (2008) reduces its price and makes the capital cheaper and encourages new investment and economic activity and consequently the consumption of fossil fuels and environmental degradation. This explanation justifies the use of **KGOV** as the independent variable.

The variables **CO_2**, **GDP_PC**, **RRE**, and **KGOV** were transformed into per capita values with the total population of each cross. This allows controlling for disparities in population growth over time and within countries (Koengkan, 2018b). The ratio of renewable energy was used as a proxy for energy transition in this chapter because this variable can capture the effect of energy transition from fossil to renewable on CO_2 emissions in the LAC countries. Table 8.2 presents the summary statistics of variables.

'Log' and 'DLog' denote variables in the natural and their first differences, 'Obs' denotes the number of observations in the model, 'Std.-dev' denotes the standard deviation, and 'Min and Max' denote minimum and maximum. These summary statistics were obtained from the command *sum* in **Stata 16.0**. The board below shows how to transform the variables in per capita values, in the natural logarithms, and the first-differences of logarithms, as well as how to obtain the summary statistics of variables.

How to do:

****Transform the variables co2, gdp_pc, rre, and kgov into per capita values****

gen co2 =(co2/population)

****Transform the variables co2, gdp_pc, rre, and kgov into natural logarithms****

gen logco2 =log(co2)

****Transform the variables co2, gdp_pc, rre, and kgov into first differences of logarithms****

gen dlogco2 =d.log(co2)

****The summary statistics****

sum logco2 loggdp_pc logrre logkgov dlogco2 dloggdp_pc dlogrre,dlogkgov

TABLE 8.2 Summary statistics of variables.

	Descriptive statistics				
Variables	Obs.	Mean	Std.-dev.	Min	Max
LogCO2	450	−6.7708	1.3936	−11.7542	−4.8785
LogGDP_PC	450	8.4056	0.7719	6.4956	9.5943
LogRRE	450	1.9695	4.5903	−2.5181	18.6579
LogKGOV	450	−12.3765	0.8855	−15.0981	−10.4755
DLogCO$_2$	432	0.0203	0.1198	−0.8107	1.0799
DLogGDP_PC	432	0.0203	0.0349	−0.1462	0.1503
DLogRRE	432	−0.0337	0.5760	−5.1653	3.5362
DLogKGOV	432	0.0174	0.0326	−0.0441	0.1813

8.2.2 Methodology

A PNARDL approach in the form of UECM will be used for the purpose of this chapter. This methodology is an asymmetric extension of the linear PARDL model in the form of UECM, which was created by Granger (1981) and Engle and Granger (1987).

The difference between the two models is that the PNARDL allows the positive and negative variations of the independent variables in the model, which has a different effect on the dependent variable (Rocher, 2017). This author also adds that the PNARDL allows the detection of the presence of asymmetric effects that the independent variables may have on the dependent variable, as well as testing the existence of cointegration in a single equation.

Indeed, the PNARDL methodology approach was used in this investigation because, in panels with long time spans (macro panels), there is generally the presence of cointegration between the variables and then endogeneity in the model. However, if the appropriate econometric techniques are not used to cope with the endogeneity and cointegration problem, it can lead to estimation errors and misinterpretation of results. Therefore, to handle the problem of endogeneity and cointegration, the literature recommends the use of PNARDL models as an econometric estimation technique that is robust enough to deal with the presence of endogeneity and cointegration between the variables. Then, the PNARDL in the form of a UECM was used to cope with endogeneity and cointegration that is expected in this investigation.

Moreover, the modelling itself should be considered in the research approach. Does this model have some advantages for this investigation? According to Rocher (2017), this model has several advantages; namely, (a) flexibility regarding the order of integration of the variables in the model, (b) the possibility of testing for hidden cointegration, (c) better performance in small samples, and (d) it permits the use of a combination of I(0) and I(1) variables. There are other justifications for opting to apply this methodology in this chapter, one of them being (e) the capacity to identify the effect of independent variables on the dependent variable in the short and long run. Indeed, as the LAC countries suffered several economic, political, and social shocks, this methodology is the best approach, and finally (f) it presents better estimations when compared to other methods.

Therefore, the PNARDL model is constructed around the following asymmetric long-run equilibrium relationship:

$$e_t = \alpha^+\gamma^+ + \alpha^-\gamma^- + \mu_t \tag{8.1}$$

where the equilibrium relationship between e and γ is divided into positive ($\alpha^+\gamma^+$) and negative ($\alpha^-\gamma^-$) effects, plus the error term (μ_t) means possible deviations from the long-run equilibrium. As shown in Eq. (8.1), the impact of the variable γ can be decomposed into two parts, positive and negative:

$$\gamma_t = \gamma_0 + \gamma_t^+ + \gamma_t^- \tag{8.2}$$

where γ_0 represents the random initial value and $\gamma_t^+ + \gamma_t^-$ denote partial sum processes which accumulate positive and negative changes, respectively, and are defined as

$$\gamma_t^+ = \sum_{j=1}^{t} \Delta\gamma_j^+ = \sum_{j=1}^{t} \max\left(\Delta\gamma_j, 0\right) \tag{8.3}$$

$$\gamma_t^- = \sum_{j=1}^{t} \Delta\gamma_j^- = \sum_{j=1}^{t} \min\left(\Delta\gamma_j, 0\right) \tag{8.4}$$

Then, the general and main PNARDL model in the form of UECM follows the specification of the following equation:

$$\begin{aligned} DLogCO2_{it} = {} & \alpha_{it} + \theta_1^+\beta_{1i1}DLogGDP\_PC_{it-1}^+ + \theta_1^- DLogGDP\_PC_{it-1}^- + \theta_2^+\beta_{2i1}DLogRRE_{it-1}^+ + \theta_2^- DLogRRE_{it-1}^- \\ & + \beta3_{i1}DLogKGOV_{it-1}^+ + \theta_3^- DLogKGOV_{it-1}^- + LogCO2_{it-1} + \theta_1^+\gamma_{1i2}LogGDP\_PC_{it-1}^+ + \theta_1^-\gamma_{1i2}LogGDP\_PC_{it-1}^- \\ & + \theta_2^+\gamma_{2i2}LogRRE_{it-1}^+ + \theta_1^-\gamma_{2i2}LogRRE_{it-1}^- + \theta_1^+\gamma_{3i2}LogKGOV + \theta_1^-\gamma_{3i2}LogKGOV_{it-1}^- + \varepsilon_{1it} \end{aligned} \tag{8.5}$$

where α_i represents the intercept, β_{ik} and γ_{ik}, with $k=1, \ldots, 4$ denoting the estimated parameters, $DLogGDP\_PC_{it-1}^+$, $DLogGDP\_PC_{it-1}^-$, $DLogRRE_{it-1}^+$, $DLogRRE_{it-1}^-$, $DLogKGOV_{it-1}^+$, $DLogKGOV_{it-1}^-$, $LogGDP\_PC_{it-1}^+$, $LogGDP\_PC_{it-1}^-$, $LogRRE_{it-1}^+$, $LogRRE_{it-1}^-$, and $LogKGOV_{it-1}^+$, $LogKGOV_{it-1}^-$ are the partial sums of positive and negative changes of variables **DLogKGOV, DLogRRE, DLogKGOV, LogGDP_PC, LogRRE**, and **LogKGOV**, respectively and ε_{it} is the error term. Consequently, before the realisation of PNARDL estimation, it is necessary to verify the proprieties of the variables that will be used in this chapter, which includes checking the cross section and time series, in addition to the existence of specificities that, when not considered in the initial verification, may produce inconsistent results and interpretation. For this, some **preliminary tests** will be computed (see e.g. Table 8.3).

Moreover, after the model regression, it is necessary also to apply **post-estimation** tests (for this analysis, we just used one test.) (see e.g. Table 8.4).

Moreover, all estimations and testing procedures will be carried out using **Stata 16.0.** This section shows the variables and methodologies and their preliminary and specification tests that will be used in our analysis. The next section will show the empirical results and discussion.

8.3 Empirical results and discussion

This section will show the empirical results of preliminary and specification tests as well as the outcomes of PNARDL estimators and the debate, as previously explained. Therefore, the first step made was to carry out the correlation matrix test developed by Snedecor and Cochran (1989) to inquire about the presence of collinearity between the variables. This test is constructed around the following equation:

$$\hat{p} = \frac{\sum_{i=1}^{n} \omega_i (x_i - \overline{x})(y_i - \overline{y})}{\sqrt{\sum_{i=1}^{n} \omega_i (x_i - \overline{x})^2} \sqrt{\sum_{i=1}^{n} \omega_i (y_i - \overline{y})^2}} \tag{8.6}$$

where ω_i are the weights, if specified, or $\omega_i = 1$ if weights are not specified $\overline{x} = (\sum \omega_i x_i)/(\sum \omega_i)$ is the mean of x, and $\overline{y}$ is similarly defined. The unadjusted significance level is calculated around the following equation:

$$p = 2{*}ttail\left(n-2, |\hat{p}|\sqrt{n-2}/\sqrt{1-\hat{p}^2}\right) \tag{8.7}$$

TABLE 8.3 Preliminary tests.

Test	Goal
Correlation matrix	To inquire about the presence of collinearity between the variables
Variance inflation factor (VIF) test	To check the existence of multicollinearity between the variables in the panel data
Pesaran CD-test	To check the existence of cross-section dependence in the panel data
Panel unit root test (CIPS-test)	To check the presence of unit roots in the variables
Hausman test	To check the presence of heterogeneity i.e. whether the panel has random effects (RE) or fixed effects (FE)
Heterogeneous estimators test	To verify the presence of heterogeneity or homogeneity in our panel data

TABLE 8.4 Post-estimation tests.

Test	Goal
Modified Wald test	To assesses the panel groupwise heteroskedasticity in the residuals of FE estimation
Wooldridge test	To assesses the autocorrelation in panel data
Pesaran test	To assesses the cross-sectional independence
Breusch and Pagan Lagrangian multiplier test	To assesses the independence for contemporaneous correlation

Therefore, the results of this test indicate that the collinearity between the variables is not a concern in the estimation, and there is only one 'high' correlation value between the variables **LogKGOV** and **LogGDP_PC** (see Table 8.A1 in Appendix). The results of the correlation matrix test were obtained from the command *pwcorr* in **Stata 16.0**. The board below shows how to carry out and obtain the results from the correlation matrix test.

How to do:

****The correlation matrix test****

```
pwcorr logco2 loggdp_pc logrre logkgov, sig
pwcorr dlogco2 dloggdp_pc dlogrre dlogkgov, sig
```

The second step in this analysis was the VIF test that informs of the presence of multicollinearity between the variables, and the CSD test for the existence of cross-sectional dependence. The VIF-test that was developed by Belsley et al. (1980) was calculated. This test is constructed around the following equation:

$$\mathrm{VIF}_i = \frac{1}{1-R_j^2}, \tag{8.8}$$

where R_j^2 is the coefficient of determination of regression of model in step one. Therefore, the results of this test indicate that the presence of multicollinearity is not a concern in the estimation, where the VIF and mean VIF values registered are lower than the usually accepted benchmark of 10 in the case of the VIF values, and six in the case of the mean VIF values (see Table 8.A2 in Appendix). The results of the VIF-test were obtained from the command *estat vif* in **Stata 16.0**. The board below shows how to carry out and obtain the results from the VIF-test.

How to do:

****The Variance Inflation Factor test****

```
reg logco2 loggdp_pc logrre logkgov
estat vif

reg dlogco2 dloggdp_pc dlogrre dlogkgov
estat vif
```

Indeed, in the presence of low multicollinearity between the variables, it is necessary to check the presence of cross-sectional dependence in the panel data. To this end, the Pesaran CD-test developed by Pesaran (2004) was calculated. This test is constructed around the following equation:

$$\mathrm{CD} = \sqrt{\frac{2T}{N(N-1)}}\left(\sum\nolimits_{i=1}^{N-1}\sum\nolimits_{j=i+1}^{N}\hat{P}ij\right) \tag{8.9}$$

The null hypothesis of this test is the nonpresence of cross-sectional dependence $\mathrm{CD} \sim N(0,1)$ for $N \rightarrow \infty$ and T is sufficiently large. The results of this test indicate that the null hypothesis is rejected in all variables in natural logarithms and some variables in first differences such as **DLogGDP_PC**, and **DLogKGOV**, leading us to the conclusion that there is a correlation between the series across countries in these variables (see Table 8.A3 in Appendix). The results of the

Pesaran CD-test were obtained from the command *xtcd* in **Stata 16.0**. The board below shows how to carry out and obtain the results from the Pesaran CD-test.

How to do:

****The Pesaran CD-test****

```
xtcd logco2 loggdp_pc logrre logkgov
xtcd dlogco2 dloggdp_pc dlogrre dlogkgov
```

The presence of cross-sectional dependence in these variables can be an indication that the countries selected in this panel data share the same characteristics. Therefore, in the presence of cross-sectional dependence in the variables, it is necessary to check the order of integration of the variables. To this end, the panel unit root (CIPS-test) developed by Pesaran (2007) was calculated. This test is constructed around the following equation:

$$CIPS(N,T) = t - bar = N^{-1}\sum_{i=1}^{N} ti(N,T) \tag{8.10}$$

where $t_i(N,T)$ is the cross-sectionally augmented Dickey-Fuller statistic for the i, the cross-section unit given by the t-ratio of the coefficient of $y_{i,\,t-1}$ in the CADF regression. Therefore, the null hypothesis of this test is that all series have a unit root. The results of this test indicate that none of the variables tested seems to be I(2). At the same time, they show that some of them may be on the borderline between the I(0) and I(1) orders of integration, as the CIPS test presents the same results; the variables **CO_2, GDP_PC, RRE** are without trend and **GDP_PC** and **KGOV** with the trend in logarithms and all variables in first differences without and with the trend are stationary too (see Table 8.A4 in Appendix). The results of the CIPS-test were obtained from the command *multipurt* in **Stata 16.0**. The board below shows how to carry out and obtain the results from the CIPS-test.

How to do:

****The CIPS-test****

```
multipurt logco2 loggdp_pc logrre logkgov,lags(1)
multipurt dlogco2 dloggdp_pc dlogrre dlogkgov,lags(1)
```

After assessing the order of integration of the variables, it is necessary to verify the presence of individual effects in the model. To this end, the Hausman test, which compares the random (RE) and fixed effects (FE), was calculated. This test is constructed around the following equation:

$$H = (\beta_{RE} - \beta_{FE})'\sum^{\wedge} - 1(\beta_{RE} - \beta_{FE}) \sim X^2(k) \tag{8.11}$$

where β_{RE} and β_{FE} are estimators of the parameter β. The null hypothesis of this test is that the difference in coefficients is not systematic, where the random effects are the most sustainable estimator. Therefore, this test indicates that the null hypotheses should be rejected (**chi2 (8) = 80.07*****, statistically significant at 1% level) (see Table 8.A5 in Appendix) and that a fixed-effects model is the most appropriate for this analysis. The results of the Hausman test were obtained from the command *hausman* with option *sigmaless* in **Stata 16.0**. The board below shows how to carry out and obtain the results from the Hausman test.

How to do:

****The Hausman test****

```
xtreg dlogco2 dloggdp_pc dlogrre dlogkgov logco2 loggdp_pc logrre logkgov,fe
estimates store fixed

xtreg dlogco2 dloggdp_pc dlogrre dlogkgov logco2 loggdp_pc logrre logkgov,re
estimates store random

hausman fixed random, sigmaless
```

So, to verify the presence of heterogeneity or homogeneity in our panel data, the mean group (MG), pooled mean group (PMG), and fixed effect (FE) estimators developed by Pesaran and Smith (1995) were performed. The MG estimator calculates the average of coefficients of all individuals, with no restrictions regarding the homogeneity of the short and long run. The PMG allows for differences in error variances in short-run coefficients, speed of adjustment and intercepts, but it imposes a homogeneity restriction on the long-run coefficients.

Moreover, the PMG estimator can combine the 'pooling' from the FE estimator with the 'averaging' from the MG estimator. Nevertheless, in the presence of panel homogeneity, in the long run, the PMG estimator is more efficient if compared with MG. Indeed for example if the ARDL is the following equation:

$$\alpha_i (L)y_{it} = b_i(L)x_{it} + d_i z_{it} + e_{it} \tag{8.12}$$

for country i, where $i=1, \ldots, N$, then the long-run parameter for country i is following equation:

$$\theta_i = \frac{b_i(1)}{d_i(1)} \tag{8.13}$$

and the MG estimation for the whole panel will be given by the following equation:

$$\theta_i = \frac{1}{N}\sum_{i=1}^{N}\hat{\theta}_i \tag{8.14}$$

Indeed, the PMG estimator follows the following equation:

$$\Delta y_{it} = \theta_i(y_{i,t-1} - \beta' x_{i,t-1}) + \sum_{j=1}^{m-1}\gamma_{ij}\Delta\gamma_{i,t-1} + \sum_{j=1}^{n-1}\gamma'_{ij}x_{i,t-j} + \mu_i + \varepsilon_{it} \tag{8.15}$$

A note should be addressed to the results for the PMG estimation when compared with the MG and FE estimations. However, the inconsistency of the MG model could raise doubts about the soundness of the research. It is most likely that the algorithm used in the MG estimator failed the starting conditions. From the information provided in Table 8. A6 in Appendix, we can conclude that there is evidence that the panel is homogeneous, with the results indicating that the FE estimator is the most appropriate. Moreover, the results also indicate that the LAC countries in this chapter return to equilibrium as quickly as expected. The results of the heterogeneous estimator's test were obtained from the command *xtpmg* with option *alleqs constant* in **Stata 16.0**. The board below shows how to carry out and obtain the results from heterogeneous estimator's test.

How to do:

****Heterogeneous estimators test****

```
qui: xtpmg dlogco2 dloggdp_pc_pos dloggdp_pc_neg dlogrre_pos dlogrre_neg dlogkgov_pos dlogkgov_neg, lr (l_logco2
l_loggdp_pc_pos l_loggdp_pc_neg l_logrre_pos l_logrre_neg l_logkgov_pos l_logkgov_neg) ec(ecm) replace mg
estimates store mg

qui: xtpmg dlogco2 dloggdp_pc_pos dloggdp_pc_neg dlogrre_pos dlogrre_neg dlogkgov_pos dlogkgov_neg, lr (l_logco2
l_loggdp_pc_pos l_loggdp_pc_neg l_logrre_pos l_logrre_neg l_logkgov_pos l_logkgov_neg) ec(ecm) replace pmg
estimates store pmg

qui: xtpmg dlogco2 dloggdp_pc_pos dloggdp_pc_neg dlogrre_pos dlogrre_neg dlogkgov_pos dlogkgov_neg, lr (l_logco2
l_loggdp_pc_pos l_loggdp_pc_neg l_logrre_pos l_logrre_neg l_logkgov_pos l_logkgov_neg) ec(ecm) replace dfe
estimates store dfe

estimates table mg pmg dfe, star(.10 0.05.01) stats(N r2 r2_a) b(%7.4f)
hausman mg pmg,alleqs constant
hausman pmg dfe,alleqs constant
hausman mg dfe,alleqs constant
```

Before the realisation of model estimation, a battery of post-estimation tests must be conducted, such as (a) the modified Wald test developed by Greene (2002) to assess the panel groupwise heteroskedasticity in the residuals of FE estimation. The null hypothesis of this test is the presence of homoskedasticity. This test is constructed around the following equation:

$$\gamma ModWald = T\left(r(\hat{\theta}) + \frac{w}{\sqrt{T}}\right)\left(\sum_{r(\theta)} + \sum_{w}\right)^{-1}\left(r(\hat{\theta}) + \frac{w}{\sqrt{T}}\right) \vec{d}\ \chi^2(J) \tag{8.16}$$

where T is the sample size, and $\chi^2(j)$ is asymptotic distribution; (b) the Wooldridge test developed by Wooldridge (2002) for autocorrelation in panel data. The null hypothesis of this test is the nonpresence of the first-order autocorrelation. This test is constructed around the following equation:

$$\begin{gathered} y_{it} - y_{it-1} = (X_{it} - X_{it-1})\beta_1 + \varepsilon_{it} - \varepsilon_{it-1} \\ \Delta_{yit} = \Delta X_{it}\beta_1 + \Delta\varepsilon_{it} \end{gathered} \tag{8.17}$$

where $\triangle$ is the first-difference operator; (c) Pesaran's test developed by Pesaran (2004) for cross-sectional independence. This test is constructed around the following equation:

$$\mathrm{CD} = \sqrt{\frac{2T}{N(N-1)}}\left(\sum_{i=1}^{N-1}\sum_{j=i+1}^{N}\hat{P}ij\right) \tag{8.18}$$

The null hypothesis of this test is the nonpresence of cross-sectional dependence CD $\sim N(0,1)$ for $N \rightarrow \infty$ and T is sufficiently large, and (d) the Breusch and Pagan Lagrangian multiplier test of independence developed by Breusch and Pagan (1980) for contemporaneous correlation. This test is constructed around the following equation:

$$LM_{BP} = T\sum_{i=1}^{n-1}\sum_{j=i+1}^{n} \check{p}_{ij}2 \tag{8.19}$$

The null hypothesis of this test is the nondependence between the residuals. The results of the specification test indicate rejection of the null hypothesis of the modified Wald and Wooldridge tests at the 1% level, indicating the presence of heteroscedasticity and first-order autocorrelation. Moreover, it cannot reject the null hypothesis of Pesaran's test, indicating the nonpresence of correlation. Regarding the Breusch and Pagan Lagrangian multiplier test, it could not be computed because the correlation matrix of residuals was singular (see Table 8.A7 in Appendix). This situation occurs because the number of crosses that are under study is less than the number of years. The results of post-estimation tests were obtained from the commands *xttest3, xtserial, xtcsd* with option *pesaran abs,* and *xttest2* in **Stata 16.0**. The board below shows how to carry out and obtain the results from the post-estimation tests.

How to do:

**** Post-estimation tests****

Modified Wald test

```
xtreg dlogco2 dloggdp_pc_pos dloggdp_pc_neg dlogrre_pos dlogrre_neg dlogkgov_pos dlogkgov_neg l_logco2 l_loggdp_pc_pos l_loggdp_pc_neg l_logrre_pos l_logrre_neg l_logkgov_pos l_logkgov_neg,fe
xttest3
```

Wooldridge test

```
xtserial dlogco2 dloggdp_pc_pos dloggdp_pc_neg dlogrre_pos dlogrre_neg dlogkgov_pos dlogkgov_neg l_logco2 l_loggdp_pc_pos l_loggdp_pc_neg l_logrre_pos l_logrre_neg l_logkgov_pos l_logkgov_neg
```

Pesaran's test

```
xtreg dlogco2 dloggdp_pc_pos dloggdp_pc_neg dlogrre_pos dlogrre_neg dlogkgov_pos dlogkgov_neg l_logco2 l_loggdp_pc_pos l_loggdp_pc_neg l_logrre_pos l_logrre_neg l_logkgov_pos l_logkgov_neg, fe
xtcsd, pesaran abs
```

Breusch and Pagan Lagrangian multiplier test

```
xtreg dlogco2 dloggdp_pc_pos dloggdp_pc_neg dlogrre_pos dlogrre_neg dlogkgov_pos dlogkgov_neg l_logco2 l_loggdp_pc_pos l_loggdp_pc_neg l_logrre_pos l_logrre_neg l_logkgov_pos l_logkgov_neg, fe
xttest2
```

Moreover, it is worth remembering that the Hausman test, MG, PMG, and DFE estimators and the post-estimation tests that were specified in Section 8.2.1 were applied in the parsimonious model. That is, insignificant variables were removed (e.g. **DLogGDP_PC_POS**, **DLogGDP_PC_NEG**, **DLogKGOV_POS**, **DLogKGOV_NEG**, **LogGDP_PC_POS**, **LogGDP_PC_NEG**, **LogKGOV_POS**, and **LogKGOV_NEG**) in previous regressions from our general model (see Eq. 8.5). The positive and negative asymmetry of these variables was not revealed in the model as expected.

After the realisation of preliminary and specification tests, the model regression can be made. Therefore, three estimations were computed in this model, the FE, FE robust standard errors (FE Robust), and FE Driscoll and Kraay (FE D.-K.). Indeed, the FE D.-K. was used due to the specification tests results, to deal with the presence of heteroscedasticity and first-order autocorrelation. This estimator can produce standard errors robust to the phenomena that were found in the sample errors. Moreover, regression dummy and shift-dummy variables were included in the model. These dummy variables were added to the model because during the period of analysis, the LAC countries suffered several shocks that, if not considered, could have produced inaccurate results that could lead to misinterpretations.

Additionally, all these dummy and shift-dummy variables following triple criteria were thus used to include (a) significant disturbances in the estimated residuals; (b) the occurrence of international events known to have disturbed the LAC region; and (c) the potential relevance of recorded economic, social, and political events at the country level. Thus, the dummy and shift-dummy variables that were added to the regression are the following: **IDBOLIVIA_2001** (Bolivia, year 2001); **IDECUADOR_1994** (Ecuador, year 1994); **IDGUATEMALA_2014** (Guatemala, year 2014); **SDHAITI_1993_1995** (Haiti, years between 1993 and 1995); **IDPANAMA_1995** (Panama, year 1995); **SDPANAMA_1996_1997** (Panama, years between 1996 and 1997).

- **IDBOLIVIA_2001**: Represents a break in the GDP of Bolivia in 2001. This break can be justified by a decrease in economic activity, where the GDP of Bolivia grew just 1.7% in 2001 (Weisbrot et al., 2009).
- **IDECUADOR_1994**: Represents a break in the GDP of Ecuador in 1994. This break can be justified by low economic growth between 1993 and 1995 (Jácome, 2004).
- **IDGUATEMALA_2014**: Represents a peak in the GDP of Guatemala in 2014. This peak can be justified by the acceleration of the country's economic activity in 2014, where the GDP of the country grew 4.2% (World Bank Open Data, 2019).
- **SDHAITI_1993_1994**: Represents two breaks in the GDP of Haiti between 1993 and 2014. These breaks can be justified by Operation Uphold Democracy that was a military intervention designed to remove the military regime installed by the 1991 *Haitian coup d'état* that overthrew the elected President Jean-Bertrand Aristide. The operation was effectively authorised in 1994 (The Carter Center, 1994).
- **IDPANAMA_1995:** Represents a break in the GDP of Panama in 1995. This break can be justified by a decrease in economic activity, where the GDP of Panama grew just 1.8% in 1995 (World Bank Open Data, 2019).
- **SDPANAMA_1996_1997**: Represents two peaks in the GDP of Panama between 1996 and 1997. These peaks can be justified by an increase in economic activity, where the GDP of Panama grew 4.1% in 1996, and 6.5% in 1997 (World Bank Open Data, 2019).

Therefore, these peaks and breaks impacted the consumption of energy and consequently, the emissions of CO_2 in these countries. Table 8.5 presents the short-run impacts, the model speed of adjustment, and the computed long-run elasticities.

In summary, the results from Table 8.5 indicate that in the short and long run, the variable Y has a positive impact of 0.5475 and 0.2186, respectively, and in the short run the variable **KGOV** has a positive impact of 0.4763 on $\mathbf{CO_2}$ emissions. However, the positive and negative asymmetry of variable **RRE** in the short and long run has a negative impact of −0.0601 **(POS)** −0.0792 **(NEG)** in the short run and −0.0281 **(POS)** and −0.0339 **(NEG)** in the long run. The results of positive and negative asymmetry are consistent because both the results are negative. Concerning the ECM term, it is negative and statistically significant at the 1% level, and the statistical significance at the 1% level of the dummy and shift-dummy variables supports the decision to include them in the model. The short-run impacts, the model speed of adjustment, and the computed long-run elasticities were obtained from the commands *xtreg* with option *fe*, *xtreg* with option *fe robust*, and *xtscc* with option *fe lag (1)* in **Stata 16.0**. The board below shows how to carry out and obtain the short-run impacts, the model speed of adjustment, and the long-run elasticities.

How to do:

****Short-run (Impacts) and long-run (elasticities)****

```
xtreg dlogco2 idbolivia2001 idecuador1994 idguatemala2014 sdhaiti19931995 idpanama1995 sdpanama19961997 dlngdp_pc
dlogrre_pos dlogrre_neg dlogkogv l_logco2 l_loggdp_pc l_logrre_pos l_logrre_neg, fe

nlcom(ratio1:-_b[l_loggdp_pc]/_b[l_logco2])
nlcom(ratio1:-_b[l_logrre_pos]/_b[l_logco2])
nlcom(ratio1:-_b[l_logrre_neg]/_b[l_logco2])
```

TABLE 8.5 Elasticities, short-run impacts, elasticities, and adjustment speed (controlling for shocks).

	Dependent variable (DLogCO2)			
Independent variables	**FE**		**FE Robust**	**FE D.-K.**
Constant	−5.2612	***	***	***
Shocks				
IDBOLIVIA_2001	−0.2468	***	***	***
IDECUADOR_1994	−0.5198	***	***	***
IDGUATEMALA_2014	0.3023	***	***	***
SDHAITI_1993_1995	−0.3676	***	***	
IDPANAMA_1995	−0.3857	***	***	***
SDPANAMA_1996_1997	0.2165	***	***	***
Short-run (impacts)				
DLogGDP_PC	0.5475	***		**
DLogRRE_POS	−0.0601	***	***	***
DLogRRE_NEG	−0.0792	***	***	***
DLogKGOV	0.4763	***		*
Long-run (elasticities)				
LogGDP_PC (-1)	0.2186	***	***	***
LogRRE_POS (-1)	−0.0281	***	***	***
LogRRE_NEG (-1)	−0.0339	***	***	***
Speed of adjustment				
ECM	−0.4452	***	***	***

Notes: ***, **, and * denote statistically significant at the 1%, 5%, and 10% levels, respectively; the ECM denotes the coefficient of the variable **LogCO2**, lagged once.

```
xtreg dlogco2 idbolivia2001 idecuador1994 idguatemala2014 sdhaiti19931995 idpanama1995 sdpanama19961997 dlngdp_pc
dlogrre_pos dlogrre_neg dlogkogv l_logco2 l_loggdp_pc l_logrre_pos l_logrre_neg, fe robust

nlcom(ratio1:-_b[l_loggdp_pc]/_b[l_logco2])
nlcom(ratio1:-_b[l_logrre_pos]/_b[l_logco2])
nlcom(ratio1:-_b[l_logrre_neg]/_b[l_logco2])

xtscc dlogco2 idbolivia2001 idecuador1994 idguatemala2014 sdhaiti19931995 idpanama1995 sdpanama19961997 dlngdp_pc
dlogrre_pos dlogrre_neg dlogkogv l_logco2 l_loggdp_pc l_logrre_pos l_logrre_neg, fe lag(1)

nlcom(ratio1:-_b[l_loggdp_pc]/_b[l_logco2])
nlcom(ratio1:-_b[l_logra_pos]/_b[l_logco2])
nlcom(ratio1:-_b[l_logra_neg]/_b[l_logco2])
```

Fig. 8.1 summarises the effect of **GDP_PC**, **RRE**, and **KGOV**, on **CO_2**.

Therefore, the possible explanation for the positive impact of economic growth on environmental degradation is due to fossil fuel sources being the primary inputs for agriculture and industry, which influences both economic growth and environmental degradation in the LAC countries (Mirza and Kanwal, 2017). Another possible clarification for this impact is that an increase in economic growth will lead to an increase in environmental degradation (CO_2) at high levels of income due to the increase in manufacturing industries.

In other words, in the early stages of development of LAC countries, environmental degradation would decrease but increase later after the economic growth exceeds the threshold parameter. Therefore, in the period of economic boom, the households and firms will have more income and consequently will increase the consumption of energy

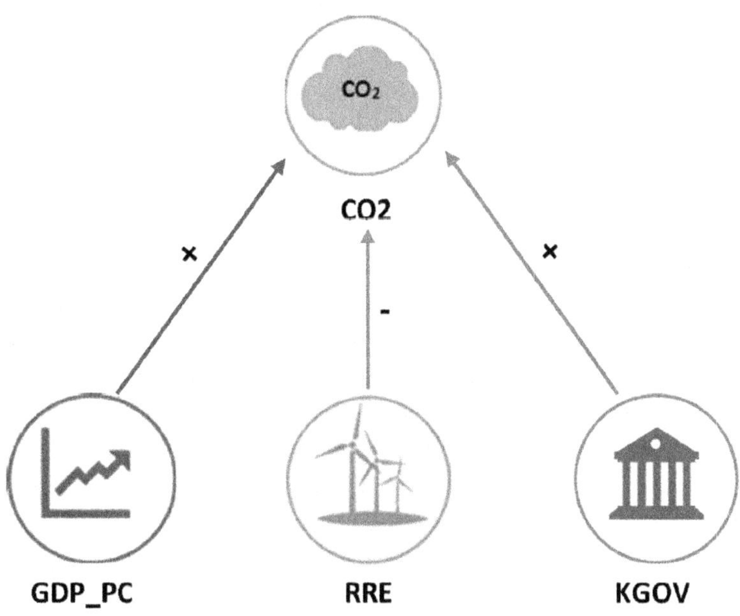

FIG. 8.1 Summary of the effect of variables **GDP_PC**, **RRE**, and **KGOV** on $\mathbf{CO_2}$.

from electric devices, transportation, appliances among others that will contribute to the increase of CO_2 emissions (Aye and Edoja, 2017). These possible explanations about the impact of economic growth on the consumption of fossil fuels were confirmed by Koengkan and Fuinhas (2020a), where it was confirmed that in Mercosur countries, economic growth increases the consumption of fossil fuels by 0.1730 and that bidirectional causality exists between the variables.

The capacity of the ratio of renewable energy, which is a proxy for energy transition, to reduce environmental degradation is probably related to the renewable energy technological efficiency that produces more clean energy and fewer emissions, as well as the increasing share of renewable energy sources in the energy matrix (Koengkan et al., 2019a). Indeed, these factors, according to Koengkan and Fuinhas (2017), are related to the globalisation process in LAC countries that exerts a positive impact on the factor of productivity and economic growth and consequently on the consumption of renewable energy and new investment in green technology. Indeed, this evidence was confirmed by Koengkan et al. (2019b), where financial openness, which is a subcomponent of globalisation, increases investment in installed capacity of renewable energy by 0.6371. This explanation also was confirmed by Koengkan and Fuinhas (2020a), where the consumption of renewable energy and the globalisation process increase economic growth, while the consumption of fossil fuels reduces it. Economic growth, consumption of fossil fuels, and globalisation increase the consumption of renewable energy. Economic growth increases the consumption of fossil fuels, while the consumption of renewable energy and globalisation reduce the consumption of energy from fossil sources. Furthermore, the consumption of energy from renewable and fossil sources increases the process of globalisation, while economic growth reduces it (Koengkan and Fuinhas, 2020a).

Shahbaz et al. (2015) add that the openness and competition brought by the globalisation process increase the environmental regulation standards regarding investment in cleaner technology. Shahbaz et al. (2016) confirm that globalisation is a way of improving economic growth and welfare by reduction of cross border restrictions on trade and investment with other countries. Therefore, this reduction in border restrictions encourages foreign firms to set up new businesses or expand their existing ones using newer and more advanced technologies that reduce the consumption of nonrenewable energy and thereby lower their overall costs. This is likely to influence the existing firms in the host country to adopt new methods of production, reducing the consumption of fossil fuels and consequently the emissions of CO_2. This idea is shared by Leitão (2014), who confirms that the process of globalisation by trade liberalisation encourages developing countries to access efficient technologies from developed countries that consequently reduce the consumption of nonrenewable energy sources and environmental degradation. This explanation was mentioned in Koengkan and Fuinhas (2020b), where the technique effect reduces energy consumption as an improvement in technology. Indeed, this improvement is due to the technology transfer that improves energy efficiency.

Therefore, as mentioned by Shahbaz et al. (2014), the technique effect is possible through trade liberalisation, which allows developing countries to import advanced technologies from developed countries. Moreover, the process of trade liberalisation, as mentioned by Zahonogo (2016), encourages the transfer of new technologies, helping technological progress and improving productivity. Henry et al. (2009) also add that this technology transfer consists of intermediated manufactured products, capital equipment, and new material that are commercialised in international markets.

Another possible explanation for this negative impact is related to the efficiency of renewable energy policies that encourage the introduction of alternative energy sources in the energy mix. For example, in the LAC countries, the most effective policies are national renewable energy targets, which provide a clear indication about the intended level of development of alternative energy sources and the timeline envisioned by governments (Fuinhas et al., 2017).

Finally, there is an indirect effect of public capital stock on environmental degradation. According to Lee and Chen (2010) and Lee et al. (2008), the abundance of capital reduces the price. It makes capital cheaper and consequently encourages new investment and economic activity and subsequently, the consumption of nonrenewable energy and environmental degradation. Moreover, Lee (2005) also adds that that capital stock can positively affect investment and industrial production, which, in consequence, leads to an increase in energy demand/consumption.

However, one doubt arises in the explanations about the impact of the energy transition on environmental degradation: can globalisation and renewable energy policies encourage energy transition in LAC countries, as mentioned in the literature? In the next section, robustness checks will be carried out to ascertain if energy transition in LAC countries is influenced by the globalisation process and renewable energy policies as mentioned before. This check is essential to confirm if the results are in line with the literature.

8.4 Robustness check

The following variables were utilised to verify if the globalisation process influences energy transition in LAC countries and renewable energy policies (Table 8.6):

The variable **GLOBA** retrieved from the KOF Globalisation index (KOF, 2018) as discussed before in Chapter 7, was used because it can present more satisfactory results than the use of other subcomponents of globalisation (e.g. capital mobility, economic integration, financial liberalisation, FDI, trade flow openness, trade openness, and trade liberalisation) when fixed-effect techniques are used (e.g. Koengkan et al., 2019; Koengkan, 2017b; Dogan and Deger, 2016). Additionally, the globalisation process, according to Iheanacho (2018), is considered one of the potential facts that encourages higher economic growth and that consequently increases the demand for and investment in energy to respond to economic growth. Koengkan et al. (2019), confirmed that this process allows countries to improve their trade and total factor productivity and raises standards of living, which consequently enhances economic growth. Therefore, this variable was also used in this robustness check for the reason that LAC countries are in a process of rapid globalisation (Koengkan and Fuinhas, 2020a). Therefore, for this reason, the inclusion of this variable is essential and indispensable for this investigation because it will evidence the influence of globalisation on the process of energy transition.

As mentioned above, the variable **RENPOL** was retrieved from the International Renewable Energy Agency (IRENA, 2016). As mentioned in Chapter 6, this variable includes all policies defined by the International Energy Agency (IRENA), namely, (a) economic instruments, (b) information and education, (c) policy support, (d) regulatory instruments, and (e) research, development, and deployment (RD&D). This variable was built in accumulated form, where each policy that was created is represented by (1) accumulated over other policies throughout its useful life or end (e.g. 1, 1, 2, 2, 2, 3, 3) to identify its effect on energy transition.

Fuinhas et al. (2017) were the first to utilise this variable to identify the impact of renewable energy policies on environmental degradation in Latin American countries. However, according to Fuinhas et al. (2017), this indicator has the shortcoming of not capturing the strength of policies, as it only registers their deployment. Zhao et al. (2013) stated that this problem is due to the fact that a precise measurement of the intensity of policies is nearly impossible because of both the unavailability of data and the diverse particularities of countries. This problem was not a severe constraint given that the objective of the use of this variable, as mentioned before, is to assess the possible effectiveness of public policies on the process of the energy transition as suggested by the literature.

The variables **RRE** and **CO_2** are the same as used in the previous model (see Table 8.1). Indeed, Table 8.A8 in Appendix presents the summary statistics of variables that were used in this robustness check. Indeed, the PNARDL model was used to carry out this check. The general PNARDL model in the form of UECM follows the specification of the following equation:

$$\begin{aligned} DLogRRE_{it} = {} & \alpha_{it} + \theta_1^+ \beta_{1i1} DLogGLOBA_{it-1}^+ + \theta_1^- DLogGLOBA_{it-1}^- + \theta_2^+ \beta_{2i1} DLogRENPOL_{it-1}^+ + \theta_2^- DLogRENPOL_{it-1}^- \\ & + \beta_{3i1} DLogCO2_{it-1}^+ + \theta_3^- DLogCO2_{it-1}^- + LogRRE_{it-1} + \theta_1^+ \gamma_{1i2} LogGLOBA_{it-1}^+ + \theta_1^- \gamma_{1i2} LogGLOBA_{it-1}^- \\ & + \theta_2^+ \gamma_{2i2} LogRENPOL_{it-1}^+ + \theta_1^- \gamma_{2i2} LogRENPOL_{it-1}^- + \theta_1^+ \gamma_{3i2} LogCO2_{it-1}^+ + \theta_1^- \gamma_{3i2} LogCO2_{it-1}^- + + \varepsilon_{1it} \end{aligned} \tag{8.20}$$

TABLE 8.6 Description of variables and source.

Description of variables		
Variable	Definition	Source
GLOBA	Globalisation index that measures the economic, social, and political dimensions of globalisation on a scale from 1 to 100	KOF Globalisation index (KOF, 2018)
RENPOL	Renewable energy policies	IRENA (International Renewable Energy Agency) (2016)

where α_i represents the intercept, β_{ik} and γ_{ik}, with $k=1, \ldots, 4$, denote the estimated parameters, $DLogGLOBA^{+}_{it-1}$, $DLogGLOBA^{-}_{it-1}$, $DLogRENPOL^{+}_{it-1}$, $DLogRENPOL^{-}_{it-1}$, $DLogCO2^{+}_{it-1}$, $DLogCO2^{-}_{it-1}$, $LogGLOBA^{+}_{it-1}$, $LogGLOBA^{-}_{it-1}$, $LogRENPOL^{+}_{it-1}$, $LogRENPOL^{-}_{it-1}$, and $LogO2^{+}_{it-1}$, $LogCO2^{-}_{it-1}$ are the partial sums of positive and negative changes of variables **DLogGLOBA**, **DLogRENPOL**, **DLogCO2**, **LogGLOBA**, **LogRENPOL**, and **LogCO2**, respectively, and ε_{it} is the error term.

The parsimonious model was used to carry out the robustness check. That is, insignificant variables were removed (e.g. **DLogGLOBA_POS, DLogGLOBA_NEG, DLogRENPOL_POS, DLogRENPOL_NEG, DLogCO2_POS, DLog-CO2_NEG, LogRENPOL_POS, LogRENPOL_NEG, LogCO2_POS, LogCO2_NEG**) in previous regressions from our general model (see Eq. 8.20). The positive and negative asymmetry of these variables was not revealed as expected in this chapter. Moreover, it should be recalled that the Hausman test, MG, PMG, and DFE estimators and the specification tests that were specified in Section 8.2.2 were applied in the parsimonious model.

The results of preliminary tests point to the existence of low collinearity between the variables, low-multicollinearity, cross-sectional dependence in all variables in the natural logarithms and the variable **GLOBA** in the first differences. Moreover, it also identified unit roots in the variables in the first differences with and without trend, and in the variables **GLOBA**, **RENPOL**, and **CO2** in natural logarithms without trend, and **RRE, GLOBA, RENPOL** with trend, as well as the presence of fixed effects in the model. These results are provided in Tables 8. A9–8.A12 in Appendix.

However, the heterogeneity/homogeneity test cannot be applied due to the MG and PMG estimators requiring a considerable number of variables, where this model has only two variables in the short run in the parsimonious model. The results of specification tests indicate rejection of the null hypothesis of modified Wald and Wooldridge tests at the 1% level, indicating the presence of heteroscedasticity and first-order autocorrelation. Additionally, we cannot reject the null hypothesis of the Pesaran and the Breusch and Pagan Lagrangian multiplier tests, indicating the nonpresence of correlation and dependence in the residuals (see Table 8.A13 in Appendix).

Dummy and shift-dummy variables were included in the model regression. The dummy and shift-dummy variables added to the regression are the following: **IDPARAGUAY_1995** (Paraguay, year 1995); **IDPARAGUAY_2000** (Paraguay, year 2000); and **SDURUGUAY_2001_2004** (Uruguay, years between 2001 and 2004).

- **IDPARAGUAY_1995**: Represents a break in the consumption of renewable energy in 1995. This break can be justified by a decrease in economic activity, where the GDP of Paraguay grew just 6.8% in 1995 (World Bank Open Data, 2019).
- **IDPARAGUAY_2000**: Represents a peak in the consumption of renewable energy in 2000, where Paraguay had 100% renewable energy in their energy matrix (e.g. hydropower) in 2000 (World Bank Open Data, 2019).
- **SDURUGUAY_2001_2004**: Represents several peaks in the consumption of renewable energy between 2001 and 2004, where between 2001 and 2003 the renewable energy sources represented 99% of Uruguay's energy matrix, and 81% in 2004 (World Bank Open Data, 2019).

TABLE 8.7 Elasticities, short-run impacts, elasticities, and adjustment speed (controlling for shocks) from robustness check.

	Dependent variable (DLogRRE)			
Independent variables	**FE**		**FE Robust**	**FE D.-K.**
Constant	−10.2464	***	***	***
Shocks				
IDPARAGUAY_1995	−2.7902	***	***	***
IDPARAGUAY_2000	2.2685	***	***	***
SDURUGUAY_2001_2004	0.9755	***	**	
Short-run (impacts)				
DLogCO2	−2.0582	***	**	***
Long-run (elasticities)				
LogGLOBA_POS (-1)	1.4870	***	***	***
LogGLOBA_NEG (-1)	1.8678	***	**	***
LogRENPOL (-1)	−0.0847	**	*	***
LogCO2 (-1)	−1.0577	**	*	**
Speed of adjustment				
ECM	−0.4612	***	***	***

Notes: ***, **, and * denote statistically significant at the 1%, 5%, and 10% levels, respectively; the ECM denotes the coefficient of the variable **LogRRE**, lagged once.

Table 8.7 presents the short-run impacts, the model speed of adjustment, and the computed long-run elasticities. The results from Table 8.7 indicate that the variable $\mathbf{CO_2}$ reduces the process of energy transition in the LAC countries in the short and long run. Moreover, the positive and negative asymmetries of the variable globalisation index have a positive effect on the proxy of energy transition in the long run. The short-run impacts, the model speed of adjustment, and the computed long-run elasticities were obtained from the commands *xtreg* with option *fe, xtreg* with option *fe robust*, and *xtscc* with option *fe lag (1)* in **Stata 16.0**. The board below shows how to carry out and obtain the short-run impacts, the model speed of adjustment, and the long-run elasticities.

How to do:

****Short-run (Impacts) and long-run (elasticities)****

```
xtreg dlogrre idparaguay1995 idparaguay2000 sduruguay20012004 dlogco2 l_logrre l_loggloba_pos l_loggloba_neg l_logren-
pol l_logco2, fe

nlcom(ratio1:-_b[l_loggloba_pos]/_b[l_logrre])
nlcom(ratio1:-_b[l_loggloba_neg]/_b[l_logrre])
nlcom(ratio1:-_b[l_logrenpol]/_b[l_logrre])
nlcom(ratio1:-_b[l_logco2]/_b[l_logrre])

xtreg dlogrre idparaguay1995 idparaguay2000 sduruguay20012004 dlogco2 l_logrre l_loggloba_pos l_loggloba_neg l_logren-
pol l_logco2, fe robust

nlcom(ratio1:-_b[l_loggloba_pos]/_b[l_logrre])
nlcom(ratio1:-_b[l_loggloba_neg]/_b[l_logrre])
nlcom(ratio1:-_b[l_logrenpol]/_b[l_logrre])
nlcom(ratio1:-_b[l_logco2]/_b[l_logrre])

xtscc dlogrre idparaguay1995 idparaguay2000 sduruguay20012004 dlogco2 l_logrre l_loggloba_pos l_loggloba_neg l_logren-
pol l_logco2, fe lag (1)

nlcom(ratio1:-_b[l_loggloba_pos]/_b[l_logrre])
nlcom(ratio1:-_b[l_loggloba_neg]/_b[l_logrre])
nlcom(ratio1:-_b[l_logrenpol]/_b[l_logrre])
nlcom(ratio1:-_b[l_logco2]/_b[l_logrre])
```

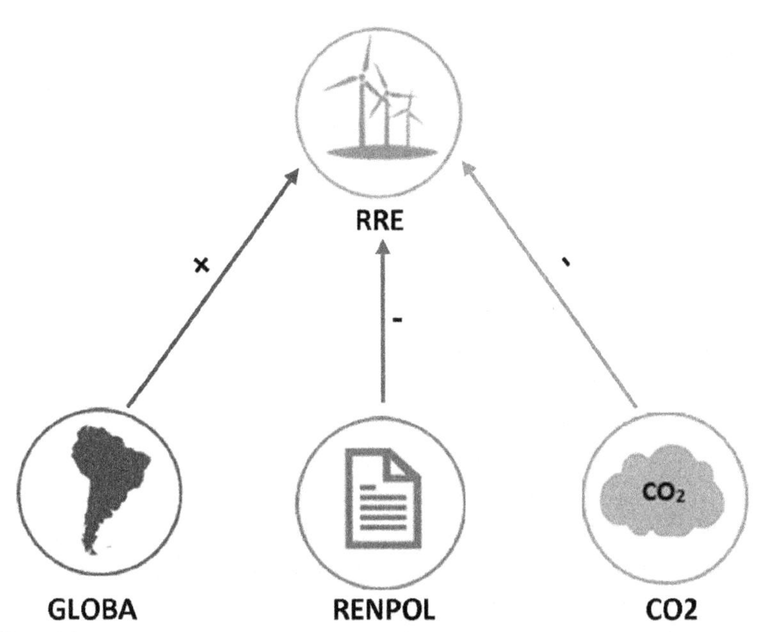

FIG. 8.2 Summary of the effect of the variables **GLOBA, RENPOL,** and **CO2** on **RRE.**

Fig. 8.2 summarises the effect of **GLOBA**, **RENPOL**, and **CO_2** on **RRE**.

The results confirm the explanations by Koengkan (2017a), Shahbaz et al. (2016), Shahbaz et al. (2015), and Leitão (2014) for example that the process of globalisation in the LAC countries influences the process of the energy transition. Therefore, the capacity of globalisation to increase the renewable energy transition is related to technology transfer. This improves energy efficiency and encourages the use of green technologies via investment and imports, as mentioned by Koengkan and Fuinhas (2020a). Indeed, these technology transfers are possible through trade and financial liberalisation, which allows developing countries to import advanced and green technologies from developed countries, as well as encouraging investment in and development of these technologies, as cited in Chapter 6.

All this helps technological progress and the improvement of productivity and subsequently an increase in economic activity and the consumption of energy from this kind of source. Koengkan et al. (2019) investigated the impact of globalisation on the development of renewable energy in the Latin American countries, and these authors confirm this explanation, whereby the globalisation process encourages investment in installed capacity of renewable energy and these new investments exert a positive impact on economic growth and subsequently on energy consumption. Moreover, the same authors also add that globalisation can allow households and firms to purchase renewable energy technology more cheaply, increasing the consumption of green energy.

However, the negative effect of renewable energy policies on energy transition is a surprise for this chapter, because this result was not expected. Indeed, the possible explanation for this impact can be related to the possible inefficiency of these policies in that it encourages the development of green energy in LAC countries or the methodology. It may also be that the construction of variable **RENPOL** is not able to reveal the real effect of this variable on energy transition. Moreover, concerning the ECM term, it is negative and statistically significant at the 1% level, and the statistical significance at the 1% level of the dummy and shift-dummy variables supports the decision to include them in the model.

After identifying that the globalisation process influences energy transition, other doubts arose in the explanations about the effect of globalisation on energy transition in the LAC countries, such as can globalisation encourage imports of advanced technology to LAC countries? Do imports of advanced technology increase the process of the energy transition? To answer these questions, it is necessary to make complementary robustness check to identify and to confirm if the results are in line with the literature, regarding the effect of globalisation on investment in renewable energy.

Therefore, to carry out this check, the following variables were used: the **ratio of renewable energy (RRE)**, **globalisation index (GLOBA)**—the same as that used in Sections 8.2 and 8.4—and the experimental variable **imports of ICT goods imports (ICT_IMPORTS)** in current US$. This variable is the multiplication of goods imports (BoP, current US$) that refers to all movable goods (including nonmonetary gold) involved in a change of ownership from nonresidents to residents by ICT goods imports (% total goods imports): information and communication technology goods imports including computers and peripheral equipment, communication equipment, consumer electronic equipment, electronic components, and other information and technology goods (miscellaneous). Both variables were retrieved from the World Bank Open Data (2019).

Therefore, we opted to use this experimental variable as a proxy for technological progress because globalisation via imports of manufactured products, capital equipment, technological goods, electronic equipment, and new materials that are commercialised in the international markets, as mentioned by Henry et al. (2009), causes technological progress and consequently encourages energy transition due to the availability of technologies. In the literature, some studies have used a similar variable (e.g. Yan et al., 2018; Mattern et al., 2010; Hilty et al., 2009; Holmgren and Thorslund, 2009). The variables **RRE** and **ICT_IMPORTS** are in per capita values.

The period between 2000 and 2014 and a group of 17 countries from the LAC region i.e. Argentina, Bolivia, Brazil, Chile, Colombia, Costa Rica, Dominican Republic, Ecuador, El Salvador, Guatemala, Mexico, Nicaragua, Panama, Paraguay, Peru, Uruguay, and Venezuela (RB), were used to carry out this 'complementary check'. This period and these countries were selected due to the availability of data for the variable **ICT_IMPORTS**.

Table 8.A14 in Appendix presents the summary statistics of variables that were used. The PVAR model estimation was used. This methodology and the preliminary and specification tests, such as panel VAR lag order selection; Granger causality Wald test; eigenvalue stability condition; forecast-error variance decomposition (FEVD); and impulse-response functions (IRF) are the same as those used in Chapter 7.

The results of preliminary tests that check the characteristics of variables point to the presence of low collinearity and multicollinearity, with cross-sectional dependence in the variables **GLOBA** and **ICT_IMPORTS** in the first differences and natural logarithms. The presence of unit roots was also identified in the variables in first differences with and without trend, except the variable **GLOBA** with trend, and in the variables **RRE**, **GLOBA**, and **ICT_IMPORTS** without and with trend in natural logarithms, as well as the presence of random effects in the model. Moreover, the results of the PVAR lag order selection point to the need to use the lag length (1) in the PVAR regression. These results can be seen in Tables 8.A15–8.A19 in Appendix.

The PVAR model indicates that globalisation increases the process of the energy transition and the import of technological goods. Moreover, it was identified that the imports of technology encourage the process of the energy transition. The results of the PVAR model are provided in Table 8.A20 in Appendix. This result confirms the possible explanation that the process of globalisation increases the technological efficiency of renewable energy by imports of technological goods and consequently increases renewable energy production.

Other authors such as Shirazi (2008) confirm that technological progress consists of learning about new technologies and materials, production processes, or organisational methods. Indeed, the indirect benefits of this process are derived from the imports of goods and services that have been developed by trade partners. Additionally, developing countries that successfully absorbed FDI inflow, particularly in the production of ICT and services (e.g. China, India, and Malaysia), have seen a variety of benefits.

Regarding the impact of imports of technological goods on the energy sector, this effect is confirmed by Yan et al. (2018); according to these authors, ICT development can stimulate economic growth with a less-than-proportionate increase in energy use i.e. improvement in energy productivity. This explication is in line with Mattern et al. (2010) and confirms that ICT reduces the consumption of resources and energy in other economic sections and consequently mitigates environmental degradation. The authors also add that ICT improves energy efficiency by the established process (i.e. increasing the ratio of a relevant target variable such as productivity or convenience to energy consumption), or by the development of new concepts to generate, allocate, distribute, share, and use energy in a resource-efficient and environmentally friendly way.

The results of specification tests indicate that the PVAR model is stable. One period after the shock, the variables themselves explained almost all the forecast error variance, and the impulse-response functions are such that all variables converge to equilibrium, supporting the fact that the variables of the model are I(1) (see Tables 8.A21 and 8.A22, and Fig. 8.A1 in Appendix).

It should be recalled that this complementary robustness check which was made is experimental in character and was used out of curiosity to discover the possible effect of globalisation on technological efficiency. Indeed, it is necessary to develop this issue further using other variables and methodology to discover the real impact. However, this verification is a kick-off for the study of this relationship.

8.5 Conclusions

The main aim of this chapter was to assess the asymmetric impact of energy transition on environmental degradation. In all, 18 LAC countries were considered, and the period from 1990 to 2014 proved to be the most appropriate. Moreover, this chapter opted to use PNARDL in the form of a UECM as the methodology.

The preliminary tests of this chapter indicated that the variables used have characteristics such as low-multicollinearity, cross-sectional dependence in all variables in natural logarithms and some variables in first differences, such as Y and PUBK, I(0)/I(1) for all variables, and the presence of fixed effects. Moreover, the specification test indicated the presence of heteroscedasticity, first-order autocorrelation, and nonpresence of cross-sectional independence. The results of these tests are essential to identify the characteristics of the countries under study as well as the possible methodologies that need to be applied.

The results of the PNARDL model estimates suggest that economic growth in the short and long run, as well as the public capital stock in the short run, have a positive effect on environmental degradation. Nevertheless, the positive and negative asymmetry of the variable ratio of renewable energy, which is a proxy for energy transition, harms the environment in the short and long run.

The capacity for the proxy for energy transition to reduce environmental degradation is probably related to the effect of globalisation on renewable energy technological efficiency that consequently produces more clean energy with fewer emissions of CO_2. It may also be due to the increasing participation of renewable energy sources in the energy matrix of these countries due to new investment and the energy demand caused by the effect of the globalisation process on economic growth. Another possible explanation for this negative impact is related to the efficiency of renewable energy policies that encourage the introduction of alternative energy sources in the energy mix.

To confirm these possible explanations the robustness checks were done, and it was identified that the positive and negative asymmetries of the variable globalisation index have a positive effect on the proxy for energy transition in the long run. However, the negative impact of renewable energy policies on the proxy of the energy transition is a surprise of this chapter. The possible explanation for this impact is related to the possible inefficiency of these policies, or that the methodology/construction of variable renewable energy policies is not able to reveal the real effect of this variable on energy transition.

Thus, based on these findings, the LAC region is recommended to make greater effort to develop policies for more efficient renewable energy that contribute to increasing growth, investment, and consumption of green energy and inversely reduce the consumption of energy from nonrenewable sources by households and industries. Regarding public capital stock, local governments should encourage public banks to support investment in renewable energy technologies or purchase technologies with higher energy efficiency that reduce the consumption of nonrenewable energy with lower interest and credit rates. Moreover, given the mistakes that were committed in the past, policymakers from the LAC region should also think about the possibility of integrating measures linked with the regulation of CO_2 emissions in their growth strategies.

This chapter makes a significant contribution to the literature for several reasons. First, it sheds light on how the process of energy transition affects environmental degradation. Second, the results of this chapter have critical consequences for local government appraisal of the relationship between economic growth, public capital stock, and environmental degradation. Finally, this study will help policymakers develop renewable energy policies more efficiently to reduce fossil fuel consumption and boost the development, investment, and use of renewable energy sources in developing countries to mitigate environmental degradation.

Appendix

TABLE 8.A1 Correlation matrix.

Variables	$LogCO_2$		LogGDP_PC		LogRRE		LogKGOV
LogCO2	**1.0000**						
LogGDP_PC	0.5330	***	**1.0000**				
LogRRE	0.2337	***	0.2763	***	**1.0000**		
LogKGOV	0.5360	***	0.7348	***	−0.0868	*	**1.0000**
Variables	**DLogCO2**		**DLogGDP_PC**		**DLogRRE**		**DLogKGOV**
DLogCO2	**1.0000**						
DLogGDP_PC	0.2266	***	**1.0000**				
DLogRRE	−0.3776	***	−0.0921	**	**1.0000**		
DLogKGOV	−0.0189		0.0448		0.0105		**1.0000**

Notes: ***, **, and * denote statistically significant at the 1%, 5%, and 10% levels.

TABLE 8.A2 VIF and Pesaran CD-tests.

Variables	VIF	1/VIF	CD-test		Average joint *T*	Mean ρ	Mean abs(ρ)
LogCO2	n.a.		33.232	***	25.00	0.54	0.58
LogGDP_PC	2.91	0.3435	43.417	***	25.00	0.70	0.84
LogRRE	2.71	0.3691	6.8782	***	25.00	0.11	0.34
LogKGOV	1.35	0.7410	12.591	***	25.00	0.20	0.68
Mean VIF	2.32						
DLogCO2	n.a.		0.262		24.00	0.00	0.17
DLogGDP_PC	1.01	0.9894	17.702	***	24.00	0.29	0.32
DLogRRE	1.01	0.9913	1.621		24.00	0.03	0.22
DLogKGOV	1.00	0.9977	18.235	***	24.00	0.30	0.42
Mean VIF	1.01						

Notes:*** denotes statistically significant at the 1% level.

TABLE 8.A3 Unit root test.

	Panel-data unit-root test			
	Panel unit root test (CIPS) (Zt-bar)			
	Without trend		With trend	
Variables	Zt-bar		Zt-bar	
$LogCO_2$	−1.719	**	−0.193	
LogGDP_PC	−2.369	***	−1.561	*
LogRRE	−2.297	**	−0.742	
LogKGOV	2.417		−2.308	**
$DLogCO_2$	−9.155	***	−8.314	***
DLogGDP_PC	−6.566	***	−5.218	***
DLogRRE	−10.036	***	−8.323	***
DLogKGOV	−2.754	***	−1.852	**

Notes:***, **, and * denote statistically significant at the 1%, 5%, and 10% levels.

TABLE 8.A4 Hausman test.

Variables	(b) Fixed	(B) Random	(b-B) Difference	Sqrt(diag(V_b-V-B)) S.E.
DLogGDP_PC	0.6396	0.6769	−0.0373	0.0609
DLogRRE_POS	−0.0623	−0.0585	−0.0038	0.0092
DLogRRE_NEG	−0.0809	−0.0881	0.0071	0.0078
DLogKGOV	0.6864	−0.1505	0.8369	0.1929
$LogCO_2$	−0.3875	−0.0041	−0.3834	0.0436
LogGDP_PC	0.1481	−0.0051	0.1533	0.0481
LogRRE_POS	−0.0297	−0.0025	−0.0272	0.0080
LogRRE_NEG	−0.0371	−0.0018	−0.0353	0.0088
Chi2 (8)	80.07***			

Notes: *** denotes statistically significant at the 1% level.

TABLE 8.A5 Heterogeneous estimators.

Independent variables	Dependent variable (DLogCO2)					
	MG		PMG		FE	
Constant	−9.7595	***	−6.2612	***	−4.3525	***
	Short-run (impacts)					
DLogGDP_PC	0.7750	***	0.8867	***	0.6396	***
DLogRRE_POS	−0.0709		−0.0732	*	−0.0623	***
DLogRRE_NEG	−0.1211	***	−0.1248	***	−0.0809	***
DLogKGOV	0.2892		0.0478		0.6864	***
	Long-run (elasticities)					
LogGDP_PC (-1)	1.6284		0.0576		0.3823	***
LogRRE_POS (-1)	0.6080		−0.0570	***	−0.0769	***
LogRRE_NEG (-1)	0.3086		−0.2416	***	−0.0958	***
	Speed of adjustment					
ECM	−0.7530	***	−0.3995	***	−0.3876	***

Notes: ***, **, and * denote statistically significant at the 1%, 5%, and 10% levels, respectively. The ECM denotes the coefficient of the variable LogCO2, lagged once; the long-run parameters are computed elasticities.

TABLE 8.A6 Heterogeneous test.

MG vs PMG	PMG vs FE	MG vs FE
Chi2(8) = 325.90***	Chi2(8) = 167.23***	Chi2(8) = 52.81***

Notes: *** denotes statistically significant at the 1% level.

TABLE 8.A7 Post-estimation tests.

Statistics	(a) Modified Wald test	(b) Wooldridge test	(c) Pesaran's test	(d) Breusch and Pagan Lagrangian multiplier test
	chi2 (18) = 6019.71***	F(1, 17) = 53.346***	1.123	n.a.

Notes: *** denotes statistically significant at the 1% level; (n.a.) denotes not available.

TABLE 8.A8 Summary statistics of variables from robustness check.

Variables	Descriptive statistics				
	Obs.	Mean	Std. dev.	Min	Max
LogRRE	450	1.9695	4.5903	−2.5181	18.6579
LogGLOBA	450	4.0277	0.1704	3.4324	4.3283
LogRENPOL	400	1.3278	1.0312	0.0000	3.8712
LogCO2	450	−6.7708	1.3936	−11.7542	−4.8785
DLogRRE	432	−0.0337	0.5760	−5.1653	3.5362
DLogGLOBA	432	0.0139	0.0266	−0.0704	0.1128
DLogRENPOL	384	0.1002	0.2266	−0.4054	1.3862
DLogCO2	432	0.0203	0.1198	−0.8107	1.0799

Notes: Obs. denotes the number of observations; Std. dev. denotes standard deviation; Min. and Max. denote minimum and maximum, respectively.

TABLE 8.A9 Correlation matrix from robustness check.

Variables	LogRRE		LogCO2		LogGLOBA		LogRENPOL
LogRRE	**1.0000**						
LogCO2	0.2337	***	**1.0000**				
LogGLOBA	0.2076	***	0.4651	***	**1.0000**		
LogRENPOL	−0.0165		0.2059	***	0.4603	***	**1.0000**
Variables	**DLogRRE**		**DLogCO2**		**DLogGLOBA**		**DLogRENPOL**
DLogRRE	**1.0000**						
DLogCO2	−0.3776	***	**1.0000**				
DLogGLOBA	−0.0233		0.0366		**1.0000**		
DLogRENPOL	0.0051		−0.0180		−0.0065		**1.0000**

Notes: *** denotes statistically significant at the 1% level.

TABLE 8.A10 VIF and CSD tests from robustness check.

Variables	VIF	1/VIF	CD-test		Average joint T	Mean ρ	Mean abs(ρ)
LogRRE	n.a.		6.872	***	25.00	0.11	0.34
LogGLOBA	1.40	0.7156	49.113	***	25.00	0.79	0.83
LogRENPOL	1.27	0.7861	47.232	***	25.00	0.68	0.68
LogCO2	1.15	0.8695	33.232	***	25.00	0.54	0.58
Mean VIF	1.27						
DLogRRE	n.a.		1.621		24.00	0.03	0.22
DLogGLOBA	1.00	0.9996	9.252	***	24.00	0.15	0.21
DLogRENPOL	1.00	0.9996	0.234		24.00	0.00	0.12
DLogCO2	1.00	0.9998	0.262		24.00	0.00	0.17
Mean VIF	1.00						

Notes: The Stata commands *estat vif* and *xtcdf* were used; *** denotes statistically significant at the 1% level.

TABLE 8.A11 Unit root test from robustness check.

	Panel-data unit-root test			
	Pesaran (2007) Panel Unit Root test (CIPS) (Zt-bar)			
	Without trend		With trend	
Variables	Zt-bar		Zt-bar	
LogRRE	−0.704		−0.074	
LogGLOBA	−2.785	***	−1.418	*
LogRENPOL	−1.490	*	−3.076	***
LogCO2	−2.148	**	0.148	
DLogRRE	−9.872	***	−8.463	***
DLogGLOBA	−7.797	***	−5.905	***
DLogRENPOL	−7.359	***	−5.313	***
DLogCO2	−8.394	***	−7.501	***

Notes: ***, **, and * denote statistically significant at the 1%, 5%, and 10% levels respectively.

TABLE 8.A12 Hausman test from robustness check.

Variables	(b) Fixed	(B) Random	(b-B) Difference	Sqrt(diag(V_b-V-B)) S.E.
DLogCO_2	−2.3429	−2.1260	−0.2168	0.1383
LogRRE	−0.4095	−0.0012	−0.4082	0.0452
LogGLOBA_POS	1.6357	0.0656	1.5701	0.4306
LogGLOBA_NEG	1.7704	0.3301	1.4402	0.8504
LogRENPOL	−0.0980	0.0005	−0.0985	0.0451
LogCO_2	−1.1074	−0.0047	−1.1026	0.2953
Chi2 (6)	81.90***			

Notes: *** denotes statistically significant at the 1% level.

TABLE 8.A13 Post-estimation tests from robustness check.

Statistics	(a) Modified Wald test	(b) Wooldridge test	(c) Pesaran's test	(d) Breusch and Pagan Lagrangian Multiplier test
	chi2 (16) = 3145.98	F(1,15) = 224.341 ***	0.371	chi2(120) = 136.911

Notes: *** denotes statistically significant at the 1% level.

TABLE 8.A14 Summary statistics of variables from complementary robustness check.

	Descriptive statistics				
Variables	Obs.	Mean	Std. dev.	Min	Max
LogRRE	255	2.0635	4.8624	−2.5181	18.6579
LogGLOBA	255	4.1280	0.0899	3.8387	4.3283
LogICT_IMPORTS	255	13,891.22	14,854.78	563.6581	66,637.91
DLogRRE	238	−0.0275	0.5424	−5.1653	3.5362
DLogGLOBA	238	0.0082	0.0227	−0.0605	0.1128
DLogICT_IMPORTS	238	823.2538	4215.752	−20,611.12	17,623.34

Notes: Obs. denotes the number of observations; Std. dev. denotes standard deviation; Min. and Max. denote minimum and maximum, respectively.

TABLE 8.A15 Correlation matrix from complementary robustness check.

Variables	LogRRE		LogGLOBA		LogICT_IMPORTS
LogRRE	**1.0000**				
LogGLOBA	0.3715	***	**1.0000**		
LogICT_IMPORTS	0.1521	*	0.4303	***	**1.0000**
Variables	**DLogRRE**		**DLogGLOBA**		**DLogICT_IMPORTS**
DLogRRE	**1.0000**				
DLogGLOBA	−0.0355		**1.0000**		
DLogICT_IMPORTS	−0.0823		0.0226		**1.0000**

Notes: *** denotes statistically significant at the 1% level.

TABLE 8.A16 VIF and CSD tests from complementary robustness check.

Variables	VIF	1/VIF	CD-test		Average joint T	Mean ρ	Mean abs(ρ)
LogRRE	n.a.		−0.202		15.00	0.00	0.44
LogGLOBA	1.00	0.9994	23.501	***	15.00	0.52	0.84
LogICT_IMPORTS	1.00	0.9994	35.669	***	15.00	0.79	0.79
Mean VIF	1.00						
DLogRRE	n.a.		−1.266		14.00	−0.03	0.29
DLogGLOBA	1.23	0.8148	11.202	***	14.00	0.26	0.31
DLogICT_IMPORTS	1.23	0.8148	17.073	***	14.00	0.39	0.41
Mean VIF	1.23						

Notes: *** denotes statistically significant at the 1% level.

TABLE 8.A17 Unit root test from complementary robustness check.

	Panel-data unit-root test			
	Pesaran (2007) Panel unit root test (CIPS) (Zt-bar)			
	Without trend		With trend	
Variables	Zt-bar		Zt-bar	
LogRRE	−1.701	**	−3.311	***
LogGLOBA	−0.458		0.078	
LogICT_IMPORTS	0.003		1.543	
DLogRRE	−6.758	***	−4.720	***
DLogGLOBA	−3.353	***	−1.174	
DLogICT_IMPORTS	−2.090	***	−2.386	***

Notes: ***, **, and * denote statistically significant at the 1%, 5%, and 10% levels.

TABLE 8.A18 Hausman test from complementary robustness check.

Variables	(b) Fixed	(B) Random	(b-B) Difference	Sqrt(diag(V_b-V-B)) S.E.
DLogGLOBA	−1.1305	−0.8029	−0.3276	0.6378
DLogICT_IMPORTS	−9.85e−0	−0.0000	6.41e−0	2.35e−0
Chi2 (1)	**0.26**			

TABLE 8.A19 PVAR lag-order selection from complementary robustness check.

Lags	CD	J	Jp-value	MBIC	MAIC	MQIC
1	0.5261	55.6071	0.4140*	−194.1414	−52.3928	−109.7917
2	−6.7212	46.8109	0.3980	−161.3128	−43.1890	−91.0213
3	−2.9161	25.7337	0.8976	−140.7653	−46.2662	−84.5321

TABLE 8.A20 PVAR model outcomes from complementary robustness check.

	Response to					
Response of	**DLogRRE$^{(t)}$**		**DLogGLOBA$^{(t)}$**		**DLogICT_IMPORTS$^{(t)}$**	
DLogRRE$_{(t-1)}$	−0.3972	***	−0.0444	***	5813.394	***
DLogGLOBA$_{(t-1)}$	8.9389	***	−0.8952	***	327,692.4	***
DLogICT_IMPORTS$_{(t-1)}$	3.07e−0	*	−2.05e−0		−0.5007	***
N. obs	102					
N. panels	17					

Notes: *** denotes statistical significance level at the 1% level.

TABLE 8.A21 Eigenvalue stability condition from complementary robustness check.

Eigenvalue			
Real	**Imaginary**	**Modulus**	**Graph**
−0.5925	−0.6147	0.8538	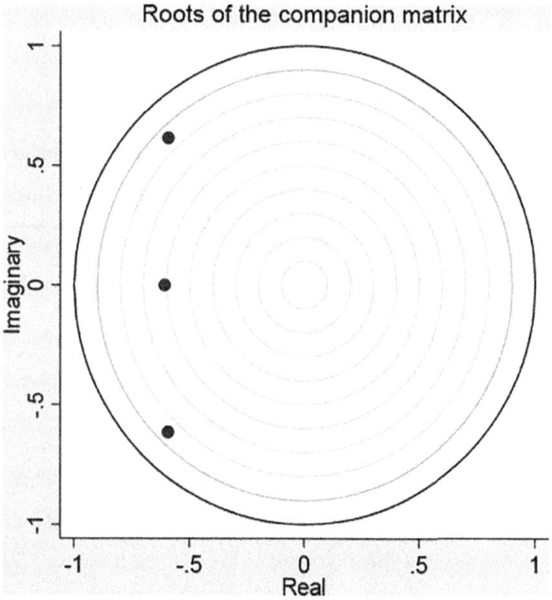
−0.5925	0.6147	0.8538	
−0.6080	0.0000	0.6080	

Roots of the companion matrix

Imaginary

Real

TABLE 8.A22 Forecast-error variance decomposition from complementary robustness check.

	Impulse variables		
Response variable and Forecast Impulse Variable Horizon	**DLogRRE**	**DLogGLOBA**	**DLogICT_IMPORTS**
DLogRRE			
0	0	0	0
1	1	0	0
5	0.6294	0.3670	0.0035
10	0.5860	0.4105	0.0034
15	0.5793	0.4171	0.0034
DLogGLOBA			
0	0	0	0
1	0.0130	0.2399	0.7469
5	0.3512	0.5725	0.0761
10	0.3620	0.6361	0.0018
15	0.3700	0.6280	0.0018
DLogICT_IMPORTS			
0	0	0	0
1	0.0130	0.2399	0.7469
5	0.3512	0.5725	0.0761
10	0.3639	0.5710	0.0650
15	0.3646	0.5726	0.0627

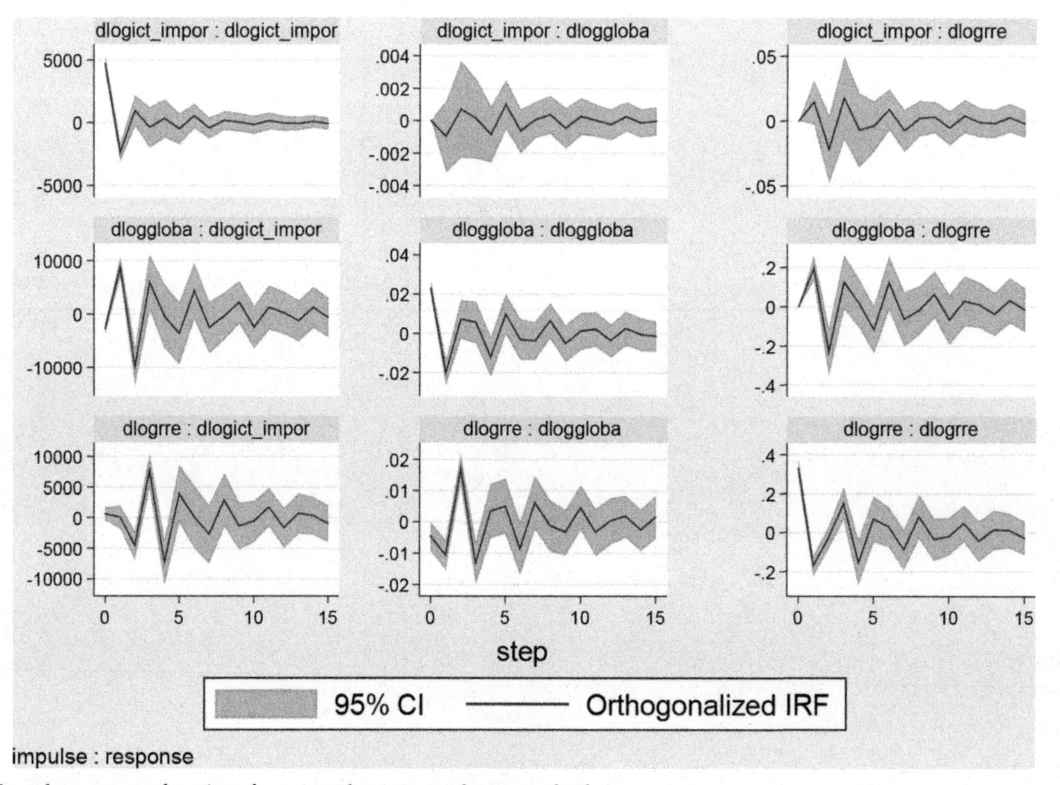

FIG. 8.A1 Impulse-response functions from complementary robustness check.

References

Akella, A.K., Saini, R.P., Sharma, M.P., 2009. Social, economical and environmental impacts of renewable energy systems. Renew. Energy 34 (2), 390–396. https://doi.org/10.1016/j.renene.2008.05.002.

Apergis, N., Payne, J.E., 2014. Renewable energy, output, CO_2 emissions, and fossil fuel prices in Central America: evidence from a nonlinear panel smooth transition vector error correction model. Energy Econ. 42, 226–232. https://doi.org/10.1016/j.eneco.2014.01.003.

Aye, G.C., Edoja, P.E., 2017. Effect of economic growth on CO_2 emission in developing countries: evidence from a dynamic panel threshold model. J. Cogent Econ. Finance 5 (1), 1–22. https://doi.org/10.1080/23322039.2017.1379239.

Balza, L.H., Espinasa, R., Serebrisky, T., 2016. Lights on? Energy Needs in Latin America and the Caribbean to 2040. Inter-American Development Bank, pp. 1–39. https://publications.iadb.org/en/publication/17053/lights-energy-needs-latin america-and-caribbean-2040.

Belsley, D.A., Kuh, E., Welsch, E.R., 1980. Regression Diagnostics: Identifying Influential Data and Sources of Collinearity. Wiley, New York, pp. 1–286, https://doi.org/10.1002/0471725153.

Bilgili, F., Koçak, E., Bulut, Ü., 2016. The dynamic impact of renewable energy consumption on CO_2 emissions: a revisited environmental Kuznets curve approach. Renew. Sust. Energy Rev. 54, 838–845. https://doi.org/10.1016/j.rser.2015.10.080.

Breusch, T.S., Pagan, A.R., 1980. The Lagrange multiplier test and its applications to model specification in econometrics. Rev. Econ. Stud. 47 (1), 239–253. https://www.jstor.org/stable/pdf/2297111.pdf.

Dogan, B., Deger, O., 2016. How globalization and economic growth affect energy consumption: panel data analysis in the sample of Brazil, Russia, India, China countries. Int. J. Energy Econ. Policy 6 (4), 806–813. ISSN: 2146-4553 https://dergipark.org.tr/tr/download/article-file/361684.

Engle, R., Granger, G., 1987. Cointegration and error correction: representation, estimation and testing. Econometrica 55, 251–276. https://www.jstor.org/stable/1913236.

Fuinhas, J.A., Marques, A.C., Koengkan, M., 2017. Are renewable energy policies upsetting carbon dioxide emissions? The case of Latin America countries. Environ. Sci. Pollut. Res. 24 (17), 15044–15054. https://doi.org/10.1007/s11356-017-9109-z.

Fuinhas, J.A., Marques, A.C., Koengkan, M., Santiago, R., Couto, A.P., 2019. The energy-growth nexus within production and oil rents context. Rev. Estudos Soc. 21 (42), 161–173. https://doi.org/10.19093/res7857.

Granger, C.W.J., 1981. Some properties of time series data and their use in econometric model specification. J. Econ. 28, 121–130. https://doi.org/10.1016/0304-4076(81)90079-8.

Greene, W., 2002. Econometric Analysis. Prentice-Hall, Saddle River, NJ.

Hauff, J., Bode, A., Neumann, D., Haslauer, F., 2014. Global Energy Transitions A Comparative Analysis of Key Countries and Implications for the International Energy Debate. World Energy Council, pp. 1–30. https://www.extractiveshub.org/resource/view/id/13542.

Henry, M., Kneller, R., Milner, C., 2009. Trade, technology transfer and national efficiency in developing countries. Eur. Econ. Rev. 53 (2), 237–254. https://doi.org/10.1016/j.euroecorev.2008.05.001.

Hilty, L.M., Coroama, V., Eicker, M.O., Ruddy, T.F., Müller, E., 2009. The Role of ICT in Energy Consumption and Energy Efficiency. Technology and Society Lab Empa, Swiss Federal Laboratories for Materials Testing and Research, St. Gallen, Switzerland, pp. 1–80. https://www.academia.edu/2686550/The_Role_of_ICT_in_Energy_Consumption_and_Energy_Efficiency.

Hollanda, L., Nogueira, R., Muñoz, R., Febraro, J., Varejão, M., Silva, T.B., 2016. Eine vergleichende Studie über die Energiewende in Lateinamerika und Europa. EKLA-KAS and FGV Energia, pp. 1–72. https://www.kas.de/web/energie-klima-lateinamerika/publikationen/einzeltitel/-/content/eine-vergleichende-studie-ueber-die-energiewende-in-lateinamerika-und-europa1.

Holmgren, S., Thorslund, E., 2009. ICT and Energy Efficiency in Sweden. Swedish Ministry of Enterprise, Energy and Communications and the Professional Association Swedish IT and Telecom Industries in Almega, pp. 1–30. https://www.government.se/49b758/contentassets/f496d0e0cc864e8fa57b22ea247a829e/report-ict-and-energy-efficiency-in-sweden.

Iheanacho, E., 2018. The role of globalisation on energy consumption in Nigeria. Implication for long run economic growth. ARDL and VECM analysis. Glob. J. Hum. Soc. Sci. 18 (1), 2–19. ISSN: 2249-460x https://pdfs.semanticscholar.org/84a2/90c7efe7cb205fd33679eebca3b94cbcb901.pdf.

IMF, 2019. Investments and Capital Stock. International Monetary Fund. https://www.imf.org/external/np/fad/publicinvestment/data/data122216.xlsx.

IRENA (International Renewable Energy Agency), 2016. Renewable Energy Market Analysis: Latin America. IRENA, pp. 1–160. ISBN 978-92-95111-49-3.

Jácome, L.H., 2004. The late 1990s financial crisis in Ecuador: institutional weaknesses, fiscal rigidities, and financial dollarization at work. IMF Working Paper: Monetary and Financial Systems Department, 4(12), pp. 1–47. https://www.imf.org/external/pubs/ft/wp/2004/wp0412.pdf.

Koengkan, M., 2017a. O nexo entre o consumo de energia primária e o crescimento econômico nos países da América do Sul: Uma análise de longo prazo. Cadernos UniFOA, Volta Redonda 12 (34), 561809–669475.

Koengkan, M., 2017b. Is the globalization influencing the primary energy consumption? The case of Latin America and the Caribbean countries. Cadernos UniFOA, Volta Redonda 12 (33), 59–69. ISSN:1809-9475.

Koengkan, M., 2018a. The decline of environmental degradation by renewable energy consumption in the MERCOSUR countries: an approach with ARDL modelling. Environ. Syst. Decis. 38 (3), 415–425. https://doi.org/10.1007/s10669-018-9671-z.

Koengkan, M., 2018b. The positive impact of trade openness on the consumption of energy: Fres evidence from Andean community countries. Energy 158 (1), 936–943. https://doi.org/10.1016/j.energy.2018.06.091.

Koengkan, M., Fuinhas, J.A., 2017. The negative impact of renewable energy consumption on carbon dioxide emissions: an empirical evidence from south American countries. Rev. Bras. Energias Renováveis 6 (5), 893–914. https://doi.org/10.5380/rber.v6i5.49252.

Koengkan, M., Fuinhas, J.A., 2020a. The interactions between renewable energy consumption and economic growth in the Mercosur countries. Int. J. Sustain. Energy 39 (6), 594–614. https://doi.org/10.1080/14786451.2020.1732978.

Koengkan, M., Fuinhas, J.A., 2020b. Exploring the effect of the renewable energy transition on CO_2 emissions of Latin American & Caribbean countries. Int. J. Sustain. Energy 39 (6), 515–538. https://doi.org/10.1080/14786451.2020.1731511.

Koengkan, M., Losekann, L.D., Fuinhas, J.A., Marques, A.C., 2018. The effect of hydroelectricity consumption on environmental degradation—the case of South America region. TAS J. 2 (2), 45–67.

Koengkan, M., Poveda, Y.E., Fuinhas, J.A., 2019a. Globalisation as a motor of renewable energy development in Latin America countries. GeoJournal, 1–12. https://doi.org/10.1007/s10708-019-10042-0.

Koengkan, M., Fuinhas, J.A., Vieira, I., 2019b. Effects of financial openness on renewable energy investments expansion in Latin American countries. J. Sustain. Finance Invest., 1–19. https://doi.org/10.1080/20430795.2019.1665379.

KOF Globalization index, 2018. https://www.kof.ethz.ch/en/forecasts-and-indicators/indicators/kof-globalisation-index.html.

Lee, C.C., 2005. Energy consumption and GDP in developing countries: a cointegrated panel analysis. Energy Econ. 27 (3), 415–427. https://doi.org/10.1016/j.eneco.2005.03.003.

Lee, C.C., Chen, P.F., 2010. Dynamic modelling of energy consumption, capital stock, and real income in G-7 countries. Energy Econ. 32 (3), 564–581. https://doi.org/10.1016/j.eneco.2009.08.022.

Lee, C.C., Chang, C.P., Chen, P.F., 2008. Energy-income causality in OECD countries revisited: the key role of capital stock. Energy Econ. 30 (5), 2359–2373. https://doi.org/10.1016/j.eneco.2008.01.005.

Leitão, N.C., 2014. Economic growth, carbon dioxide emissions, renewable energy and globalization. Int. J. Energy Econ. Policy 4 (3), 391–399. ISSN: 2146-4553 https://core.ac.uk/download/pdf/70620032.pdf.

Mardani, A., Streimikiene, D., Cavallaro, F., Loganathan, N., Khoshnoudi, M., 2019. Carbon dioxide (CO_2) emissions and economic growth: a systematic review of two decades of research from 1995 to 2017. Sci. Total Environ. 649, 31–49. https://doi.org/10.1016/j.scitotenv.2018.08.229.

Marland, G., Boden, T.A., Andres, R.J., 2011. Global, Regional, and National Fossil-Fuel CO_2 Emissions. Carbon Dioxide Information Analysis Center, Oak Ridge National Laboratory, U.S. Department of Energy, Oak Ridge, TN, USA, https://doi.org/10.3334/CDIAC/00001_V2011.

Mattern, F., Staake, T., Weiss, M., 2010. ICT for Green: How Computers Can Help Us to Conserve Energy. Institute for Pervasive Computing, ETH Zurich, pp. 1–10. https://www.vs.inf.ethz.ch/publ/papers/ICT-for-Green.pdf.

Mirza, F.M., Kanwal, A., 2017. Energy consumption, carbon emissions and economic growth in Pakistan: dynamic causality analysis. Renew. Sust. Energy Rev. 72, 1233–1240. https://doi.org/10.1016/j.rser.2016.10.081.

OECD Environment Directorate, 2008. OECD Key Environmental Indicators. OECD, pp. 1–38. https://www.oecd.org/env/indicators-modelling-outlooks/37551205.pdf.

Pesaran, M.H., 2004. General Diagnostic Tests for Cross Section Dependence in Panels. Cambridge Working Papers in Economics, N. 0435, The University of Cambridge, Faculty of Economics. http://ftp.iza.org/dp1240.pdf.

Pesaran, M.H., 2007. A simple panel unit root test in the presence of cross-section dependence. J. Appl. Econ. 22 (2), 265–312. https://doi.org/10.1002/jae.951.

Pesaran, M.H., Smith, R., 1995. Estimating long-run relationships from dynamic heterogeneous panels. J. Econ. 68 (1), 79–113. https://doi.org/10.1016/0304-4076(94)01644-F.

Rocher, C.L., 2017. Linear and Nonlinear Relationships Between Interest Rate Changes and Stock Return: International Evidence. Universidad de Valencia. Working Paper nº 017/016 https://www.uv.es/bfc/TFM2017/16%20Carlos%20Lopez%20Rocherpdf.

Sadorsky, P., 2009. Renewable energy consumption, CO_2 emissions and oil prices in the G7 countries. Energy Econ. 31 (3), 456462. https://doi.org/10.1016/j.eneco.2008.12.010.

Shafiei, S., Salim, R.A., 2014. Non-renewable and renewable energy consumption and CO_2 emissions in OECD countries: a comparative analysis. Energy Policy 66, 547–556. https://doi.org/10.1016/j.enpol.2013.10.064.

Shahbaz, M., Nasreen, S., Ling, C.H., Sbia, R., 2014. Causality between trade openness and energy consumption: what causes what in high, middle and low-income countries. Energy Policy 70, 126–143. https://doi.org/10.1016/j.enpol.2014.03.029.

Shahbaz, M., Bhattacharya, M., Ahmed, K., 2015. Growth-globalisation-emissions Nexus: the role of population in Australia. Department of Economics, 23(15):1–32. ISSN: 1441-5429.

Shahbaz, M., Mallick, H., Mahalik, M.H., Sadorsky, P., 2016. The role of globalization on the recent evolution of energy demand in India: implications for sustainable development. Energy Econ. 55, 52–68. https://doi.org/10.1016/j.eneco.2016.01.013.

Shirazi, F., 2008. The impact of foreign direct investment and trade openness on ICT expansion. In: PACIS 2008 Proceedings. vol. 148, pp. 1–22. https://aisel.aisnet.org/pacis2008/148.

Smil, V., 2010. Energy Transitions: History, Requirements, Prospects. Praeger Publishers, Santa Barbara, CA. ISBN-10: 0313381771.

Snedecor, G.W., Cochran, W.G., 1989. Statistical Methods, eighth ed. Iowa State University Press, Ames, IA.

Tavares, F.B., 2017. Energy transition enablers in Latin American countries. In: 6ELAEE, 2017, pp. 1–2. https://www.iaee.org/en/.../proceedingsabstractpdf.aspx?id=13999.

The Carter Center, 1994. President Carter Leads Delegation to Negotiate Peace With Haiti. https://www.cartercenter.org/news/documents/doc218.html.

United Nations Environment Programme (UNEP), 2001. Climate Change: Information Kit. United Nations Environment Programme, pp. 1–63. https://unfccc.int/resource/iuckit/cckit2001en.pdf.

Weisbrot, M., Ray, R., Johnston, J., 2009. Bolivia: The Economy During the Morales Administration. Center for Economic and Policy Research, pp. 1–31. http://www.cepr.net/documents/publications/bolivia-2009-12.pdf.

Wooldridge, J.M., 2002. Econometric Analysis of Cross Section and Panel Data. The MIT Press, Cambridge, Massachusetts/London, England.

World Bank Open Data, 2019. http://www.worldbank.org/.

Yan, Z., Shi, R., Yang, Z., 2018. ICT development and sustainable energy consumption: a perspective of energy productivity. Sustainability 10, 1–15. https://doi.org/10.3390/su10072568.

Zahonogo, P., 2016. Trade and economic growth in developing countries: evidence from sub-Saharan Africa. J. Afr. Trade 3 (1–2), 41–56. https://doi.org/10.1016/j.joat.2017.02.001.

Zhao, Y., Tang, K.-K., Wang, L., 2013. Do renewable electricity policies promote renewable electricity generation? Evidence from panel data. Energy Policy 62, 887–897. https://doi.org/10.1016/j.enpol.2013.07.072.

CHAPTER

9

The capacity of energy transition to decrease deaths from air pollution: Empirical evidence from Latin America and the Caribbean countries

JEL codes E00, I15, N16, Q43

9.1 Introduction

Air pollution is considered as one of the world's largest environmental and public health problems in the present day. According to the Institute for Health Metrics and Evaluation (IHME) in its Global Burden of Disease study (2018), this problem is responsible for more than 5 million deaths each year in the world. Air pollution is a combination of outdoor and indoor (household) air pollution particulate matter that can contain gases such as ammonia (NH_3), carbon monoxide (CO_2), sulphur dioxide (SO_2), nitrous oxides (NO_x), methane (CH_4), chlorofluorocarbons (CFCs), particulate matter, and lead (Romieu et al., 1990). These substances are risk factors for many of the leading causes of death such as stroke, heart disease, lower respiratory infections, diabetes, lung cancer, and chronic obstructive pulmonary disease (COPD) (Ritchie and Roser, 2020; Institute for Health Metrics and Evaluation (IHME), 2018). More than 80% of these deaths are due to strokes and ischemic heart disease, 14% due to lower respiratory infections or chronic obstructive pulmonary disease, and 6% due to lung cancer and diabetes (Riojas-Rodríguez et al., 2016).

The process of urbanisation and growing industrialisation in addition to the industrial processes in large cities and consumption of fossil fuels by households, firms, and industries is responsible for the increased air pollution problem and consequently these deaths (Romieu et al., 1990). Indeed, the electricity and heat production sector accounted for 25% of the total air pollution in the world in 2010, the agriculture, forestry, and other land use (AFOLU) sector for 24%, the industry sector 21%, the transport sector 14%, other energies 10%, and the building sector accounted for 6% (see Fig. 5.4).

More than 80% of people living in these urban areas are exposed to air quality levels that exceed the World Health Organisation (WHO) guideline limits, with low- and middle-income countries suffering from the highest exposures of outdoor and indoor air pollution (WHO, 2020). Latin America and the Caribbean (LAC) region is no exception, with countries in this region registering a growth in air pollution exposure between 1990 and 2014 (see Fig. 9.1).

This growth was caused by high urbanisation trends, where 79% of the population of the region live in urban areas, and by economic liberalisation (e.g. trade and financial) that intensified during the periods 1989–92 and 2004–14. This consequently encouraged a rapid increase in the demand for energy and in air pollution (Koengkan and Fuinhas, 2020a, b; Riojas-Rodríguez et al., 2016). Indeed, in the LAC region, the electricity and heat production sector accounted for 48% of total air pollution in 2014, the AFOLU sector for 23%, the industry sector 4%, and waste 6% of air pollution emissions (see Fig. 5.8).

However, as can be seen in Fig. 9.2, air pollution in the region has been decreasing since 2015 due to the biggest insertion of new energy sources. Moreover, the death rates from indoor and outdoor air pollution that measure the number of deaths per 100,000 population were 67.62 in 1990, and this value dropped to 33.84 in 2016.

Indeed, the slow growth in air pollution exposure and the significant decrease in deaths from air pollution between 1990 and 2016 coincides with the rapid intensification in the process of energy transition in the region

https://doi.org/10.1016/B978-0-12-824429-6.00007-3

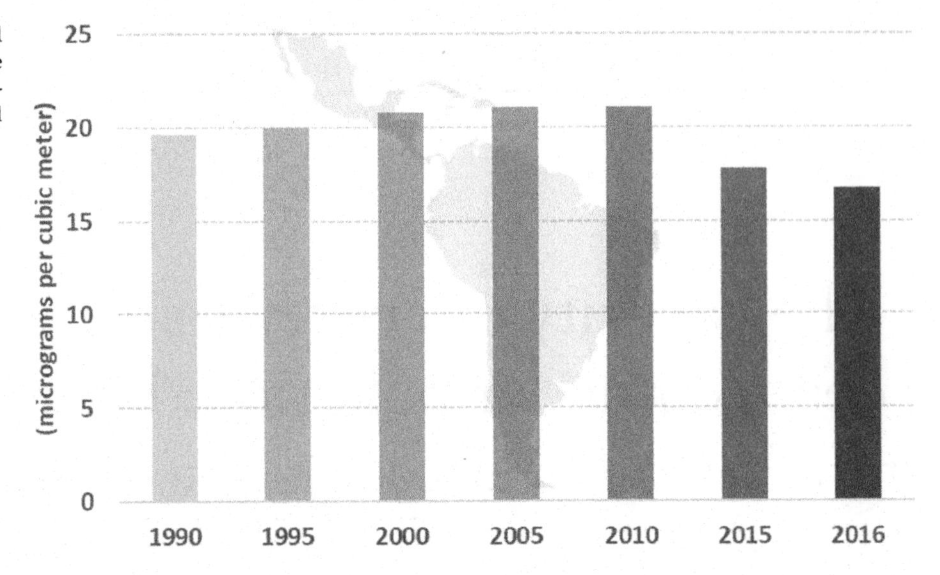

FIG. 9.1 PM2.5 air pollution, mean annual exposure (micrograms per cubic meter) in the LAC region between 1990 and 2016. This figure was created by the author and was based on the World Bank Open Data (2020).

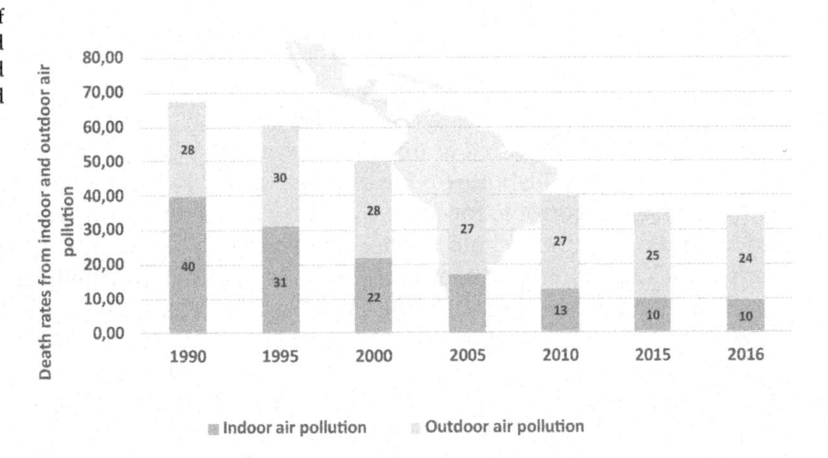

FIG. 9.2 Death rates are measured as the number of deaths per 100,000 population from both outdoor and indoor air pollution in the LAC region between 1990 and 2016. This figure was created by the author and was based on the Our World in Data (2020).

(Koengkan et al., 2019a). This rapid process of the energy transition was caused by the rapid development of new renewable energy sources (e.g. biomass, geothermal, solar, photovoltaic, waste, and wind) during the 1990s and up to 2016, where this kind of sources comprised only 5.03% of total installed capacity of energy in the region in 2016. Wind energy comprised 2.12%, solar energy 0.40%, and other renewable energy that includes geothermal, biomass, and waste comprised 2.51%. Large hydropower dam plants also underwent rapid growth in this period, with the share of this kind of energy in the energy matrix reaching a value of 23.57% (see Fig. 5.14). However, investments in large hydropower dam plants have been declining in the LAC region in recent decades due to investments in new renewable energy sources.

However, in the main countries of the LAC region such as Mexico, the renewable energy sources accounted for 5.91% of the total energy matrix in 2016, 10.74% in Argentina, and 18.56% in Chile, while it accounted for an incredible 36.40% of the total energy matrix in Brazil (see Fig. 9.3).

Moreover, between 1990 and 2016, the installed capacity of these energy sources was 360 terawatt-hours (TWh) in 1990 and reached a value of 814 TWh in 2016. This increase is related to large investments in biomass and waste and by wind, geothermal, and also solar and large hydropower dam plants which make up most of this growth (see Fig. 9.4).

This increase in the installed capacity of renewable energy in the region is a result of high investment in renewable energy technologies that was US$1.6 billion in 2004, and reached US$8.7 billion in 2016 (see Fig. 6.5). That is, in the period between 2000 and 2013, investment in renewable energy technologies grew 13% (Koengkan et al., 2019b).

This rapid expansion of renewable energy in the LAC region is also associated with the fast process of economic liberalisation, as mentioned earlier, which began in the 1990s after the implementation of neoliberal policies during the

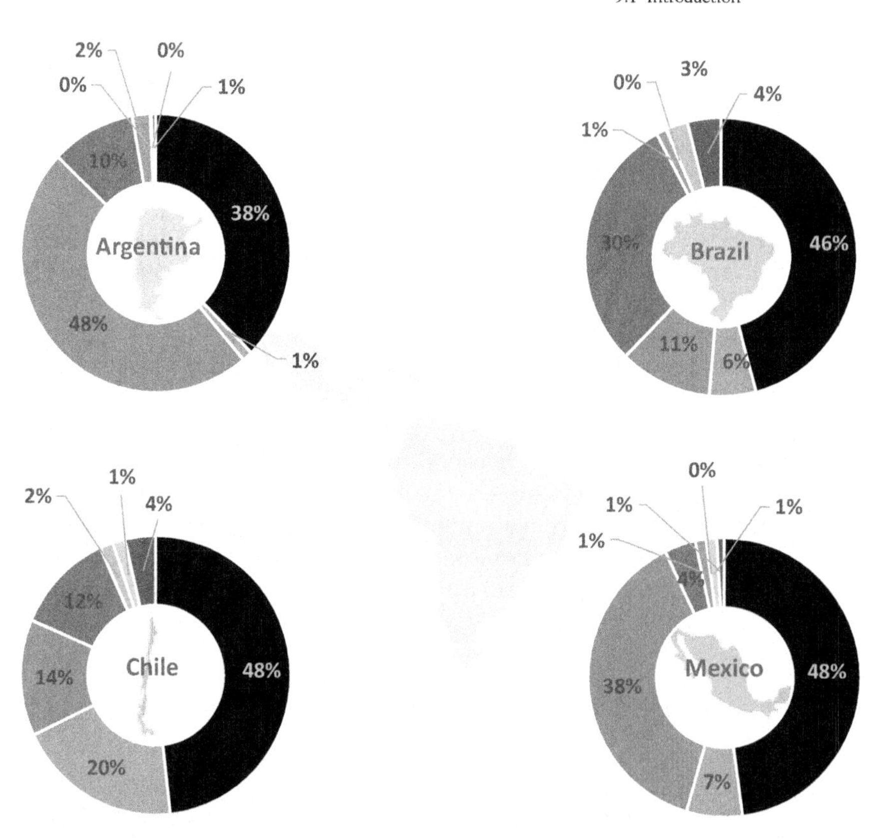

FIG. 9.3 Share of energy sources in Argentina, Brazil, Chile, and Mexico in 2016. Other renewables include geothermal, biomass, and waste energy. This figure was created by the author and was based on the Our World in Data (2020).

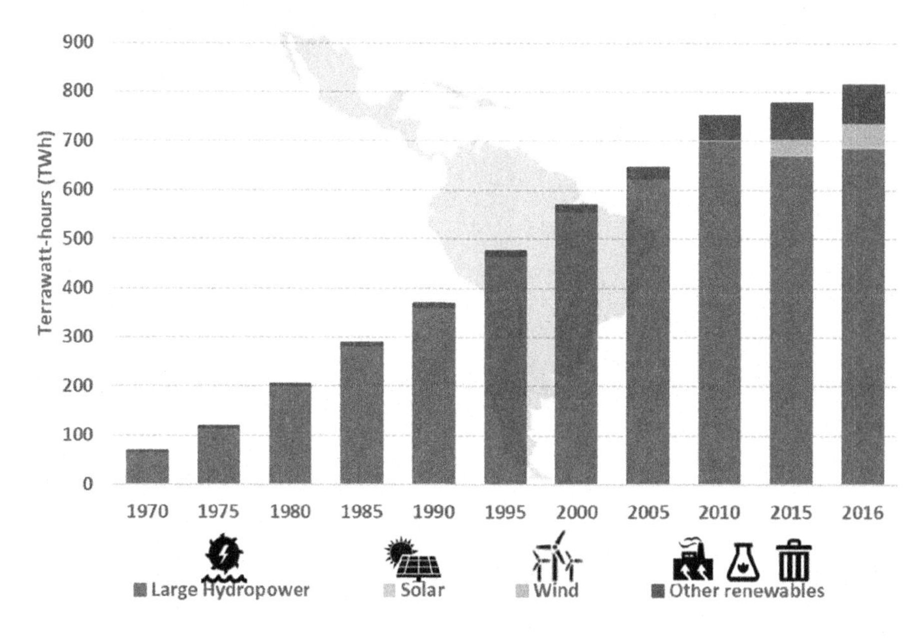

FIG. 9.4 Installed capacity of renewable energy in the LAC region energy matrix between 1970 and 2016. Other renewables include geothermal, biomass, and waste energy. The installed capacity is measured in terawatt-hours (TWh). This figure was created by the author and was based on the Our World in Data (2020).

process of the 'Washington Consensus', which is a combination of measures in order to promote 'macroeconomic adjustment', as well as by the 'Brady Plan', which was an external debt restructuring plan (Koengkan and Fuinhas, 2020a, b). This process of macroeconomic adjustment occurred between 1989 and 1992, where several countries from the region adopted this strategy such as Costa Rica and Mexico in 1989, Venezuela (RB) in 1990, Uruguay in 1991, and Argentina and Brazil in 1992 (Koengkan and Fuinhas, 2020a). According to Aizenman (2005) and Vásquez (1996), all these countries underwent an in-depth process of trade and financial liberalisation, in addition to the process of privatisation of significant portions of the public sector.

This process of liberalisation in the region intensified with the 'commodities boom' that occurred between 2004 and 2014. The region had an average growth rate of 7.40% (Koengkan and Fuinhas, 2020a). That is the commodities boom that occurred in the LAC region impacted the degree of economic openness, with the countries in this region creating a dependence on external demand (Carneiro, 2012). This process of liberalisation during 1989–92 and between 2004 and 2014, as mentioned before, exerted a positive impact on economic growth, and the LAC's GDP per capita growth (annual %) had an average annual growth rate of approximately 2.67% between 1989 and 2016 (see Fig. 9.5).

This was also the case with energy consumption, where the electric power consumption in kilowatt-hours (kWh) per capita in the region was 1175.97 kWh in 1990, and consumption had reached a value 2155.70 kWh in 2014 (see Fig. 9.6).

However, to meet energy demand, new investments in renewable energy technologies were necessary (Koengkan et al., 2019b). In addition, this same process of liberalisation also facilitated the introduction of new renewable energy technologies via trade liberalisation that consequently contributed to an increase in installed capacity and consumption of this kind of energy in the region (Koengkan et al., 2019c).

The relevance of the events described has inspired the main research question of this investigation, namely: Can energy transition decrease deaths from air pollution in Latin America and the Caribbean? The main objective of this investigation is to study the effect of the energy transition on deaths from air pollution in a group of 19 countries from the LAC region between 1995 and 2016. To this end, the panel vector autoregressive or panel (VAR) model that was developed by Holtz-Eakin et al. (1988) will be used as the methodological approach to identify the existence of this possible phenomenon.

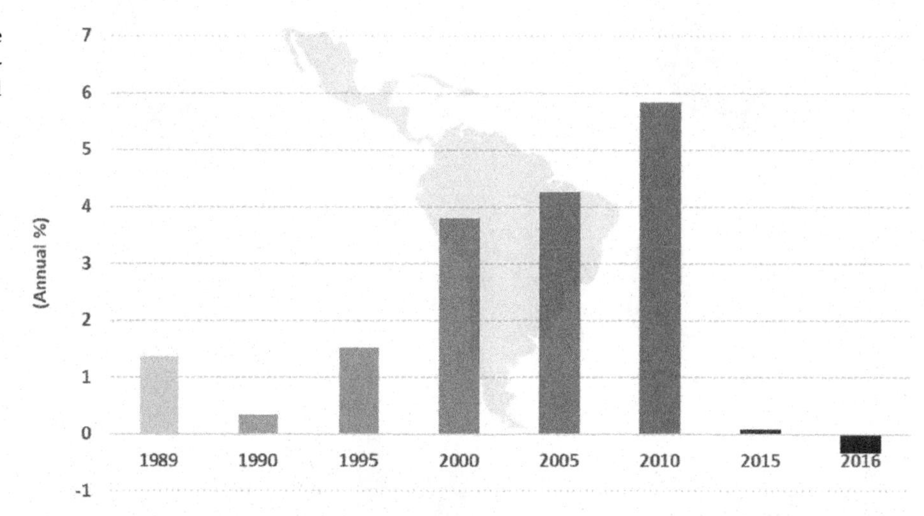

FIG. 9.5 GDP growth (annual %) in the LAC region between 1989 and 2016. This figure was created by the author and was based on the World Bank Open Data (2020).

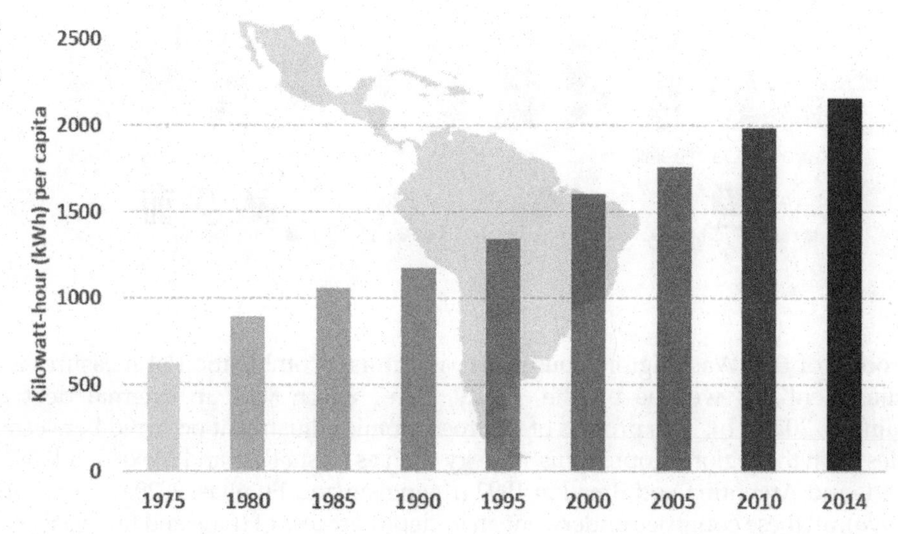

FIG. 9.6 Electric power consumption (kWh per capita) in the LAC region. This figure was created by the authors and was based on the World Bank Open Data (2020).

The current literature does not indicate that the process of the energy transition can decrease the deaths caused by air pollution, much less in the LAC region, because this topic of study has received little attention from scholars. The literature is focused on the reduction of CO_2 emissions or air pollution by renewable energy sources, where there are several studies, leaving aside the possible consequence of this reduction on deaths from stroke, heart disease, lower respiratory infections, diabetes, lung cancer, and COPD.

Therefore, as mentioned earlier, there are no studies that approach the effect of the renewable energy transition on deaths caused by air pollution directly; however, there do exist close studies, for example, that approach the effect of renewable energy consumption and energy transition on CO_2, which is one of the main components of air pollution (e.g. Koengkan and Fuinhas, 2020a; Charfeddine and Kahia, 2019; Koengkan and Fuinhas, 2018; Zoundi, 2017; Bilgili et al., 2016).

Koengkan and Fuinhas (2020a) investigated the effect of the ratio of renewable energy that includes renewables (e.g. biomass, biofuels, geothermal, solar, wind, hydropower, waste, and wave), which are a proxy for the renewable energy transition on CO_2 emissions in the LAC region between 1990 and 2014. The authors found that the process of energy transition mitigates emissions of CO_2 by −0.0675 and −0.031 in the short and long run. The explanation for this effect is related to technological efficiency of renewable energy caused by globalisation, which consequently produces more clean energy with fewer emissions, and, in addition, increases the share of this kind of energy sources in the energy mix in the LAC region.

Charfeddine and Kahia (2019) studied the impact of renewable energy that includes all renewable energy sources on CO_2 emissions in the Middle East and North Africa (MENA) region from 1980 to 2015. This partly complements the explanation provided by Koengkan and Fuinhas (2020a). The expansion and development of financial activities caused by financial liberalisation (that is a component of globalisation) boost the renewable energy sector via capital lending and provide equity financing and consequently increases the share of renewable energy sources in the total energy consumption.

Indeed, the idea that CO_2 emissions decrease with greater insertion of renewable energy sources in the energy matrix is also shared by Zoundi (2017). This author examined the effect of renewable energy including biomass, biofuels, geothermal, solar, wind, hydropower, waste, and wood, on CO_2 emissions for 25 African countries, over the period from 1980 to 2012. The insertion of renewable energy sources contributes to the improvement of air quality. Indeed, an increase of 1% of renewable energy consumption mitigates CO_2 emissions by 0.13%.

However, for the development and insertion of renewable energies to occur, resulting in a decrease in CO_2, adequate and consistent policies are necessary. Therefore, the evidence that the success of renewable energy in mitigating environmental degradation or CO_2 emissions is related to renewable energy policies was found by Koengkan and Fuinhas (2018), who investigated the impact of renewable energy consumption, including biomass, biofuels, geothermal, solar, wind, hydropower, waste, and wave, on CO_2 emissions in the South America region, over the period between 1980 and 2012. The authors found that an increase of 1% of renewable energy consumption mitigates emissions of CO_2 in the region by 0.0420%. Indeed, this result is possible due to the efficiency of renewable energy policies that encourage the development of green technologies, given the enormous biodiversity and the abundance of renewable sources.

Indeed, Bilgili et al. (2016) identified the capacity of renewable energy to decrease CO_2 emissions in 17 countries of the Organisation for Economic Co-operation and Development (OECD) over the period 1977–2010, agreeing with Koengkan and Fuinhas (2018). The increase of production and consumption of renewable energy sources is only possible with effective short-, mid-, and long-term policies.

Other researchers have explored the effect of renewable energy sources on the air pollution problem that is also caused by CO_2 emissions (e.g. Moorkens and Dauwe, 2019; Alvarez-Herranz et al., 2017; Zheng et al., 2015; Tsilingiridis et al., 2011). According to a report developed by Moorkens and Dauwe (2019), who approach the effect of renewable energy on air pollution in European Union countries, it was found that renewable energy sources (e.g. solar, wind, and hydropower energy sources) prevent air pollution, while the higher relative increase in the use of solid renewable fuels (e.g. biomass and waste) increases the air pollution implicitly.

Alvarez-Herranz et al. (2017) investigated the impact of renewable energy consumption (include all renewable energy sources) on air pollution levels in 17 OECD countries from 1990 to 2012 and found that the renewable energy consumption, regardless of the source, helps to improve air pollution levels. Indeed, according to the authors, the capacity of renewable energy sources to mitigate air pollution is related to the promotion of renewable sources by energy policies and public budget on energy innovation that reduces CO_2 emissions. This vision is also shared by Zheng et al. (2015), who studied the impact of clear energy production (e.g. solar, wind, hydropower, biomass, and waste) on air pollution control in 26 provinces and four centrally controlled municipalities in China over 10 years from 2002 to 2011. The authors found that clean energy production encouraged by environmental policies improved air pollution levels. This explanation is in keeping with Koengkan and Fuinhas (2018), Alvarez-Herranz et al. (2017), and Bilgili et al. (2016).

Therefore, if renewable energy sources can mitigate CO_2 emissions and air pollution, then the process of the energy transition can reduce deaths from air pollution by reducing health problems caused by air pollution as mentioned by Buonocore et al. (2015). Energy efficiency and renewable energy sources can benefit public health and the climate by displacing emissions from fossil-fuelled generation of electricity.

However, some authors disagree that some renewable energy sources can benefit public health and the climate. For example, Rinne et al. (2006) investigated the relationship of pulmonary function among women and children to indoor air pollution from biomass use in rural Ecuador. The following authors found that users of biomass fuels for domestic energy increase the exposure to indoor air pollution. Moreover, many respiratory diseases in women and children have been associated with biomass smoke. However, it is not clear what relationship exists between biomass use and pulmonary diseases. This evidence was also found by Smith and Mehta (2003), who examined the burden of diseases from indoor air pollution in the developing countries. The authors confirmed that household solid fuel use increases respiratory infections, chronic obstructive pulmonary disease, tuberculosis, asthma, lung cancer, ischaemic heart disease, and blindness, as well as deaths.

Moreover, burning waste as a solid household fuel for cooking or heating is also linked to various health problems and causes of death. According to Finkelman (2004), who investigated the impacts of burning coal beds and waste banks on health, the burning of coal beds and waste banks contributes to indoor and outdoor air pollution and several diseases. These findings are in agreement with Moorkens and Dauwe (2019), who indicate that an increase in the use of solid renewable fuels such as biomass and waste encourages an increase in air pollution. Indeed, the increase in health problems and deaths from biomass and waste is related to energy poverty, which is the use of firewood or charcoal and waste as fuel for cooking or heating due to the lack of access to electricity or gas (World Energy Council, 2006).

This chapter is innovative in the literature as (i) it investigates the effect of the energy transition on deaths from air pollution, an underexplored theme, opening a new topic of study in the literature related to energy transition and health; (ii) it uses the ratio of renewable energy as a proxy for energy transition; (iii) it utilises the panel-VAR model as the methodological approach, as this method fits this type of investigation due to its flexibility and the fact that is useful in the presence of little theoretical information about the relationship between the variables; (iv) it uses a macroeconomic approach to identify the possible effect of the energy transition on deaths from air pollution; and (v) it addresses the countries from the LAC region, bearing in mind that this region is not outlined in the literature in general.

Additionally, this chapter is essential and will contribute to the literature for several reasons such as: (i) the empirical results of this chapter can help policymakers develop more initiatives to accelerate the process of energy transition in the region to reduce the consumption of fossil fuels, and consequently CO_2 emissions and air pollution; (ii) the empirical finds of this chapter will open a new topic of study in the literature regarding the link between energy and health; and (iii) the results of this chapter can lead to the realisation of new research regarding the effect of energy transition on other diseases related to air pollution such as stroke, heart disease, lower respiratory infections, diabetes, lung cancer, and COPD.

Finally, the motivation that drives the realisation of this chapter is related to the necessity to extend or expand knowledge regarding the effect of the energy transition in other scientific areas such as health and provide evidence that the process of the energy transition goes far beyond of that of changing the energy matrix, where energy and health are inextricably linked. That is, this process of energy transition can save the lives of people from low- and middle-income countries, the case of the LAC region, who suffer from diseases caused by air pollution, and consequently can improve the life expectancy in these countries.

This chapter is organised as follows: Section 9.2 presents the methodology approach and data; Section 9.3 presents the results and a brief discussion; and Section 9.4 presents the conclusions and policy implications.

9.2 Methodological approach and data

This section will demonstrate the methodological approach and data that this chapter will use. Therefore, to help to answer the central question of this study, the panel-VAR model that was developed by Holtz-Eakin et al. (1988) and improved by Love and Zicchino (2006) will be used. This methodology is used in several research fields, being most commonly used by macroeconomists working with data for many countries and with a long period of time (Kroop and Korobilis, 2016). According to Antonakakis et al. (2017), Abrigo and Love (2015), and Canova and Ciccarelli (2009), this methodology is excellent at showing how shocks are transmitted across countries; it treats all variables as endogenous in the presence of restrictions; and it is useful in the presence of little theoretical information about the relationship between the variables. That is, this methodology is flexible and suitable for this chapter.

The specification of the equation for the first-order panel-VAR model, according to Love and Zicchino (2006), can be seen in the following equation is expressed as:

$$y_{it} = \Gamma_0 + \Gamma_1 y_{it-1} + f_i + d_{c,t} + u_t \tag{9.1}$$

where y_{it} is a vector of variables in the first differences, where the panel-VAR model requires that all variables must be I(0). Γ_0 is the vector of constant, $\Gamma_1 y_{it-1}$ in equation designates the matrix polynomial, the fixed effects in the model regression f_i, the effects of time are represented for $d_{c,t}$, and the term of random errors u_t.

However, since the fixed effects are correlated with the regressions due to delays of the dependent variable, the average differentiation procedure commonly used to eliminate the presence of fixed effects would create biased coefficients in the model. Therefore, to avoid this problem in the model regressions, this investigation will apply the technique called the 'Helmert procedure' developed by Arellano and Bover (1995).

Before the realisation of panel-VAR regression, some **preliminary tests** need to be applied, for example, as summarised in Table 9.1.

Moreover, after the panel-VAR regression, some **specification tests** also need to be applied for example as summarised in Table 9.2.

All Stata commands of these tests will be available in this chapter. As mentioned before, 19 countries from the LAC region (e.g. Argentina, **Bolivia, Brazil, Chile, Colombia, Costa Rica, Dominican Republic, Ecuador, El Salvador, Guatemala, Haiti, Jamaica, Mexico, Nicaragua, Panama, Paraguay, Peru, Uruguay**, and **Venezuela**) will be used in carrying out this investigation. Additionally, this study used annual data that were collected from 1995 to 2016. The variables which were chosen to perform this investigation are presented in Table 9.3.

TABLE 9.1 Preliminary tests.

Test	Goal
Variance inflation factor (VIF) test	To check the existence of multicollinearity between the variables in the panel data
Pesaran CD-test	To check the existence of cross-section dependence in the panel data
Panel unit root test (CIPS-test)	To check the presence of unit roots in the variables
Hausman test	To check the presence of heterogeneity i.e. whether the panel has random effects (RE) or fixed effects (FE)
Panel VAR lag-order selection	To report the overall model coefficients of determination

TABLE 9.2 Specification tests.

Test	Goal
The eigenvalue stability condition test	To indicate that the panel-VAR model is stable
The panel Granger causality Wald test	To analyse the causal relationship between the variables of the model
Forecast-error variance decomposition (FEVD) test	To show how a variable responds to shocks in specific variables
Impulse-response function (IRF)	To indicate the impulse-response function of variables of the model

TABLE 9.3 Description of variables, source, and summary statistic.

Description of variables			Summary statistic				
Variable	**Definition**	**Source**	**Obs**	**Mean**	**Std.-dev**	**Min**	**Max**
DLogDRAP	Death rates from air pollution (DRAP) measures the number of deaths per 100,000 population from both outdoor and indoor air pollution	Our World in Data (2020)	399	0.0233	1.7433	−4.6624	4.6576
DLogRREC	The ratio of renewable energy consumption (RREC) from biomass, hydropower, solar, photovoltaic, wind, wave, and waste in (kWh) per capita divided by the fossil fuel consumption from oil, gas and coal sources in (kWh) per capita. This variable is a proxy for energy transition	World Bank Open Data (2020)	399	−0.0188	0.1480	−1.1321	1.5385
DLogGDP_PC	Gross domestic production (GDP_PC) in constant local currency unity (LCU) and expressed per capita	World Bank Open Data (2020)	399	0.0199	0.0346	−0.1261	0.1506
DLogURBA	Urban population rate (URBA), which refers to people living in urban areas as defined by national statistical offices. This variable is a proxy for urbanisation	World Bank Open Data (2020)	399	0.0059	0.0052	−0.0035	0.0234
DLogITU	International tourism (ITU) measures the overnight visitors who travel to a country whose main purpose in visiting is not commercial. This variable is a proxy for tourism	Our World in Data (2020)	399	0.0506	0.1242	−0.7962	0.7847
DLogKOFEcGI	Economic globalisation index (KOFEcGI) that measure the de facto trade and financial globalisation. Trade globalisation is determined based on trade in goods and services, and financial globalisation includes foreign investment in various categories	KOF Globalization Index (2020)	399	0.0063	0.0498	−0.1983	0.2092

'DLog' denotes variables in the first differences of logarithms, 'Obs' denotes the number of observations in the model, 'Std.-dev' denotes the standard deviation, and 'Min and Max' denote minimum and maximum. The variable **GDP_PC** was transformed into per capita values with the total population of each cross. These summary statistics were obtained from the command *sum* in **Stata 15.0**. The board below shows how to transform the variables into per capita values, natural logarithms, and first differences of logarithms, as well as how to obtain the summary statistics of variables.

How to do:
****Transform the variable gdp_pc into per capita values****
gen gdp_pc =(gdp/population)
****Transform the variables drap, rrec, gdp_pc, urba, itu, and kofecgi into natural logarithms****
gen logdrap =log(drap)

Transform the variables drap, rrec, gdp_pc, urba, itu, and kofecgi into first differences of logarithms
gen dlogdrap =d.log(drap)
The summary statistics
sum dlogdrap dlogrrec dloggdp_pc dlogurba dlogitu dlogkofecgi

9.3 Results and discussion

As previously explained in the introduction, this section will present the empirical results of preliminary tests, the panel-VAR estimator, and post-estimation tests in addition to a brief discussion of results. Then, to identify the level of multicollinearity between the variables in the panel's data, the VIF test that was developed by Belsley et al. (1980) was calculated. This test is constructed around the following equation:

$$VIF_i = \frac{1}{1 - R_j^2} \tag{9.2}$$

where R_j^2 is the coefficient of determination of regression of model in step one. The results of VIF test indicate that the values are lower than the usually accepted benchmark of 10 in the case of the VIF values, and 6 in the case of the mean VIF values (see Table 9.A1 in Appendix). The results of VIF test were obtained from the command *estat vif* in **Stata 15.0**. The board below shows how to carry out and obtain the results from the VIF test.

How to do:
**** The Variance Inflation Factor test****
reg dlogdrap dlogrrec dloggdp_pc dlogurba dlogitu dlogkofecgi
estat vif

Moreover, to identify the presence of cross-sectional dependence (CSD) in the panel data, the Pesaran CD test developed by Pesaran (2004) was calculated. This test is constructed around the following equation:

$$CD = \sqrt{\frac{2T}{N(N-1)}}\left(\sum_{i=1}^{N-1}\sum_{j=i+1}^{N}\hat{P}ij\right) \tag{9.3}$$

The null hypothesis of this test is the nonpresence of cross-sectional dependence $CD \sim N(0,1)$ for $N \rightarrow \infty$ and T is sufficiently large. The results of CSD test indicate that variables such as **DLogITU**, **DLogURBA**, **DLogGDPPC**, and **DLogKOFEcGI** have the presence of cross-sectional dependence (see Table 9.A2 in Appendix). The results of Pesaran CD test were obtained from the command *xtcd* in **Stata 15.0**. The board below shows how to carry out and obtain the results from the Pesaran CD test.

How to do:
The Pesaran CD-test
xtcd dlogdrap dlogrrec dloggdp_pc dlogurba dlogitu dlogkofecgi

The presence of CSD in these variables can be an indication that the selected countries of this study share the same characteristics and shocks (Fuinhas et al., 2017). However, the nonpresence of cross-sectional dependence in the variable **DLogDRAP** can be related to different death rates from air pollution in the LAC region, making the identification of the same death rates for all countries in the region impossible. Regarding the variable **DLogRREC**, the nonpresence of cross-sectional dependence is due to each country of the LAC region having its own characteristics of generation of renewable energy sources (Fuinhas et al., 2017).

In the presence of CSD, it is necessary to verify the order of integration of the variables that will be used in the panel-VAR regression. To this end, the panel unit root (CIPS test) developed by Pesaran (2007) was calculated. This test is constructed around the following equation:

$$\text{CIPS}(N,T) = t - bar = N^{-1}\sum_{i=1}^{N} ti(N,T) \tag{9.4}$$

where $t_i(N,T)$ is the cross-sectionally augmented Dickey-Fuller statistic for i, the cross-section unit given by the t-ratio of the coefficient of $y_{i,\,t-1}$ in the CADF regression. Therefore, the null hypothesis of this test is that all series have a unit root. The results from the CIPS test obtained indicate that most of the variables are of order 1, which is a precondition for the use of a panel-VAR estimator, as mentioned before. Indeed, most of the variables appear to be somewhere between stationarity and integration of order 1 with and without trend, except the variable **DLogURBA** without trend (see Table 9.A3 in Appendix). The results of CIPS test were obtained from the command *multipurt* in **Stata 15.0**. The board below shows how to carry out and obtain the results from the CIPS test.

How to do:
****The CIPS-test****

```
multipurt dlogdrap dlogrrec dloggdp_pc dlogurba dlogitu dlogkofecgi,lags(1)
```

The nonstationarity of the variable urban population was expected, and this result also was found by Koengkan and Fuinhas (2020a), who used the same variable in their investigations. The next step of this investigation is to identify the presence of individual effects in the model. To this end, the Hausman test, which compares the random (RE) and fixed effects (FE), was calculated. This test is constructed around the following equation:

$$H=(\beta_{RE}-\beta_{FE})'\hat{\sum}-1(\beta_{RE}-\beta_{FE})\sim X^2(k) \tag{9.5}$$

where β_{RE} and β_{FE} are estimators of the parameter β. The null hypothesis of this test is that the difference in coefficients is not systematic, where the random effects are the most suitable estimator (Koengkan and Fuinhas, 2020a). The results of this test indicate that the null hypothesis should be rejected [**chi2 (5) = 25.96*****, statistically significant at the 1% level] (see Table 9.A4 in Appendix). In this case, the FE model is the most appropriate for the realisation of this analysis. The results of the Hausman test were obtained from the command *hausman* with option *sigmaless* in **Stata 15.0**. The board below shows how to carry out and obtain the results from the Hausman test.

How to do:
****The Hausman test****

```
xtreg dlogdrap dlogrrec dloggdp_pc dlogurba dlogitu dlogkofecgi,fe
estimates store fixed
xtreg dlogdrap dlogrrec dloggdp_pc dlogurba dlogitu dlogkofecgi,re
estimates store random
hausman fixed random, sigmaless
```

Next, the overall model coefficients of determination need to be reported. To this end, the panel-VAR lag-order selection test developed by Abrigo and Love (2015) was used. This test is constructed around the following equation:

$$\begin{aligned} MMSC_{BIC,n}(k,p,q)&=J_n(k^2p,k^2q)-(|q|-|p|)k^2\ \ln n\\ MMSC_{AIC,n}(k,p,q)&=J_n(k^2p,k^2q)-2k^2(|q|-|p|)\\ MMSC_{HQIC,n}(p,q)&=J_n(k^2p,k^2q)-Rk^2(|q|-|p|)\ \ln\ \ln n, R>2 \end{aligned} \tag{9.6}$$

where $J_n\ (k,p,q)$ is the J statistic of overidentifying restriction for a k-variate PVAR of order p and moment conditions based on q lags of the dependent variables with sample size n. By construction, the above MMSC is available only when $q>p$. As an alternative criterion, the overall coefficient of determination (CD) may be calculated even with just-identified GMM models. Suppose we denote the $(k\times k)$ unconstrained covariance matrix of the dependent variables by Ψ. CD captures the proportion of variation explained by the PVAR model as

$$CD=1-\frac{\det\left(\sum\right)}{\det(\Psi)} \tag{9.7}$$

The results of the panel-VAR lag-order selection test points to the use of 1 lag or 2 lags in the panel-VAR model (see Table 9.A5 in Appendix). This selection was based on the MAIC criterion, which is supported by Serena and Perron (2001). The results of the panel-VAR lag-order selection test were obtained from the command *pvarsoc* in **Stata 15.0**. The board below shows how to carry out and obtain the results from the PVAR Lag-order selection test.

TABLE 9.4 Panel-VAR model regression.

	1/6 instrumental lags			1/6 instrumental lags		
	1 Lag			2 Lags		
Response of DLogDRAP to:	**Coefficient**	**Heteroskedasticity adjusted *t*-statistics**		**Coefficient**	**Heteroskedasticity adjusted *t*-statistics**	
DLogRREC	3.8999	2.81	***	−3.7174	−2.29	**
DLogGDP_PC	−53.0628	−8.04	***	−58.4877	−6.22	***
DLogURBA	488.8492	4.90	***	855.3862	3.04	***
DLogITU	0.1161	0.09		3.2326	1.96	**
DLogKOFEcGI	−26.8585	−7.69	***	−1.9878	−0.53	
Test of overidentifying restriction: Hansen's *J*						
	Chi2(180) = 168.79923*			Chi2(144) = 137.33849*		

Note: Instruments: 1 (1/6) and 2 (1/6) were used.

How to do:
The PVAR Lag-order selection test

```
pvarsoc dlogdrap dlogrrec dloggdp_pc dlogurba dlogitu dlogkofecgi, maxlag (4) pvaropts (instl(1/6))
```

After the realisation of preliminary tests, it is necessary to carry out the panel-VAR regression. Table 9.4 presents the results of the first equation from the panel-VAR regression model. The results of this equation will answer the central question of this investigation. The lag length (1) and (2) indicated by the panel-VAR lag-order selection test was used in this estimation.

The results indicate the existence of endogeneity in the variables in the models that used 1/6 lags on the instruments and 1 and 2 lags included in the dependent variables list. Moreover, the variables in the model that used 1/6 lags on the instruments and 1 lag included points out that the variables are statistically significant at the 1% and 5% levels. The Hansen's J test of this model indicates the null hypothesis is not rejected, which states that the overidentification restrictions are not valid, thus making the specification valid. The model that used 1/6 lags on the instruments and 2 lags included also indicates that the variables are statistically significant at 1% and 5% levels and that the Hansen's J test of this model indicates the null hypothesis is not rejected. However, although these models indicate that both specifications are valid, this investigation opted to use the model with 2 lags included in the dependent variables list due to the model having a loss of observations much smaller (144) than (180) observations from the model with 1 lag.

Moreover, the model with 2 lags presented better results in the panel Granger causality Wald test (see Table 9.5) than the model with 1 lag. The results are obtained from the command *pvar* in Stata with the option *gmmst overid,* in **Stata 15.0**. The board below shows how to carry out and obtain the results from the PVAR model regressions.

How to do:
Panel-VAR model regression

```
pvar dlogdrap dlogrrec dloggdp_pc dlogurba dlogitu dlogkofecgi, lags(1) instl (1/6) gmmst overid
pvar dlogdrap dlogrrec dloggdp_pc dlogurba dlogitu dlogkofecgi, lags(2) instl (1/6) gmmst overid
```

After the realisation of panel-VAR regression, some specification tests need to be applied to verify the robustness of the model that was chosen. The first test that needs to be computed is the eigenvalue stability condition. The results of this test obtained from the command *pvarstable* in Stata indicate that the panel-VAR model is stable (see Table 9.A6 in Appendix), because all eigenvalues are inside the unit circle, satisfying the stability condition of the test (Lutkephol, 2005; Hamilton, 1994).

The next specification test that needs to be computed is the panel Granger causality Wald test. This test is important to analyse the causal relationship between the variables of the model. To find this causal relationship between the variables, it is necessary to use all equations of panel-VAR regression. Table 9.5 presents the outputs of the panel Granger causality Wald test.

TABLE 9.5 Panel Granger causality Wald test.

Equation	Excluded	Chi2	Df	Prob > chi2	
DLogDRAP	DLogRREC	5.284	2	0.071	*
	DLogGDP_PC	40.658	2	0.000	***
	DLogURBA	14.694	2	0.001	***
	DLogITU	3.841	2	0.147	
	DLogKOFEcGI	18.323	2	0.000	***
	All	54.772	10	0.000	***
DLogRREC	DLogDRAP	2.337	2	0.311	
	DLogGDP_PC	17.386	2	0.000	***
	DLogURBA	19.487	2	0.000	***
	DLogITU	0.517	2	0.772	
	DLogKOFEcGI	22.839	2	0.000	***
	All	57.485	10	0.000	***
DLogGDP_PC	DLogDRAP	6.692	2	0.035	**
	DLogRREC	7.867	2	0.020	**
	DLogURBA	5.157	2	0.000	***
	DLogITU	5.157	2	0.076	*
	DLogKOFEcGI	42.359	2	0.000	***
	All	71.772	10	0.000	***
DLogURBA	DLogDRAP	5.343	2	0.069	*
	DLogRREC	14.936	2	0.000	***
	DLogGDP_PC	19.048	2	0.000	***
	DLogTU	4.605	2	0.100	
	DLogKOFEcGI	50.024	2	0.000	***
	All	70.297	10	0.000	***
DLogITU	DLogDRAP	1.599	2	0.449	
	DLogRREC	5.157	2	0.076	*
	DLogGDP_PC	48.747	2	0.000	***
	DLogURBA	7.945	2	0.019	**
	DLogKOFEcGI	8.113	2	0.017	**
	All	60.818	10	0.000	***
DLogKOFEcGI	DLogDRAP	4.462	2	0.107	
	DLogRREC	3.780	2	0.151	
	DLoGDP_PC	20.102	2	0.000	***
	DLogURBA	14.420	2	0.000	***
	DLogITU	2.918	2	0.232	
	All	53.149	10	0.000	***

Notes: ***, **, and * denote statistical significance level at the 1%, 5%, and 10% levels, respectively.

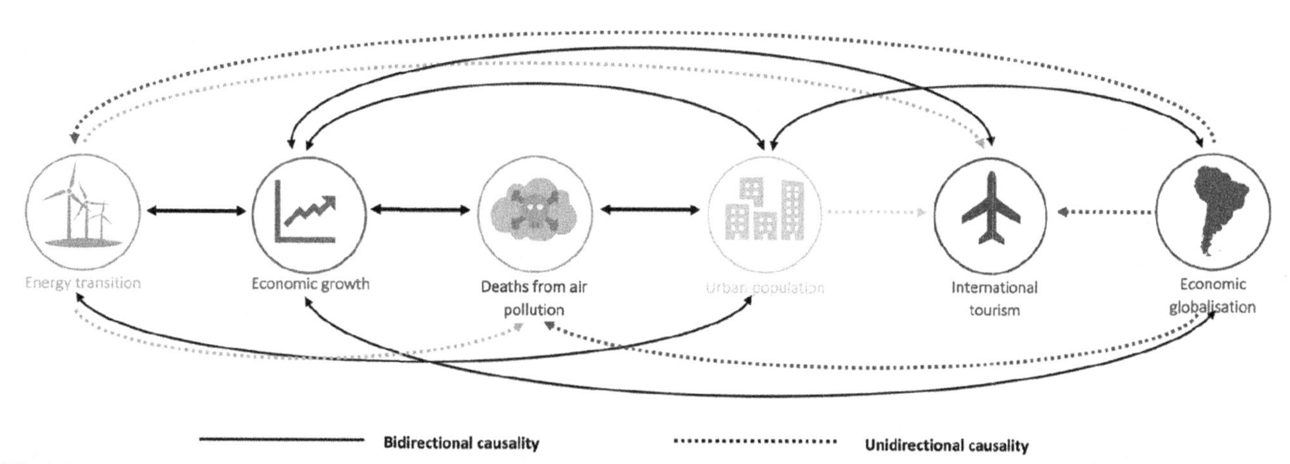

FIG. 9.7 Summary of causality of the variables. This figure was created by the authors.

The results of this test obtained from the command *pvargranger* in Stata indicate that there is a unidirectional and bidirectional causality among the variables of the model, with a statistic significance at 1%, 5%, and 10% levels. Fig. 9.7 summarises the causalities between the variables. This figure was based on results from the panel Granger causality Wald test (see Table 9.5).

After the realisation of the Granger causality Wald test, it is necessary to calculate the forecast error variance decomposition. This test represents how a variable responds to shocks in specific variables (e.g. Koengkan et al., 2020; Koengkan et al., 2019d). This test is constructed around the following equation:

$$Y_{it+h} - E(Y_{it+h}) = \sum_{i=0}^{h-1} e_{i(t+h-i)}\Phi_i \tag{9.8}$$

where Y_{it-h} is the observed vector at time $t+h$ and $E(Y_{it+h})$ is the h-step ahead predicted vector made at time t.Then, the results of this test obtained from the command *pvarfevd* in Stata are consistent with the Hausman test and the impulse-response functions. However, when we look at the outcomes of the first equation from the FEDV test in detail, we come across exciting situations. Two periods after a shock on the variable **DLogDRAP** explain 90% of forecast error variance; the variable **DLogRREC** 10 periods after a shock explains 2%. The variable **DLogGDP_PC** explains 29%, **DLogURBA** explains 11%, and **DLogITU** explains 13%, while the variable **DLogKOFEcGI** two periods after a shock explains 7% (see Table 9.A7 in Appendix).

This investigation computed the impulse-response function (IRF) to finalise the realisation of the specification tests that were mentioned before. The results of this test, obtained from the command *pvarirf* in Stata, indicate that in the long run, all variables converge to equilibrium supporting that the variables are stationary (see Fig. 9.A1 in Appendix). Moreover, the impulse-response functions are in concordance with the FEDV test (see Table 9.A7 in Appendix).

The panel-VAR regression model using 1/6 lags on the instruments and 2 lags included in the dependent variable list indicates that the ratio of renewable energy consumption (**DLogRREC**), which is a proxy for energy transition and economic growth (**DLogGDP_PC**), decreases deaths from air pollution (**DLogDRAP**). In contrast, the urban population (**DLogURBA**), which is a proxy for urbanisation and international tourism (**DLogITU**), increases them.

Therefore, the capacity of energy transition to decrease deaths from air pollution is related to the capacity of renewable energy consumption in decreasing air pollution in the LAC region and of its components such as CO_2 emissions and other gases. This capacity is related to the rapid expansion of renewable energy technology in the LAC region that began in the 1970s and was intensified between 1990 and 2012, influenced by economic development and liberalisation. This evidence was confirmed in Chapter 6.

Buonocore et al. (2015) further added that energy efficiency and renewable energy can improve the climate and consequently public health by reducing emissions from burning fossil fuels. The capacity of renewable energy to reduce the consumption of fossil fuels and consequently CO_2 emissions are described in Chapters 7 and 8. Indeed, the evidence that energy transition mitigates emissions of CO_2 and other gases in the LAC region was found by several authors (e.g. Koengkan and Fuinhas, 2018, 2020a; Dong et al., 2018; Fuinhas et al., 2017).

Moreover, the efficiency of renewable energy policies may also be related to the capacity of the energy transition to reduce deaths from air pollution. These policies will encourage the development and consumption of clean energy sources and consequently will decrease air pollution (e.g. Koengkan and Fuinhas, 2018; Alvarez-Herranz et al., 2017; Bilgili et al., 2016).

The capacity of economic growth to decrease these deaths could be related to the existence of better health systems in the LAC countries, which were influenced by economic growth. Indeed, this improvement in the health system reduces the mortality rate from air pollution in the region. The capacity of economic growth to improve the health system in the LAC region is related to macroeconomic adjustments which occurred between 1989 and 1992, where several countries in the region adopted this strategy (Koengkan and Fuinhas, 2020a).

The health sector was also impacted by these macroeconomic reforms. At the end of the 1980s, the health sector was included in many cases of social and health sector reforms to create macroeconomic stability and alleviate the poverty caused by economic crisis (Atun et al., 2015). Indeed, many countries in the LAC region (e.g. Argentina, Brazil, Chile, Colombia, Mexico, Peru, and Uruguay) took advantage of the good performance of economic growth after the macroeconomic reforms, and implemented conditional cash transfer schemes to expand access to health and education and reduce poverty and malnutrition (Koengkan and Fuinhas, 2020a; Atun et al., 2015).

Moreover, this process of improvement of the health system in the LAC region is accelerated in the period of rapid economic expansion that occurred between 2004 and 2014 with the 'commodities boom'. This created a period of sustained economic growth and enabled the implementation of policies with supply side to strengthen the health systems and expand access to the most vulnerable populations, as well as introducing universal health coverage (Koengkan and Fuinhas, 2020a; Atun et al., 2015). That is, the process of economic growth in the LAC region from 1989 to 2014 provided a fiscal space and budgetary flexibility to introduce improvements in the health system.

Furthermore, the period of accelerated economic expansion in the region also reduced energy poverty. The reduction of this problem is visible in access to electricity in rural and urban areas; in 1992, 63.4% of the rural population had access to electricity, while in 2016 this value reached 93.6% of the rural population. In urban areas, 97.5% of the urban population had access to electricity, and this value reached 99.5% in 2016 (World Bank Open Data, 2020). Regarding access to clean fuels and technologies for cooking, 78.1% of the population in the LAC region in 2000 had access to clean fuel technologies for cooking, with this value reaching 87% of the population in the region in 2016 (World Bank Open Data, 2020). That is, it is evident that the process of economic development in the region reduced the problem of energy poverty that is responsible for indoor air pollution deaths caused by burning solid fuel sources—such as firewood, crop waste, and dung—for cooking and heating.

The capacity of urbanisation in the LAC region positively influences deaths from air pollution, according to Romieu et al. (1990), and is due to the rapid process of urbanisation that occurred during the 20th century. Indeed, the LAC region is the most urbanised area in the less developed world. The total urban population increased by more than 400%, from 68 million inhabitants in 1950 to 466 million by 2000. The region has four large urban agglomerations with more than 2 million inhabitants, such as Mexico City (Mexico), Sao Paulo (Brazil), Bueno Aires (Argentina), and Rio de Janeiro, (Brazil). This rapid growth in urbanisation in the region is linked with economic development in the last 30 years caused by various economic reforms. Indeed, this process of development caused the introduction of new agricultural technologies and the industrialisation process, which led to a restructuring of rural economies in most Latin American countries (Koengkan et al., 2019e).

Therefore, it is natural that urbanisation causes deaths from air pollution, where economic growth causes urbanisation and consumption of energy from fossil fuels and consequently increases emissions of CO_2 and other gases. The evidence that economic growth causes urbanisation and consequently, the consumption of fossil fuels and emissions of CO_2 in the LAC region was found by Koengkan et al. (2019e).

Moreover, the positive influence of international tourism on deaths from air pollution is due to tourism has a positive on economic growth and consequently causes a negative effect on the environment in the form of air pollution (Paramati et al., 2016). According to the same authors, tourism activities require energy for direct use for scenic flights, jet boating, or air travel or indirectly for hotels, events, museums, or experience centres. Most of this consumption of energy is often generated from oil, coal, and natural gas. Paramati et al. (2016) add that the tourism sector is responsible for 5% of CO_2 emissions in the world, with most of these emissions coming from transportations, accommodation, and other tourism activities.

In the LAC region, the flow of inbound tourism has grown in recent decades; the region registered 15.3 million visitors in 2000, and in 2012 it received 26.9 million tourists, 4.2% more than in 2011 (WTM, 2020). Indeed, this growth in the tourism sector was driven by rapid economic growth in the 1980s–1990s after the economic reforms of liberalisation that facilitated the entrance of visitors, capital in flows, goods and services, and by global economic expansion in 2000–08 which helped to create a period of sustained economic growth and stability political in the region. Moreover, this process of expansion of the tourism section impacted the economic growth of the region (e.g. Andreu-Boussut and Salin, 2015; Wilson, 2008; Eugenio-Martin et al., 2004) and consequently positively affected the consumption of energy and the environment (e.g. Wilson, 2008).

9.4 Conclusions and policy implications

The main objective of this investigation was to assess the impact of energy transition on deaths from air pollution in 19 LAC countries in the period 1995–2016. Moreover, this investigation opted to use panel-VAR as the methodology. The results from preliminary tests indicated the presence of low-multicollinearity, cross-sectional dependence, stationarity in the variables, fixed effects in the model, and the need to use the lag length (1) or (2) in the regressions.

The results of the panel-VAR model estimates suggest that the energy transition and economic growth decrease deaths from air pollution, while urbanisation and international tourism increase them. The results from post-estimation tests indicated the existence of unidirectional and bidirectional causality among the variables of the model, that the model estimation is stable, and that two periods after a shock, the variables themselves explained almost all the forecast error variance. The impulse-response functions of all variables converge to equilibrium, supporting that the variables of the model are stationary.

Therefore, the capacity of energy transition to decrease these deaths is related to the rapid growth of investment in renewable energy technologies that occurred between 1990 and 2012, and the consumption of energy from these technologies and by the efficiency of renewable energy policies that consequently decrease the air pollution in the LAC region and its components such as emissions of CO_2 and other gases. This evidence demonstrates that the LAC region is on the right path in the process of energy transition. However, more initiatives are necessary for the development of renewable energy technologies, such as encouraging the public and private banks to support investments in renewable energy technologies or purchase technologies with higher energy efficiency. Policies should also be created that facilitate access to renewable energy by families, farms, and industries. All these will reduce the consumption of nonrenewable energy and consequently air pollution and deaths.

Moreover, the capacity for economic growth to decrease the deaths from air pollution is related to the existence of better health systems in the LAC countries caused by rapid economic growth in the last 30 years. This rapid economic development in the region provided a fiscal space and budgetary flexibility in most countries of the region (e.g. Argentina, Brazil, Chile, Colombia, Mexico, Peru, and Uruguay) to introduce health-system improvements. This evidence demonstrated that the countries from the LAC region took advantage of the economic boom that occurred in the region and correctly implemented social policies, welfare reforms, and cash transfer schemes to reduce poverty and expand access to health.

Moreover, the period of an economic boom in the LAC region also reduced the energy poverty related to the use of firewood or charcoal as fuel for cooking/or heating due to the lack of access to electricity or gas. However, although energy poverty has been decreased in the LAC region, it is necessary to create more policies, such as price support measures as an essential complement (e.g. social tariff, tax reduction, and others) to increase access for the urban and rural population to clean fuels and technologies for cooking or heating, as 13% of the population in the region does not have access to these technologies.

The capacity of urbanisation to increase deaths is due to the process of urbanisation caused by economic growth that consequently increases the consumption of energy from nonrenewable sources and air pollution. Indeed, that is an indicator that the process of urbanisation caused by the process of economic development in the LAC still causes several impacts on the environment and health. Therefore, in the case of the LAC region particularly, it is necessary to create policies that reduce the consumption of nonrenewable energy sources in urban cities, where individual transportation can be reduced by the introduction of a better urban public transportation network. Moreover, it is necessary to create fiscal policies to support green construction and encourage private firms to engage and expand the scale of this type of construction.

Additionally, the increase of deaths from air pollution by international tourism in the LAC region is due to the tourism sector requiring energy for direct use for scenic flights, jet boating, or air travel or indirectly for hotels, events, museums, or experience centres. Indeed, most of this consumption of energy is often generated from oil, coal, and natural gas that consequently increases air pollution. The tourism sector in the LAC region is based on the consumption of fossil fuels for the realisation of their activities. In this case, it is necessary to create more policies that encourage the consumption of renewable energy sources and the acquisition of green technologies by hotels, museums, or experience centres, as well as more policies that encourage the development of green technologies that replace the dirty technologies from aircraft, jets, and boats, and also other types of transport in tourism activities.

Appendix

TABLE 9.A1 VIF-test.

Variables	VIF	1/VIF
DLogRREC	1.01	0.9949
DLogGDP_PC	1.18	0.8509
DLogURBA	1.02	0.9827
DLogITU	1.20	0.8323
DLogKOFEcGI	1.04	0.9642
Mean VIF	1.09	

TABLE 9.A2 Pesaran CD-test.

Variables	CD-test	*P*-value		Corr	Abs (corr)
DLogDRAP	−1.15	0.252		−0.019	0.206
DLogRREC	−0.39	0.696		−0.007	0.186
DLogGDP_PC	22.58	0.000	***	0.377	0.393
DLogURBA	17.74	0.000	***	0.296	0.570
DLogITU	9.41	0.000	***	0.157	0.214
DLogKOFEcGI	10.61	0.000	***	0.177	0.252

Notes: *** denotes statistical significance at the 1% level.

TABLE 9.A3 Panel unit root test (CIPS-test).

	Panel unit root test (CIPS) (Zt-bar)				
		Without trend		With trend	
Variables	Lags	Zt-bar		Zt-bar	
DLogDRAP	1	−13.919	***	−11.559	***
DLogRREC	1	−6.670	***	−4.322	***
DLogGDP_PC	1	−3.158	***	−2.557	***
DLogURBA	1	1.799		−2.355	***
DLogITU	1	−2.955	***	−2.484	***
DLogKOFEcGI	1	−5.587	***	−2.807	***

Notes: *** denotes statistically significant at the 1% level.

TABLE 9.A4 Hausman test.

Variables	(b) Fixed	(B) Random	(b-B) Difference	Sqrt(diag(V_b-V-B)) S.E.
DLogRREC	0.2159	0.0145	0.2013	0.2777
DLogGDP_PC	0.5938	4.3719	−3.7781	0.9130
DLogURBA	−50.7604	−4.3007	−46.4596	32.6903
DLogITU	0.2284	−0.3408	0.5693	0.1667
DLogKOFEcGI	−0.5681	−0.3408	−0.7558	0.5248
Chi2 (5)	25.96***			

Notes: *** denotes statistically significant at the 1% level.

TABLE 9.A5 PVAR Lag-order selection test.

Lag	CD	J	Jp-value	MBIC	MAIC	MQIC
1	**0.9050**	**162.443**	**0.8216**	**−842.5863**	**−197.557**	**−456.6904**
2	**0.9684**	**121.1267**	**0.9172**	**−682.8968**	**−166.8733**	**−374.1801**
3	0.9929	99.6329	0.7050	−503.3846	−116.367	−271.8471
4	0.9910	58.7433	0.8695	−343.2684	−85.2566	−188.91

Note: The overall coefficient of determination (CD), Hansen's *J* statistic (*J*), *P*-value (*JP*-value), MMSC-Bayesian information criterion (MBIC), MMSC-Akaike information criterion (MAIC), and MMSC-Hannan and Quinn information criterion (MQIC) were computed.

TABLE 9.A6 Eigenvalue stability condition test.

Eigenvalue			Graph
Real	Imaginary	Modulus	
−0.8560	0.0000	0.8560	
0.8011	0.0000	0.8011	
0.5400	0.0000	0.5400	
−0.2463	0.4561	0.5184	
−0.2463	−0.4561	0.5184	
0.3966	−0.2461	0.4668	
0.3966	0.2461	0.4668	
−0.0406	0.3808	0.3830	
−0.0406	−0.3808	0.3830	
−0.2031	−0.1081	0.2301	
−0.2031	0.1081	0.2301	
0.1162	0.0000	0.1162	

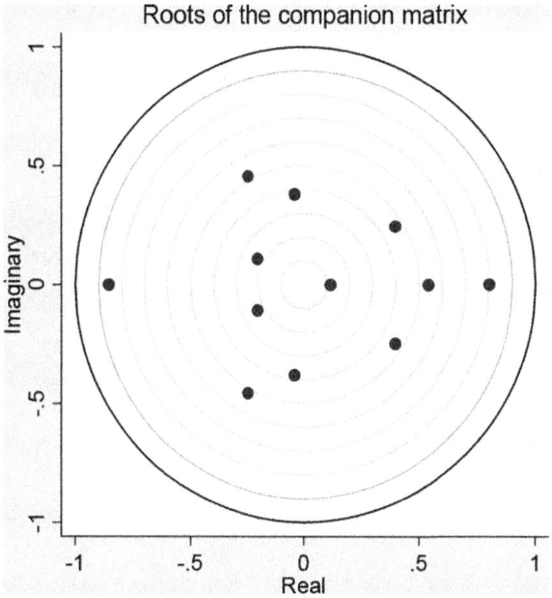

Roots of the companion matrix
Imaginary
Real

TABLE 9.A7 Forecast-error variance decomposition (FEVD).

Response variable and Forecast horizon	Impulse variables					
	DLogDRAP	DLogRREC	DLogGDP_PC	DLogURBA	DLogITU	DLogKOFEcGI
Dlogdrap						
1	1	0	0	0	0	0
2	0.8934	1.44e-06	0.0071	0.0227	0.0019	0.0746
5	0.5451	0.0179	0.2670	0.0999	0.0104	0.0594
10	0.5095	0.0201	0.2879	0.1145	0.0126	0.0553

Continued

TABLE 9.A7 Forecast-error variance decomposition (FEVD)—cont'd

Response variable and Forecast horizon	Impulse variables					
	DLogDRAP	**DLogRREC**	**DLogGDP_PC**	**DLogURBA**	**DLogITU**	**DLogKOFEcGI**
Dlogrrec						
1	0.0285	0.9714	0	0	0	0
2	0.0491	0.7217	0.0470	0.1381	0.0019	0.0420
5	0.0775	0.5856	0.1010	0.1417	0.0048	0.0892
10	0.0945	0.5518	0.1173	0.1441	0.0057	0.0863
Dloggdp_pc						
1	0.0203	0.0011	0.9018	0.0000	0.0767	0
2	0.0208	0.0034	0.8143	0.0072	0.1134	0.0406
5	0.0503	0.0092	0.7511	0.0401	0.1088	0.0401
10	0.0613	0.0104	0.7251	0.0568	0.1041	0.0420
Dlogurba						
1	0.0063	0.0073	0	0.9779	0.0083	0
2	0.0047	0.0058	0.0000	0.9309	0.0059	0.0523
5	0.0058	0.0145	0.0134	0.8708	0.0061	0.0891
10	0.0063	0.0145	0.0145	0.8584	0.0069	0.0991
Dlogitu						
1	0.0588	0.0147	0.9263	0	0	0
2	0.0479	0.0136	0.1262	0.0650	0.7427	0.0043
5	0.0606	0.0183	0.1364	0.0752	0.6860	0.0230
10	0.0658	0.0186	0.1406	0.0785	0.6723	0.0239
Dlogkofecgi						
1	0.0118	0.0189	0.0012	0.0094	0.0585	0.8999
2	0.0230	0.0302	0.0518	0.0296	0.0528	0.8123
5	0.0978	0.0302	0.1013	0.0515	0.0481	0.6708
10	0.1207	0.0305	0.1276	0.0665	0.0457	0.6087

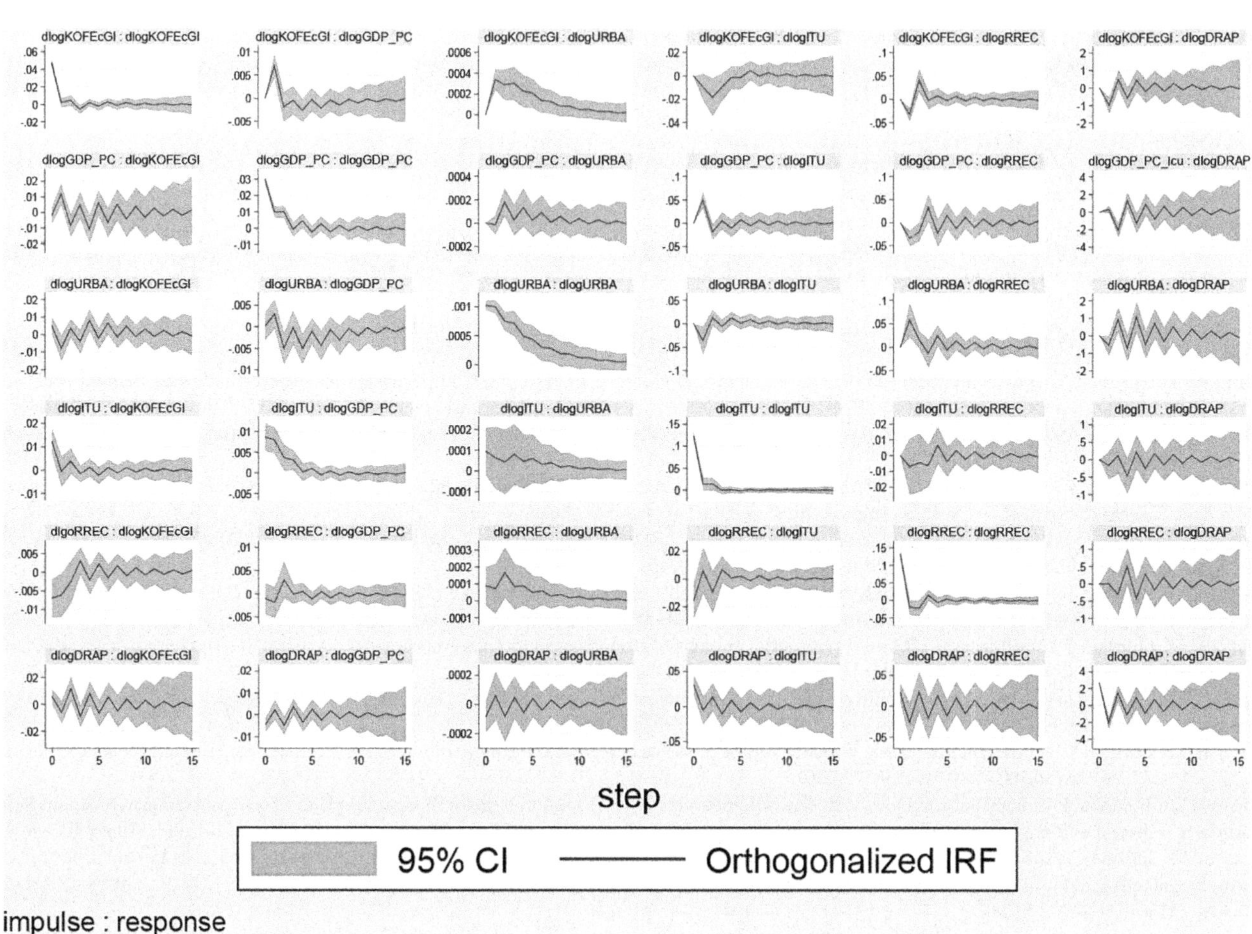

FIG. 9.A1 Impulse-response functions.

References

Abrigo, M.R.M., Love, I., 2015. Estimation of Panel Vector Autoregression in Stata: A Package of Programs. The University of Hawai'i at Mānoa Department of Economics. http://www.economics.hawaii.edu/research/working.papers/WP 16-02.pdf.

Aizenman, J., 2005. Financial Liberalizations in Latin America in the 1990s: A Reassessment. NBER Working Paper Series, 11145, pp. 1–30. https://www.nber.org/papers/w11145.

Alvarez-Herranz, A., Balsalobre-Lorente, D., Shahbaz, M., Cantos, J.M., 2017. Energy innovation and renewable energy consumption in the correction of air pollution levels. Energy Policy 105, 386–397. https://doi.org/10.1016/j.enpol.2017.03.009.

Andreu-Boussut, V., Salin, E., 2015. O turismo nas Américas: territórios, experiências e novosdesafios? IdeAs. pp. 1–9. http://journals.openedition.org/ideas/5691.

Antonakakis, N., Chatziantoniou, L., Filis, G., 2017. Energy consumption, CO_2 emissions, and economic growth: an ethical dilemma. Renew. Sust. Energy Rev. 68 (1), 808–824. https://doi.org/10.1016/j.rser.2016.09.105.

Arellano, M., Bover, O., 1995. Another look at the instrumental variable estimation of error components models. J. Econ. 68, 29–51. https://doi.org/10.1016/0304-4076(94)01642-D.

Atun, R., Andrade, L.O.M., Almeida, G., Cotlear, D., Dmytraczenko, T., Frenz, P., Garcia, P., Dantés, O.G., Knaul, F.M., Muntaner, C., Paula, J.B., Rígoli, F., Serrate, C.-F., Wagstaff, A., 2015. Health-system reform and universal health coverage in Latin America. Lancet 385 (9974), 1230–1247. https://doi.org/10.1016/S0140-6736(14)61646-9.

Belsley, D.A., Kuh, E., Welsch, E.R., 1980. Regression Diagnostics: Identifying Influential Data and Sources of Collinearity. Wiley, New York, https://doi.org/10.1002/0471725153.

Bilgili, F., Koçak, E., Bulut, Ü., 2016. The dynamic impact of renewable energy consumption on CO_2 emissions: a revisited environmental Kuznets curve approach. Renew. Sust. Energy Rev. 54, 838–845. https://doi.org/10.1016/j.rser.2015.10.080.

Buonocore, J.J., Luckow, P., Norris, G., Spengler, J.D., Biewald, B., Fisher, J., Levy, J.I., 2015. Health and climate benefits of different energy-efficiency and renewable energy choices. Nat. Clim. Change 6 (1), 100–105. https://doi.org/10.1038/nclimate2771.

Canova, F., Ciccarelli, M., 2009. Estimating multi-country VAR models. Int. Econ. Rev. 50, 929–959. https://doi.org/10.1111/j.1468-2354.2009.00554.x.

Carneiro, R.M., 2012. Commodities, choques externos e crescimento: reflexões sobre a América Latina. 117 CEPAL, pp. 1–47. ISSN: 1680-8843 http://www.eco.unicamp.br/cecon/images/arquivos/observatorio/Commodities_choques_externos_crescimento.pdf.

Charfeddine, L., Kahia, M., 2019. Impact of renewable energy consumption and financial development on CO_2 emissions and economic growth in the MENA region: a panel vector autoregressive (PVAR) analysis. Renew. Energy 139, 198–213. https://doi.org/10.1016/j.renene.2019.01.010.
Dong, K., Hochman, G., Zhang, Y., Sun, R., Li, H., Liao, H., 2018. CO_2 emissions, economic and population growth, and renewable energy: empirical evidence across regions. Energy Econ. 75, 180–192. https://doi.org/10.1016/j.eneco.2018.08.017.
Eugenio-Martin, J.L., Martín, M.N., Scarpa, R., 2004. Tourism and Economic Growth in Latin American Countries: A Panel Data Approach. FEEM Working Paper No. 26, https://doi.org/10.2139/ssrn.504482.
Finkelman, R.B., 2004. Potential health impacts of burning coal beds and waste banks. Int. J. Coal Geol. 59 (1–2), 19–24. https://doi.org/10.1016/j.coal.2003.11.002.
Fuinhas, J.A., Marques, A.C., Koengkan, M., 2017. Are renewable energy policies upsetting carbon dioxide emissions? The case of Latin America countries. Environ. Sci. Pollut. Res. 24 (17), 15044–15054. https://doi.org/10.1007/s11356-017-9109-z.
Hamilton, J., 1994. Time Series Analysis. Prentice-Hall, New Jersey, pp. 837–900.
Holtz-Eakin, D., Newey, W., Rosen, H., 1988. Estimating vector autoregressions with panel data. Econometrica, 1371–1395. doi: 0012-9682(198811) 56:6<1371:EVAWPD>2.0.CO;2-V.
Institute for Health Metrics and Evaluation (IHME), 2018. Global, Regional, and National Comparative Risk Assessment of 84 Behavioural, Environmental and Occupational, and Metabolic Risks or Clusters of Risks for 195 Countries and Territories, 1990–2017: A Systematic Analysis for the Global Burden of Disease Study 2017. http://www.healthdata.org/research-article/global-regional-and-national-comparative-risk-assessment-84-behavioral-0.
Koengkan, M., Fuinhas, J.A., 2018. The impact of renewable energy consumption on carbon dioxide emissions – the case of South American countries. Rev. Bras. Energ. Renováveis 7 (2), 1–21. https://doi.org/10.5380/rber.v7i2.58266.
Koengkan, M., Fuinhas, J.A., 2020a. Exploring the effect of the renewable energy transition on CO_2 emissions of Latin American & Caribbean countries. Int. J. Sustain. Energy, 1–24. https://doi.org/10.1080/14786451.2020.1731511.
Koengkan, M., Fuinhas, J.A., 2020b. The interactions between renewable energy consumption and economic growth in the Mercosur countries. Int. J. Sustain. Energy 39 (6), 594–614. https://doi.org/10.1080/14786451.2020.1732978.
Koengkan, M., Fuinhas, J.A., Santiago, R., 2019a. Asymmetric impacts of globalisation on CO_2 emissions of countries in Latin America and the Caribbean. Environ. Syst. Decis., 1–13. https://doi.org/10.1007/s10669-019-09752-0.
Koengkan, M., Fuinhas, J.A., Vieira, I., 2019b. Effects of financial openness on renewable energy investments expansion in Latin American countries. J. Sustain. Finance Invest., 1–19. https://doi.org/10.1080/20430795.2019.1665379.
Koengkan, M., Poveda, Y.E., Fuinhas, J.A., 2019c. Globalisation as a motor of renewable energy development in Latin America countries. GeoJournal, 1–12. https://doi.org/10.1007/s10708-019-10042-0.
Koengkan, M., Fuinhas, J.A., Losekann, L.D., 2019d. The relationship between economic growth, consumption of energy, and environmental degradation: renewed evidence from Andean community nations. Environ. Syst. Decis. 39 (1), 95–107. https://doi.org/10.1007/s10669-018-9698-1.
Koengkan, M., Fuinhas, J.A., Santiago, R., 2019e. The relationship between CO_2 emissions, renewable and non-renewable energy consumption, economic growth, and urbanisation in the Southern Common Market. J. Environ. Econ. Policy, 1–19. https://doi.org/10.1080/21606544.2019.1702902.
Koengkan, M., Fuinhas, J.A., Vieira, I., 2020. Effects of financial openness on renewable energy investments expansion in Latin American countries. J. Sustain. Finance Invest. 10 (1), 65–82. https://doi.org/10.1080/20430795.2019.1665379.
KOF Globalization Index, 2020. https://www.kof.ethz.ch/en/forecastsand indicators/indicators/kof-globalisation-index.html.
Kroop, G., Korobilis, D., 2016. Model uncertainty in panel vector autoregressive models. Eur. Econ. Rev. 81, 115–131. https://doi.org/10.1016/j.euroecorev.2015.09.006.
Love, I., Zicchino, L., 2006. Financial development and dynamic investment behaviour: evidence from panel VAR. Q. Rev. Econ. Finance 46 (2), 190–210. https://doi.org/10.1016/j.qref.2005.11.007.
Lutkephol, H., 2005. New Introduction to Multiple Time Series Analysis. Springer-Verlag, Berlin Heidelberg, pp. 21–764. ISBN: 978-3-540-27752-1.
Moorkens, I., Dauwe, T., 2019. Impacts of Renewable Energy on Air Pollutant Emissions: Calculation of Implied Emission Factors Based on GAINS Data and Estimated Impacts for the EU-28. pp. 1–47. https://www.eionet.europa.eu/etcs/etc-cme/products/etc-cmereports/impacts-of-renewable-energy-on-air-pollutantemissions/@@download/file/ETCCME%20technical%20report%202019_2.pdf.
Our World in Data, 2020. Air Pollution. https://ourworldindata.org/air-pollution.
Paramati, S.R., Alam, M.S., Chen, C.-F., 2016. The effects of tourism on economic growth and CO_2 emissions: a comparison between developed and developing economies. J. Travel Res. 56 (6), 712–724. https://doi.org/10.1177/0047287516667848.
Pesaran, M.H., 2004. General Diagnostic Tests for Cross-Section Dependence in Panels. Cambridge Working Papers in Economics, N. 0435, The University of Cambridge, Faculty of Economics, https://doi.org/10.17863/CAM.5113.
Pesaran, M.H., 2007. A simple panel unit root test in the presence of cross-section dependence. J. Appl. Econ. 22 (2), 256–312. https://doi.org/10.1002/jae.951.
Rinne, S.T., Rodas, E.J., Bender, B.S., Rinne, M.L., Simpson, J.A., Gale-Unti, R., Glickman, L.T., 2006. Relationship of pulmonary function among women and children to indoor air pollution from biomass use in rural Ecuador. Respir. Med. 100 (7), 1208–1215. https://doi.org/10.1016/j.rmed.2005.10.020.
Riojas-Rodríguez, H., Silva, A.S., Texcalac-Sangrador, J.S., Moreno-Banda, G.T., 2016. Air pollution management and control in Latin America and the Caribbean: implications for climate change. Rev. Panam. Salud Publica 40 (3), 150–159. https://www.scielosp.org/pdf/rpsp/v40n3/1020-4989-RPSP-40-03-150.pdf.
Ritchie, H., Roser, M., 2020. Air Pollution. Published Online at OurWorldInData.org https://ourworldindata.org/air-pollution.
Romieu, I., Weitzenfeld, H., Finkelman, J., 1990. Urban air pollution in Latin America and the Caribbean. J. Air Waste Manage. Assoc. 41 (9), 1166–1171. https://doi.org/10.1080/10473289.1991.10466910.
Serena, N., Perron, P., 2001. Lag length selection and the construction of unit root tests with good size and power. Econometrica 69 (6), 1519–1554. http://www.jstor.org/stable/2692266.
Smith, K.R., Mehta, S., 2003. The burden of disease from indoor air pollution in developing countries: comparison of estimates. Int. J. Hyg. Environ. Health 206 (4), 279–289. https://doi.org/10.1078/1438-4639-00224.

Tsilingiridis, G., Sidiropoulos, C., Pentaliotis, A., 2011. Reduction of air pollutant emissions using renewable energy sources for power generation in Cyprus. Renew. Energy 36 (12), 3292–3296. https://doi.org/10.1016/j.renene.2011.04.030.

Vásquez, I., 1996. The Brady plan and market-based solutions to debt crises. Cato J. 16 (2), 1–11. https://www.cato.org/sites/cato.org/files/serials/files/cato-journal/1996/11/cj16n2-4.pdf.

Wilson, T.D., 2008. Introduction: the impacts of tourism in Latin America. Lat. Am. Perspect. 35 (3), 3–20. https://doi.org/10.1177/0094582X08315760.

World Bank Open Data, 2020. http://www.worldbank.org/.

World Energy Council, 2006. Alleviating Urban Energy Poverty in Latin America. pp. 1–69. https://www.worldenergy.org/assets/downloads/PUB_Alleviating_Urban_Energy_Poverty_in_Latin_America_2006_WEC.pdf.

World Health Organization (WHO), 2020. Air Pollution. https://www.who.int/health-topics/air-pollution.

World Travel Market (WTM), 2020. Viagens e Turismo na América Latina. https://latinamerica.wtm.com/pt-br/Blogs/Viagens-e-Turismo-na-America-Latina/.

Zheng, S., Yi, H., Li, H., 2015. The impacts of provincial energy and environmental policies on air pollution control in China. Renew. Sust. Energy Rev. 49, 386–394. https://doi.org/10.1016/j.rser.2015.04.088.

Zoundi, Z., 2017. CO_2 emissions, renewable energy and the environmental Kuznets curve, a panel cointegration approach. Renew. Sust. Energy Rev. 72, 1067–1075. https://doi.org/10.1016/j.rser.2016.10.018.

Conclusion

This book has analysed the physical capital development and energy transition in Latin America and the Caribbean (LAC) countries. From the analysis conducted in the previous chapters, we are now capable of presenting several strategies and policies that could contribute to the LAC region's sustainable development. In this book, the conclusions are divided into two sections. In the first section, we present strategies and policy recommendations for the LAC physical capital stock development, whereas in the second section we do the same but for the case of the regional energy transition process.

Is the state of Latin America and Caribbean capital stock affecting the development of the region?

In the first four chapters of this book, we analysed the impact of public and private capital stocks on the development of the LAC region, with an emphasis on their effects on economic growth, income inequality, and energy intensity. First, looking at the historical evolution of the LAC region's physical capital since the 1970s, we saw that its public and private physical capital investments were constantly affected by the region's economic booms and busts. In fact, we saw that this region's propensity affected by external shocks always exerted a strong influence on the LAC economic strategies and, subsequently, on its economic performance. More precisely, we observed that in the 1970s, the beginning of the 1990s, and in the 'commodity boom' that occurred in the period between 2003 and 2014, the LAC countries experienced periods of considerable growth. In contrast, in the 1980s, and from 1998 to 2002, these countries experienced periods of severe crises and economic depression.

Therefore, we were able to note considerably different trends in the evolution of public and private capital in the LAC depending on the economic conjuncture of the period that was being analysed. For example, we saw that in the 1970s, the LAC countries took advantage of the favourable economic conditions to increase the region's public and private capital, while in the 1980s, due to the debt crisis that hit the region and which led to private capital flight and the accentuated reduction on public investment, the public and private capital investment sharply dropped. With the macroeconomic stabilisation and liberalisation plans implemented in the region after the 1980s crisis, private capital returned to the region in the 1990s. This return was influenced mainly by the opening of several economic sectors, such as the infrastructure sector, to private participation. However, due to the macroeconomic stabilisation and fiscal consolidation strategies, the LAC public capital investment continued to be very cautious in the 1990s. Indeed, the region showed relatively low public investment levels in this same decade.

During the period 1998–2002, the LAC passed through another crisis, which negatively affected its public and private capital investment. Then, after this challenging period, the commodity prices boom came (2003–14) and, driven by this favourable condition, the public and private investment grew again until the global financial crisis of 2008. Although the drop in private investment after the crisis hit the region, the LAC governments did not drop their investment levels, following a countercyclical strategy. Powered by the high commodity prices and the increased connection to China, the region overcame the effects of the global financial crisis in an impressive (and surprisingly fast) way. In 2010 the region seemed to be already recovered from the impacts of the crisis (at least looking at its growth rate). After this period, the LAC public investment only decreased again with the end of the commodity prices boom in 2014. However, the private investment, which returned to a growing trend after 2010, never returned to the levels of the years before the financial crisis.

Overall, despite the rising and falling trends, during the period 1970–2017, the LAC public investment was always relatively low. In fact, only in 1979, 1980, and 1981, it has crossed the fair value of 6% of the GDP. Regarding the private investment evolution, we noted that it was much more volatile, with picks (in good economic times) and breaks (in times of crisis) of considerable magnitude. If we look at the LAC public and private capital stocks as percentages of GDP, we see that, although they had increased from 1970 to 2017, their evolution was always relatively slow and, in some decades, it was nearly constant. By adding to this observation that some of the highest shares were registered in years of the economic deceleration, we are left with a greater conviction on the idea that there really seems to exist a lack of physical capital investment in the LAC region. This lack of investment (especially in infrastructure) is worrisome. It can prevent the LAC region from being more competitive and from achieving the desired sustainable growth path.

Through the observations drawn from our analysis, and given that capital stock is one primary input for production, representing various types of physical capital and, in a great deal, a country's infrastructure assets (e.g. roads, bridges, railroads, airports, tunnels, etc.), we are in accordance with the idea that was already highlighted by some previous authors and institutions that, in fact, there seems to exist an 'infrastructure gap' in LAC, especially when compared with similar developing (emerging) countries, as the East Asian (EA) or East Asia and Pacific (EAP) countries. This can be seen for example in the marked differences between LAC and EAP public investment shares since the 1970s.

The second chapter was dedicated to the analysis of the relationship between public capital stock, private capital stock, and economic growth in the LAC recurring to annual data from 1970 to 2014 for 30 countries from this region. In this chapter's empirical analysis, the panel vector autoregression (PVAR), panel dynamic ordinary least squares (PDOLS), and panel fully modified ordinary least squares (PFMOLS) methodologies were used to unveil the variables short- and long-run relationships.

Starting with the overall PVAR model results, they support the idea that economic growth, indeed, contributes to the increase of both public and private capital stocks in this region. This result is far from unexpected given that an enhancement in a country's economic output is expected to lead to a rise in the economy's degree of investment, i.e. to an increase in both public and private capital investment. For example, we can cite the study of Blomstrom et al. (1996) who had already reached a similar conclusion that positive changes in economic growth enhance the capital formation rates.

Conversely, it was also expected that increases in both types of capital stocks (public and private) would enhance growth, as the neoclassic growth models postulate (e.g. Solow, 1956). However, following the PVAR outcomes, only the private capital stock seems to positively affect these countries' economic growth in the short run, with the public capital stock showing to have a depressing effect on both growth and private capital. This could be linked with the Agenor and Moreno-Dodson (2006) theory that, in the short run, the public capital stock may have harmful effects on growth if it produces a crowding-out effect on private investment (in our estimation, we identified this effect, with the public capital showing a negative unidirectional causal relation with private capital).

Given that an increase in public capital should raise the returns to private capital (Aschauer, 1989), one would possibly think that the increases in public capital would have a different effect on private capital (a crowd-in effect) and, in the end, would contribute to increase the economic growth of these countries. However, this result seems to point to the possibility that the public and private capital act as substitutes in the LAC rather than acting as complements (Erden and Holcombe, 2006). In addition, there is a set of critical factors which can further explain this negative relationship between public and private investment (e.g. Bahal et al., 2018; Presbitero, 2016; Cavallo and Daude, 2011), namely (i) institutions' quality and strength; (ii) how easily a country has access to finance; (iii) and the degree of absorptive capacity. Overall, this means that countries with weak institutions, with difficult access to finance, and limited absorptive capacity, are more prone to revealing a crowding-out effect. All these characteristics seem to be usually associated with almost all the countries from the LAC region, meaning that they could probably explain why we achieved such a result.

Now, regarding the reasons for the negative effect that public capital stock seems to also have on economic growth in the short run, additionally to the negative effect that public capital has on private capital and which can be part of the explanation, there is also the fact of the LAC's insufficient level of public capital stock, with a particular emphasis on the recognised case of the regional infrastructure shortfalls (Faruqee, 2016). In addition, the inaccurate public investment strategies that these countries sometimes follow could also be an additional explanation for the verified effect (Gupta et al., 2014). Moreover, this adverse effect could also be exacerbated by factors usually linked to this group of countries, such as corruption, political instability, and 'white elephants' (e.g. Pritchett, 2000). Lastly, we should clarify that public capital stock is not always centred on profit. It is often primarily aimed at increasing social welfare (acting in areas which are not attractive to the private initiative and where it is difficult to make a profit), it can lead to situations where the positive economic effects are not immediate (i.e. may not be felt in the short run).

Turning to the results of the PDOLS and PFMOLS estimations, we see that, contrary to the short run, in the long run, both public and private capital stock positively impact the economic growth of LAC countries. Concerning the turn in the public capital stock effect, following Agenor and Moreno-Dodson (2006) and Erenburg and Wohar (1995), we can say that the crowding-out effects of public capital stock are, in general, observed in the short run and that as we move forward in time, this negative effect usually vanishes. The achieved outcome leads us to believe that this situation occurred in the LAC and, probably, the public capital stock starts to crowd-in private capital in the long-run. In addition, the alteration of the public capital stock effect, in the long run, can also be associated with the large marginal returns produced by the increases on both public capital stock levels (e.g. Fournier, 2016) and public capital efficiency (e.g. Berg et al., 2019), especially given the situation of the LAC's countries, where public capital stock is relatively low and where past investments were inefficient.

As in the short run, in the long run, the private capital stock continues to positively influence the economic growth of these countries, supporting the idea that higher investment rates lead to higher output levels (e.g. Solow, 1956). Moreover, it should be stressed that private capital demonstrates a higher coefficient than public capital, as it was already observed in past studies (e.g. Arslanalp et al., 2010). Devadas and Pennings (2018) inclusively mention that private investment is one of the primary factors responsible for the increase in the output of developing countries. As in the case of public capital stock, this effect is also linked with the low levels of private capital in these countries, which produces relatively high returns to private investment.

Even though economic growth is an indicator, which shows (in a great deal) if a certain country is on the right development path, being a tool that the countries can use to achieve various macroeconomic objectives, there are more standards that the countries/regions need to attain in order to consider them as developed ones. In the case of LAC countries, there is the case of their high inequality levels, which could prevent them from entering a desired sustainable development path if nothing is done. Thus, after the analysis of the relationship between LAC capital stock and the regional economic growth, in the third chapter, we focused our analysis on the effects that the LAC capital stock has had on the income inequality levels of the countries from this region. To conduct our analysis, we collected annual data for a panel of 18 countries of the region from 1995 to 2017. Two models were built based on a PARDL methodology: Model I with public capital stock as the interest variable, and Model II with the private capital stock as the interest variable. As control variables, we used the gross domestic product, human development index, trade-in percentage of the gross domestic product, tax revenue as a percentage of gross domestic product, and the unemployment rate in percentage of the total labour force.

From our estimations we saw that economic growth seems to be a powerful tool to reduce the LAC income inequality levels, given that gross domestic product showed a depressing effect on income inequality in all models, both in the short and long run. This result suggests that the economic performance of these countries can influence their income inequality levels and that they are making some advances in the promotion of inclusive growth policies. The combination of growth-enhancing policies with measures focused on promoting a more equitable society allows for the positive effects of their economic performance on all populations. In our view, this strategy should continue to be followed in order to grant that the gains from their growth will not only be channelled to the highest strata of the population. Finally, we should stress that this result seems to be in line with the ones from past studies, for example the one from Tsounta and Osueke (2014) who used a sample similar to ours.

Conversely, to economic growth, unemployment (proxied by the unemployment rate) showed an augmenting effect on these countries income inequality levels, both in the short and long run. This means that the increased unemployment contributes to enlarge these countries income gap with a persistent effect that extends over time. It also means that to tackle income inequality, LAC governments should concentrate a substantial part of their efforts on the development of policies to encourage job creation and on the promotion of measures that guarantee enlarged job opportunities for all. From this observed outcome, we conclude that fighting against unemployment is also fighting against income inequality. As in the previous case, this result is also in line with the ones from past literature findings (e.g. Hacibedel et al., 2019).

Concerning the effects from the human development index, which incorporates information related to population health, education, and standard of living, we observed that, as expected, it can contribute to the reduction of the LAC countries income inequality levels in the long run. This is in line with the general view that policies aimed at improving the standard of living of populations contribute to an equal society (e.g. Martínez-Vázquez et al., 2012). Some examples of these policies can be the ones linked with the public investment in education and health or with social policies as the development of social protection programmes. Therefore, our suggestion is that if the governments of LAC countries want to achieve greater social cohesion and more equalitarian income distribution, they should continue to invest in the well-being of their populations (especially of the lower-income groups).

In the same line, tax revenues also decreased the income inequality of LAC countries in the long run. This means that taxation if adequately done can generate a redistributive effect in these countries and contribute to lower their income gap. One example of this redistributive effect is that tax revenues could be used to promote social welfare and social protection programmes in these countries. In sum, they could help these governments to support their public expenditure policies, namely the ones focused on income inequality reduction. However, as most of these countries have low tax revenue levels (see e.g. Martorano, 2018), it could be essential to improve their tax schemes with for example more progressive taxation. This outcome also seems to be validated by some previous literature (e.g. Martorano, 2018; Balseven and Tugcu, 2017).

Now, regarding the effects of trade on the income inequality of LAC countries, we can say that, although the literature on this theme has found mixed results, our outcomes support the view that trade has a reducing effect on income inequality in the long run (e.g. Cerdeiro and Komaromi, 2017). Despite the fact that the LAC region has seen an increase in the income inequality levels in the 1980s and 1990s, which coincided with the increasing integration of the region in the global economy, at the beginning of the new millennium, this trend was reversed. This happened mainly due to the stabilisation of the trade liberalisation process in the region and the policy reforms that the LAC made to promote growth and control inequality (Székely and Sámano-Robles, 2014; Cornia, 2011). In our opinion, LAC countries should continue to pursue their integration process, given the positive effects that it seems to have had on their economic output (e.g. Santiago et al., 2020) at the same time as they continue to develop policies aimed at extending the gains from trade to all population layers.

Finally, concerning the effects of our interest variables, we saw that both the public capital stock and private capital stock have an enhancing effect on income inequality levels of these countries. Indeed, both variables seem to have contributed to the deterioration of the income distribution in these countries, due to private capital stock being sparingly higher than the one from public capital stock. However, we should mention that these effects were only observed in the short run. In the long run, none of these variables showed a statistically significant effect on income inequality. Moreover, we should also mention that these results held when we corrected the models for the presence of outliers, suggesting that even with this correction i.e. with the inclusion of dummy variables, the previous inferences remained accurate.

In sum, our findings revealed that these countries physical capital investments strategies must be rethought. In fact, following the outcomes of our estimations, we see that contrary to what is intended, public capital stock fails to contribute to the progress of these countries in terms of equality. Indeed, it seems that public capital stock has contributed to the increase in income inequality levels of these countries in the short run. This result probably means that the public investment in physical capital (e.g. roads, railways, bridges, schools, hospitals, sanitation and water systems, telecommunications and energy systems, and public transportation) is being made already more prosperous/wealthiest areas, where there is a proven economic dynamism, rather than being channelled to the impoverished/undeveloped areas (see e.g. Lopez, 2003).

As the public capital stock, the private capital stock also demonstrated an enhancing effect on income inequality in the short run. The justifications for this result probably match those of the previous case. However, as the private interest is majorly driven by profit, in the absence of government incentives, it is natural that they invest in areas where higher profits are guaranteed (generally in the most developed areas). In addition, we must also consider the possible barriers that the private control of for example energy, infrastructure, and transport services, can generate to the population's lowest income groups. As private enterprises usually charge higher prices for their services compared to public enterprises, they could reduce the access and affordability of these services by the most disadvantaged strata of the population. These additional assumptions can also help to explain why the magnitude of the effect of private capital on income inequality is greater than the one from public capital.

Finally, regarding the absence of a long-run statistically significant effect of both variables i.e. public and private capital stocks, on income inequality, we can say that it can come from the possibility that governments of these countries try to correct the detected negative impact that was verified in the short run, over a period of time, through the investment in the less developed areas and through the creation of incentives to the private sector to also invest in these areas. However, the level of investment does not seem to be large enough to reduce income inequality in the long run. Overall, this result points to problems related to the lack of investment.

In the fourth and last chapter, we focused our attention on the downward trend of the LAC's energy intensity, investigating if the region's physical capital had contributed to this tendency. The analysis conducted in this chapter was connected with the fact that, although a country needs to use energy to support its production, there is an increased worry about the energy demand and the energy security of LAC countries. The demand for energy efficiency is another critical subject for the region's development. To achieve the goals of this chapter, we used the PARDL methodology, the log t regression test method, the club clustering algorithm, and the ordered logit model, to investigate the relationship between public and private capital stocks and energy intensity in a group of 21 LAC countries, from 1970 to 2014. In addition to the interest variables, in the PARDL model, we used the gross domestic product per capita, CO_2 emissions per capita, and energy (commodities) prices as the control variables. Moreover, it must be said that two different specifications of the PARDL were estimated: a nonparsimonious version (with the statistically significant and nonstatistically significant variables) and a parsimonious version (only with the statistically significant variables).

Overall, the results from both PARDL specifications were very similar. As in the case of the Jimenez and Mercado (2014) study, the results from our PARDL models supported the hypothesis that income is negatively related to energy intensity. However, this effect only seems to occur in the short run. This result supports the idea that the income effect on energy intensity fades away when countries reach a certain income level (Deichmann et al., 2019). Besides, according to Deichmann et al. (2019), when a certain income level is reached, the development and application of energy efficiency policies become much more significant for reducing energy intensity than the income effect alone.

According to our PARDL models, another factor contributing to the decrease in energy intensity is CO_2 emissions, with this effect being felt only in the long run. This result seems to transmit the idea that environmental pressure (in this case, proxied by CO_2 emissions) can be a catalyst for the development of environmental and energy policies (e.g. energy efficiency policies) and the adoption of more environmentally friendly (and more energy efficient) technologies and innovations (Khan et al., 2019, 2020). Directly or indirectly, these actions can lead to a decrease in energy intensity. However, as this process can take a sizeable amount of time, from the instant that the problems are felt until the instant that these actions start having results, it is natural that this effect only arises in the long run.

Moreover, from our PARDL models results, we also saw that the prices of energy commodities contributed to the increase in LAC energy intensity either in the short or long run. This result is very similar to the one found by Samargandi (2019) between oil prices and energy intensity in the OPEC (Organization of the Petroleum Exporting Countries). Following a similar theory of the one from Samargandi (2019) for OPEC, as a significant portion of the LAC countries has abundant energy commodities, higher energy commodities prices can induce the LAC countries to increase their rents, leading them to higher energy consumption levels which, eventually, leads to the increase of their energy intensity.

Finally, regarding the interest variables, both specifications from the PARDL model point to public and private capital stocks that did not have contributed to the LAC energy intensity decreasing trend. In fact, by the results, we see that both present an enhancing effect on the energy intensity of this group of countries in the long run. This can be a sign of the lack of investment in the LAC physical capital (Faruqee, 2016), more appropriately, of the lack of investment in new and more energy-efficient capital (Araújo et al., 2016). This result indicates the necessity to upgrade the LAC physical capital in the public and private sectors, from equipment to infrastructure. As the LAC physical capital still seems to be very energy intensive, it can prevent the region from achieving an even lower energy intensity level. Moreover, as the acceleration of these economies was not accompanied by appropriate investment in new physical capital, over time, the effect of the lack of investment in more energy-efficient physical capital on this region's energy intensity becomes more noticeable and significant. This is why this effect is only noticed in the long run.

After the PARDL models estimation, through the log t regression test method, we found that our sample of LAC countries does not converge to the same steady-state equilibrium in terms of energy intensity. Additionally, through the club clustering algorithm, we could identify four convergence clubs and one divergent group i.e. a group composed of countries that do not converge to any club. The composition of the convergence clubs is as follows: 'Club 1' was composed of Haiti and Honduras; 'Club 2' composed of Argentina, Bolivia, Brazil, Guatemala, Nicaragua, Paraguay, Uruguay, and Venezuela; 'Club 3' composed of Barbados, Chile, Costa Rica, El Salvador, Grenada, Mexico, and Peru; and 'Club 4' composed of Ecuador and Panama. Regarding the so-called divergent group, it was composed of Colombia and the Dominican Republic.

Then, through analysis of the four convergence clubs' descriptive statistics in terms of energy intensity, we found that the clubs were ordered according to their energy intensity levels. More precisely, we observed that 'Club 1' was the most energy-intensive club, followed by 'Club 2', 'Club 3', and 'Club 4' (the least energy-intensive club). By analysing the descriptive statistics and the clubs' average transition paths, it was easy to perceive that within the LAC region, there are countries that need to make some additional efforts to reduce their energy intensity (and promote energy efficiency) when compared with others. By the results from our analysis, it seems that these countries are the ones from 'Club 1' and 'Club 2', given that they have presented considerably higher energy intensity levels and transition paths that raise doubts on their effective energy intensity reduction efforts. Conversely, in addition to their lower energy intensity levels, 'Club 3' and 'Club 4' also presented transition paths that showed a strong and clear decreasing trend in energy intensity since the 1970s.

Despite the importance of the previous analysis and its respective inferences, we must not forget that this chapter's primary purpose was to examine the role of public and private capital stocks in the LAC energy intensity. Accordingly, through the identification of the LAC convergence clubs, we were able to build an ordinal response variable that we can use as the dependent variable in an ordered logit regression model, representing the club to which a country belongs. Using the public and private capital stocks as the independent variables of the ordered logit regression model, we were able to investigate if the capital stocks were determinant factors for the formation of LAC energy intensity convergence clubs. This means that we were able to understand if public and private capital stocks decrease (increase) a country's probability of moving to a low (high) energy intensity club. According to previous literature, we estimated two models, one with the averages of the public and private capital stocks as a percentage of the GDP between 1970 and 2014 and the other with the annual % averages of public capital stock and private capital stock over the period 1970–2014. In both cases, the control variable was the initial energy intensity, representing the countries' energy intensities in the first year of the analysis.

By the results from the ordered logit regression model, we saw that, contrary to what is verified with initial energy intensity (a higher level of initial energy intensity increases the probability that a certain country will belong to a high energy intensity club), public and private capital stocks are not determinants of the convergence clubs' formation i.e. their effects were not statistically significant. This result can come from the evolution of both types of capital. Their respective effects on energy intensity can be similar in all the clubs under analysis (the difference between these clubs may come from other factors). Overall, it seems that the results from the ordered logit regression model support the assumption that the investment in newer and more energy-efficient capital should be extended to all the countries of the region, regardless of their club.

After this brief summary of the results, we can now try to answer the question which is presented in the title of Part II: 'Is the state of Latin America and Caribbean capital stock affecting the development of the region?' From our estimations, we can say that it is and that to the LAC region be able to achieve higher growth rates and inclusive and sustainable development, some changes need to be addressed in its public and private physical capital investment strategies. Hence, according to results from Part II chapters, we will now give some policy recommendations to help the LAC policymakers develop future physical capital investments and guarantee that they are done focused on the growth, cohesion, and sustainability of the LAC region. Starting by Chapter 1, in which we analysed the public and private capital evolutions in the LAC since the 1970s recurring to data from the IMF 'Investment and Capital Stock Dataset', we can say that, as we suspected, the available data clearly supports the idea that there is a lack of physical capital investment in the region. In our view, and according to some previous authors and international institutions (as the IMF), it would be of benefit to the region if governments of their countries were able to increase their pubic investment levels, namely in infrastructure development and that, at the same time, they were able to create conditions to attract (and maintain) private investment in this region.

Turning to Chapter 2, where we analysed the relationships of public and private capital stocks with economic growth, we can say that, in our view, LAC governments should continue to promote their public investment in physical capital and simultaneously create or improve the conditions to encourage private physical capital investment, given that both public and private capital stocks demonstrated a positive impact on long-run economic growth. As we already stressed, besides the increase in the LAC public and private capital, it is essential to enhance the regional economic growth. It is also indispensable to boost the competitiveness of the LAC region countries regarding their main competitors, as the emerging Asian countries. However, given the adverse short-run effects of the public capital on private capital and economic growth, we think it is imperative to improve the selection, evaluation, and management of the public investment projects in the LAC region. An example is that the governments could sometimes retrieve better outcomes with the investment on the maintenance and upgrade of the existing capital instead of moving directly to new capital investments. Moreover, due to the problems associated with this region's macroeconomic stability, the LAC governments must weigh their fiscal space and ensure that no negative fiscal costs will be associated with their investment, ending up in situations where public capital investment hampers their growth. In the most problematic situations, it could be vital to improve public financing instruments or search for new funding sources, as the development of public–private partnerships (PPPs). This can also be a solution to increase the cooperation between the public and private sectors and to ensure that public capital does not crowd-out private capital. Public and private capital should act as complements and not as substitutes. However, this type of contracts must be very well planned so that the investment can be sustainable and meet the desired quality standards. It is also a priority to ensure that it does not originate harmful effects, as barriers to access these services or with the emergence of for example rent-seeking situations. In sum, LAC governments should guarantee that public spending on physical capital will raise their countries' social well-being and the marginal productivity of the private capital.

Regarding Chapter 3, where we investigated the impact of the LAC public and private capital stocks on income inequality, we must say that it seems that LAC governments should rethink their physical capital investment strategies, given that, according to our results, it appears that the investment is being concentrated in areas where there is already a certain level of development/richness. Suppose no changes are made in this field, and they continue to forget the areas where the investment in physical capital is essential i.e. in the most impoverished/undeveloped areas. In that case, cohesion of these countries will continue to be threatened. As in the previous chapter, we once again think that the governments of these regions should improve the management and the selection criteria of their public investments, but this time we think this should be done with special attention to the development of the most impoverished/rural areas. An effort should be made to link these areas to the more prosperous ones, where there is a more thriving economic activity, which ultimately, will allow an increased income convergence. As the adverse effects detected in the short run seem to have vanished in the long run, maybe LAC governments are trying to correct this undesirable situation. However, it appears that investment is not yet enough to have a depressing effect on income inequality, which means that it still needs to be increased.

Moreover, given the low degree of economic slack that many of these countries face, it could also be important that the regional governments create incentives to promote the private initiative to invest in these areas, which otherwise, is unlikely to happen. Still, as in Chapter 2, we think that the private initiative should be intensively examined and discussed by the public entities in order to guarantee that it does not neglect the low-income layers of the population. In sum, the LAC governments must continue to increase the connection between the rural/more underdeveloped areas and the developed areas, where there is a higher degree of economic dynamism and opportunities, with increased long-term and territorially balanced physical capital investment.

Moving on to Chapter 4, where we analysed the impacts from the LAC capital stock (public and private) on the region's energy intensity, we can say that, by the outcomes from our estimations, we think that the LAC countries should invest in new and more energy-efficient physical capital, given the enhancing effect that both types of capital (public and private) seem to have had on the region's energy intensity. This investment should be focused on improving the region's energy efficiency, which will subsequently lead to the region's energy intensity decrease. Moreover, we think that this should be done on their infrastructures and the equipment and machines used in the LAC production process (with particular attention to the most energy-intensive economic sectors). Again, as in the previous chapter, to achieve a suitable and satisfactory investment level in this field, the LAC governments will probably need to improve their public financing instruments and, if necessary, call upon institutional investors. Besides, in order to improve the energy efficiency of the physical capital, LAC governments should also improve their laws and their regulatory framework regarding this matter. As examples we can stress the establishment of (i) energy efficiency targets for the several economic sectors (especially for the most energy-intensive ones); (ii) the deployment of energy efficiency norms and regulation for equipment and appliances; and (iii) the creation and promotion of efficiency standards and energy audits for buildings/infrastructures. Although the improvement in the public sector energy efficiency could influence the private sector to follow a similar tendency, some additional measures should also be developed in order to accelerate this process. Some examples of these measures can be the development of new financing schemes and financial incentives (e.g. fiscal or subsidies) for energy efficiency investments in the private sector, the creation of loans and lines of credit to energy efficiency investments, and the grant of tax reductions and tax credits for the private investment in energy efficiency projects. It is also advisable that the financial sector complements the governments' efforts in this field. Finally, as the LAC energy transition will depend on the shift to renewable energy and improvements in the energy efficiency of the region, the discussion on how the LAC countries can follow to attain higher energy efficiency levels and lower energy intensity levels should be increasingly present in the agendas of regional organisations as the Economic Commission for Latin America and the Caribbean (ECLAC) or the Latin American Energy Organisation (OLADE). This type of organisations can, and should, serve as a stage for creating and promoting energy efficiency measures for the LAC region, contributing to the creation of a more homogeneous energy efficiency promotion plan among the several countries from this region.

Essays on the Latin America and Caribbean energy transition process

The LAC energy transition process was analysed in the last five chapters of this book. The first one analysed the initiatives and challenges of the energy transition in the LAC region. The analysis focused on difficulties in implementing the energy transition in three major countries (Brasil, México, and Argentina) and the case of success (Uruguay). The second one was dedicated to analysing the role of public, private, and PPP capital stock on the expansion of renewable energy investment in the LAC region. The third one was committed to analysing the effect of the energy transition on economic growth and consumption of nonrenewable energy sources of LAC countries. The fourth was devoted to an overview of the consequences of the energy transition on environmental degradation of LAC countries. The last chapter was dedicated to analysing the energy transition's capacity to decrease air pollution deaths in LAC countries.

Chapter 5 was committed to analysing the initiatives and challenges of the energy transition in the LAC region. The countries of LAC region have common and unique characteristics that arise challenges to the energy transition process. What are the challenges to the process of the energy transition in the LAC region? The challenges of LAC countries will be briefly discussed.

In **Brazil**, we have several initiatives that encourage the process of energy transition in the country. However, these initiatives come across the country's structural problems. First, energy expansion in Brazil does not have targets.

It made long-run projections for a possible composition of the energy matrix in the country. Therefore, the goals to mitigate climate change are not an objective for Brazil's energy expansion energy planner.

This lack of targets affects the organisation and dialogue between the numerous economic agents involved. As identified by the FGV Energia (2016), exist in Brazil, inefficient communication between the principal planner and the energy transition agenda. When referring to the principal planner, we have in mind the Ministry of the Environment (ME), and the Ministry of Mines and Energy (MME). Consequently, we are thinking about their regulatory agencies, such as *Agência Nacional de Energia Elétrica* (ANEEL), *Instituto Brasileiro do Meio Ambiente e dos Recursos Naturais Renováveis* (IBAMA), and *Agência Nacional do Petróleo, Gás Natural e Biocombustíveis* (ANP). However, as identified by this same report, it does not exist between the programs and central coordination.

Another challenge that we can evidence in Brazil is related to the no parallel between the various mechanisms of planning, development, and financing of cleaner energy sources, with the transport sector. No parallel between renewable energy sources and the transport sector is visible. The widespread use of fossil fuels in the transport sector is a significant source of CO_2 emissions.

The evidence shows that the transport sector contributes 11% of the total CO_2 emissions, which is caused by energy consumption. That is, this contradicts the Brazilian leadership regarding the development of biofuel technology. Please note, that Brazil is the only country that produces cars with technology flex engines, where consumers can interchangeably use gasoline and ethanol.

Indeed, high participation of the transport sector in CO_2 emissions is related to discovering the presalt layer in the middle of the 2000s. This discovery made the government's resources, and the private sector attention focused on the potential of oil and the production of its by-products in the country. The investment in renewable energy sources agenda was put aside. There is a lack of engagement and public planning to promote renewable energy sources in Brazil. Moreover, this lack is correlated with the transport sector deficit, where this sector has a small rail network and uses very little of its coastal shipping potential. As the low economic activity in the last 5 years, other factors have been presenting a substantial slowdown in renewable energy projects.

Even with some ambitious proposals of the energy transition in **Argentina**, their capacity for achieving them is questioned due to the weak economic growth in the last years. Indeed, the search for social and economic stability that is in fall in the last years has been overshadowing the energy transition agenda in the country. This lack of attention for the renewable energy agenda reflects the structured planning for the energy sector. According to the Institute of Americas (2016), the government of Argentina planned that renewable energy consumption would increase from 6.6% today to 14.6% in 2025. The oil consumption will decrease from 32.6% to 23.7%, and natural gas from 51.1% to 49.6%. However, these intentions have not yet been formalised due to low economic growth that consequently impacted the investment in renewable energy sources.

Indeed, the annual investment in renewable energy from 2003 to 2012 was on average 550 MW, well half of the 1000 MW annual investment target for this kind of energy, as well as far from reaching the goal of 10 GW of installed capacity in 2025 (FGV Energia, 2016). This low investment has contributed to the government's declaration of emergency in the electricity sector in 2015, due to fuel shortages. This state of emergency in the electricity sector lasted until the end of 2017. This evidence shows the Argentine energy plan's weakness and possible obstacles/or challenges for the country's energy transition process.

Another challenge for Argentina's energy transition process is allocating natural resources and the energy structure based on fossil fuels energy sources, 64% of energy production in Argentina in 2013 coming from fossil fuels. Therefore, it is necessary to engender great effort in planning and investment to change the situation. Moreover, the transport sector is not in the same stage of development as the energy sector. In Argentina, the transport sector is responsible for around 12% of its total CO_2 emissions.

Despite being the third-largest emitter of these gases, the sector has no mitigation targets, plans to increase energy efficiency, or plans to change the transport sector's fuels. It reflects the poor coordination between the environmental planners and the energy industry planners in the country. Additionally, according to FGV Energia (2016), establishing an environmental and energy transition agenda was hampered by several institutional, governance, and macroeconomic barriers.

The environmental agenda in **Mexico** is regarded as a matter of international relations, where there is no integration with domestic planning bodies as also identified by FGV Energia (2016). This lack of integration between the environmental agenda and the local government was reflected lack of consistency in Mexico's renewable energy plan. This inconsistency is also related to the official definition of clean energy by the Mexican government. It comprises renewable sources, nuclear energy, hydropower, natural gas, and high-efficiency cogeneration process.

However, considering natural gas as renewable energy by the government of Mexico raises questions from environmentalists. Indeed, this kind of energy source in energy consumption in the country is about 44.6%.

Three factors cause the substantial participation of natural gas in energy use in Mexico. The first factor is the significant reduction of CO_2 emissions in the energy sector caused by thermal oil replacement of natural gas. The second factor is related to the international environmental debate indicating that natural gas emits fewer pollutants than oil and its derivates. The third factor is related to the low cost of gas and lobbying by the industries, which encourages the use of this source. All these diminish the appeal of the inclusion of renewable energy sources in the energy matrix, such as wind and solar, where the country has immense potential for these two energy sources.

Another area that is not widely covered in the energy transition and environmental planning is the transport sector. Like other LAC countries, such as Argentina and Brazil, Mexico's transport sector is the second-largest CO_2 emitter representing about 20%.

The transport sector's high CO_2 emissions in Mexico contribute to severe air pollution problems that repeatedly occur in Mexico City. An energy transition to a more efficient one in Mexico's transport sector, particularly from the energy point of view, could solve this problem in Mexico's big cities. Nevertheless, the discussion on energy transition for transport is inevitably left aside. It occurred at a moment, where the oil price is low. In the last years, the Mexican government has been reducing the tax revenues from the exploration and production of oil. It makes this kind of energy more economically competitive than the renewable energy source.

We can highlight a case of the success of energy transition in the LAC region. This success comes by making clean energy sources available equally to the entire population. It can be achieved by (i) the implementation of well-structured legal and regulatory processes that encourage the development of renewable energy technologies; (ii) the robustness of institutions and economic development; and (iii) the development of strategies between private and public companies for the development of clean energy sources. The country with all these feats is **Uruguay**.

The Uruguayan energy transition is a case of success. Uruguay's capacity to make available clean energy sources equally by developing strategies between private and public companies. Uruguay has been benefited from well-structured regulatory and legal processes. The robustness of institutions and economic growth since 2004 has encouraged the development of renewable energy technologies in the country.

According to FGV Energia (2016), these factors have also attracted private foreign investment, which has helped incorporate new business models for the country's energy sector. Moreover, the development and implementation of a clean energy transition plan with long-term energy goals, such as the 'National Energy Policy 2005–2030' plan, was designed by the Uruguayan government (IRENA, 2015) to attract these investments. This plan has, as objective diversity, the country's energy matrix, reduces dependency on nonrenewable energy sources, improves energy efficiency, and increases renewable energy sources. Indeed, this plan sets a target of 50% primary energy from renewable energy sources for electricity generation, industrial, domestic heat, and transport sector by 2015 (IRENA, 2015). Therefore, this energy transition plan is evidence of a well-structured plan that contributes to the realisation of a successful energy transition.

As a result of this well-structured energy transition plan, renewable energy sources accounted for about 93% of Uruguay's energy consumption in 2015. Hydropower electricity accounted for about 58% of energy consumption and, biofuels used in transport sector accounted for about 18% and other renewable energy sources that include solar, wind, and solar photovoltaic, accounted for 15.9%.

In 2012, the country ranked first among the countries that most invested in renewable energy technologies per unit of GDP. Additionally, the country is considered the second green energy leader in the LAC region. The existence of a successful renewable energy plan focused on the promotion of renewable energy technologies and energy efficiency (FGV Energia, 2016). Given a successful energy agenda focused on energy transition, Uruguay was also considered a successful study case at COP 21. It established a reduction of 25% in CO_2 emissions from electricity consumption and heat production by 2030, based on 1990 values. Indeed, Uruguay's success in reducing these emissions.

Indeed, it is essential to clarify that Uruguay is a small country, with less diversity and political and economic complexity than their neighbours, such as Argentina and Brazil that have been suffering from severe economic and political problems in the last years. Moreover, worth remembering that the country's electrical integration is less complicated than other LAC countries, contributing to the development and diffusion of renewable energy technologies.

Chapter 6 was devoted to analysing the role of public, private, and PPP capital stock on the expansion of renewable energy investment in the LAC region. We analysed the effect of public, private, and PPP capital stock on the installed capacity of renewable energy that is a proxy of renewable energy investment in a panel of 18 countries from the LAC region from 1990 to 2015 Quantile via Moment's methodology was used.

The preliminary tests results revealed that we have low multicollinearity in the variables, cross-sectional dependence, stationarity in some variables, fixed effect in all estimated models, and presence of serial correlation up to second order. Therefore, the Quantile results via Moment's methodology indicated that the public and PPP capital stock positively affect renewable energy's installed capacity. In contrast, the private capital stock does not cause any effect on the installed capacity of renewable energy. The positive effect of public and PPP capital stock on the installed capacity of renewable energy is the high investment and maintenance costs, complex construction issues, and economic returns that are not always high in renewable energy projects. Actually, during the initial process of developing these projects, the initial access to capital can be challenging. This kind of energy is often supported by public and PPP capital, which are cheaper than private capital support. Moreover, the nonimpact of private capital stock is related to the high investment and maintenance costs of renewable energy projects and the low private capital supply, so increasing the financing costs that discourage private participation in investment in renewable energy technologies. This explanation agrees with Koengkan (2020) that investigated the effect of capital development on renewable energy sources.

It is necessary to reduce the cost of capital and the barriers to renewable energy investment in the LAC region. Indeed, the financial sector has an essential role in scaling up the finance level available for renewable energy technologies investment, significantly reducing the risks between public and private investors and increasing private sector participation. For this, it is necessary to catalyse private capital finance. It is necessary to create concrete renewable energy policies, instruments to reduce financial risk, and develop well-structured finance mechanisms. Enable technological progress, where the public financing institutions need to support the initial development of renewable energy technologies in the LAC region. In this stage, public financing institutions' participation is due to the high risks in the deployment stage of renewable energy technologies. The private sector may be more reticent in committing resources in the development of renewable energy technologies.

The LAC region should promote energy innovation and entrepreneurship. Several utilities in the region are spending substantial financial resources for energy efficiency and renewable research and development programmes. Indeed, many of these programmes can be turned into commercial projects and create new markets for renewable energy technologies. Finally, accelerate the learning curve, wherein the LAC region must disseminate information regarding the information on effective financing mechanisms for renewable energy technologies. That can facilitate access to these finance sources by local project developers and increase the access to specific credit lines for renewable energy technologies. Indeed, this dissemination of information includes knowledge by policymakers, which will reduce the costs of acquiring information and local renewable energy markets that consequently can contribute to attracting foreign investment in the sector.

Chapter 7 analysed the effect of the energy transition on economic growth and consumption of nonrenewable energy sources of LAC countries. The effect of the energy transition on economic growth and consumption of nonrenewable energy is investigated. Five Mercosur countries, from 1981 to 2014, were analysed. The PVAR model was used as a methodology. The preliminary tests results indicated low multicollinearity between the model's variables, cross-sectional dependence, the stationarity of all variables in the first differences of logarithms and the need to use the lag length (1) in the PVAR regression.

The PVAR model results indicated the consumption of renewable energy that is a proxy of energy transition increase economic growth and decrease the consumption of fossil fuels in the Mercosur countries. Moreover, the Granger causality Wald test confirmed the existence of a bidirectional relationship between energy consumption (renewable and fossil sources) and economic growth. The countries are dependent on fossil fuels to grow due to the bidirectional relationship between the consumption of fossil fuels and economic growth. The existence of substitutability between renewable and fossil sources in periods of drought in the reservoirs was found. Indeed, hydropower was substituted by thermoelectric plants that are powered by oil or gas. The process of globalisation in the countries harms renewable energy and consumption of fossil fuels. Thus, the dependency on fossil fuels for growth and the substitutability between renewable and fossil reveals low energy source diversification in the Mercosur countries. The low energy diversification in these countries is due to low public and private green energy investment to supply the growing and future demand.

More public policies and incentives should be created to attract more investment in renewable energy and increase the consumption of this source. Policies should be advanced that encourage households and firms to purchase appliances with a high energy efficiency standard to reduce energy consumption. Policies should be developed that encourage public and private banks to support investment in renewable energy technologies or the purchase of technologies that reduce energy consumption and environmental degradation by firms and households with low-interest rates and credit. The bureaucracy that discourages the renewable energy foreign investment should be reduced, as should the political lobby between governments and large fossil fuels producers.

These policies need to be implanted to reduce the dependency of Mercosur countries on fossil fuels and reduce environmental degradation by increasing renewable energy consumption. It is also advisable to promote economic growth and take advantage of the enormous abundance of renewable energy sources in Mercosur countries. Finally, this study's empirical findings help advance the existing literature and warrant governments' and policymakers' attention. Moreover, the Mercosur countries need to realise a deep process of integrating their economies with the rest of the world and reducing the imports and capital barriers to facilitate access to renewable energy technologies and encourage investments in this energy source. As Koengkan et al. (2020) mentioned, the process of trade and financial liberalisation is a motor to renewable energy development in the LAC economies.

Chapter 8 made an overview of the consequences of the energy transition on the environmental degradation of LAC countries. This chapter's main aim was to assess the asymmetric impact of the energy transition on environmental degradation. In 18 LAC countries over the period from 1990 to 2014 a PNARDL in a UECM form was used as the methodology.

The preliminary tests of this chapter indicated that the variables have the following characteristics (i) low-multicollinearity; (ii) cross-sectional dependence in all variables in natural logarithms and some variables in first-differences, such as Y and PUBK; (iii) I(0)/I(1) for all variables; (iv) and the presence of fixed effects. Moreover, the specification test indicated the presence of (i) heteroscedasticity; (ii) first-order autocorrelation; (iii) and nonpresence of cross-sectional independence. Results of these tests are essential to identify the characteristics of the countries under study and the possible methodologies that need to be applied.

The results of the PNARDL model estimates suggest that economic growth in the short and long run and the public capital stock in the short run, have a positive effect on environmental degradation. Nevertheless, the positive and negative asymmetry of the variable ratio of renewable energy, which is a proxy of the energy transition, hurts the environment in the short and long run.

The capacity for the proxy of the energy transition to reduce environmental degradation is probably related to the effect of globalisation on renewable energy technological efficiency that consequently produces more clean energy with fewer CO_2 emissions, as well as being due to the increasing participation of renewable energy sources in the energy matrix of these countries due to the new investment and the energy demand caused by the effect of the globalisation process on economic growth (see e.g. Fuinhas et al., 2017). Another possible explanation for this negative impact is the efficiency of renewable energy policies that encourage alternative energy sources in the energy mix.

Indeed, to confirm these possible explanations the robustness check was made, and it was identified that the positive and negative asymmetries of the variable globalisation index, in the long run, have a positive effect on the proxy of the energy transition. However, the negative impact of renewable energy policies on the energy transition's proxy is a surprise of this chapter. The possible explanation for this impact can be related to the inefficiency of these policies. The methodology/construction of variable renewable energy policies cannot reveal this variable's real effect on energy transition.

Thus, based on these findings, the LAC region is recommended to develop policies for more efficient renewable energy that contribute to increasing growth, investment, and consumption of green energy and inversely reducing energy consumption from nonrenewable sources by the households and industries. Regarding the public capital stock, local governments should encourage public banks to support renewable energy technologies or purchase technologies with higher energy efficiency that reduce nonrenewable energy consumption with lower interest and credit rates. Given the mistakes committed in the past, the LAC region policymakers should consider integrating measures of regulation of CO_2 emissions in their growth strategies.

Chapter 9 analyses the capacity of energy transition to decrease air pollution deaths in LAC countries. The main objective of this investigation was to assess the impact of the energy transition on deaths from air pollution in 19 LAC countries from 1995 to 2016, using the PVAR methodology. The preliminary tests results indicated low-multicollinearity, cross-sectional dependence, stationarity in the variables, the fixed effects in the model, and the need to use the lag length (1) or (2) regressions.

The PVAR model estimates suggest that the energy transition and economic growth decrease air pollution deaths, while urbanisation and international tourism increase them. The postestimation tests indicated unidirectional and bidirectional causality among the variables of the model. The model estimation is stable, and that two periods after a shock, the variables themselves explained almost all the forecast error variance. The impulse–response functions of all variables converge to equilibrium, supporting that the model variables are stationary.

The energy transition's capacity to decrease these deaths is related to the rapid investment growth in renewable energy technologies from 1990 to 2012. The energy consumption from these technologies and the efficiency of renewable energy policies decrease air pollution in the LAC region and their components such as CO_2 emissions and other gases. This evidence proves that the LAC region is in the right way, in the energy transition process, as identified by Koengkan et al. (2021). However, it is necessary for more initiatives for the development of renewable energy technologies. For example, encouraging the public and private banks to support renewable energy technologies or purchase technologies with higher energy efficiency and create policies that facilitate access to renewable energy by families, farms, and industries. All these will reduce the consumption of nonrenewable energy and consequently, air pollution and deaths.

The capacity of economic growth to decrease air pollution deaths is related to better health systems in the LAC countries caused by rapid economic growth in the last 30 years. This rapid economic development in the region provided a fiscal space and budgetary flexibility in most countries of the region (e.g.. Argentina, Brazil, Chile, Colombia, Mexico, Peru, and Uruguay) to introduce health-system improvements. This evidence demonstrated that the LAC region countries took advantage of the economic boom in the region and implemented correctly social policies, welfare reforms, and cash transfer schemes to reduce poverty and expand access to health.

The LAC region's economic boom also reduced the energy poverty related to firewood or charcoal as fuel for cooking or heating due to the lack of electricity or gas access. However, although energy poverty has been decreased in the LAC region, it is necessary to create more policies. For example, price support measures as an essential complement (e.g. social tariff, tax reduction, and others) to increase the access of the urban and rural population to the clean fuels and technologies for cooking or heating, where 13% of the population in the region does not have access to these technologies.

The process of urbanisation caused by economic growth has induced an increase in deaths. It, consequently, had increased the consumption of energy from nonrenewable sources that intensified air pollution. Indeed, that is an indicator that urbanisation, caused by economic development in the LAC, still causes several impacts on the environment and health. Therefore, it is necessary to create policies that reduce nonrenewable energy sources in urban cities in the LAC region. Personal transportation can be reduced by introducing a more urban public transportation network. Moreover, it is necessary to create fiscal policies to support green construction and encourage private firms to engage and expand the scale of this type of construction.

Additionally, the increase of deaths from air pollution by international tourism in the LAC region is due to the tourism sector require energy from the direct use for scenic flights, jet boating or air travel or indirectly for hotels, events, museums or experience centres. Indeed, most of this energy consumption is often generated from oil, coal, and natural gas, consequently increasing air pollution. That is the tourism section in the LAC region based on fossil fuel consumption to realise their activities. In this case, it is crucial to create more policies that encourage renewable energy sources and the acquisition of green technologies by hotels, museums, or experience centres. It is necessary to have more policies that encourage green technologies that replace old technologies from aircraft, jets, boats, and other transport types in tourism activities.

References

Agenor, P.R., Moreno-Dodson, B., 2006. Public Infrastructure and Growth: New Channels and Policy Implications. Policy Research Working Paper, No. 4064, World Bank, Washington, DC, https://doi.org/10.1596/1813-9450-4064.

Araújo, J.T., Vostroknutova, E., Wacker, K.M., Clavijo, M., 2016. Understanding the Income and Efficiency Gap in Latin America and the Caribbean. Directions in Development-Countries and Regions. World Bank Publications, Washington, DC, https://doi.org/10.1596/978-1-4648-0450-2.

Arslanalp, S., Bornhorst, F., Gupta, S., Sze, E., 2010. Public Capital and Growth. IMF Working Paper, No. 10/175, International Monetary Fund, Washington, DC, https://doi.org/10.5089/9781455201860.001.

Aschauer, D.A., 1989. Does public capital crowd out private capital? J. Monet. Econ. 24 (2), 171–188. https://doi.org/10.1016/0304-3932(89)90002-0.

Bahal, G., Raissi, M., Tulin, V., 2018. Crowding-out or crowding-in? Public and private investment in India. World Dev. 109, 323–333. https://doi.org/10.1016/j.worlddev.2018.05.004.

Balseven, H., Tugcu, C.T., 2017. Analyzing the effects of fiscal policy on income distribution: a comparison between developed and developing countries. Int. J. Econ. Financ. Issues 7 (2), 377–383. Available from: https://www.econjournals.com/index.php/ijefi/article/view/4235.

Berg, A., Buffie, E.F., Pattillo, C., Portillo, R., Presbitero, A.F., Zanna, L.F., 2019. Some misconceptions about public investment efficiency and growth. Economica 86, 409–430. https://doi.org/10.1111/ecca.12275.

Blomstrom, M., Lipsey, R.E., Zejan, M., 1996. Is fixed investment the key to economic growth? Q. J. Econ. 111 (1), 269–276. https://doi.org/10.2307/2946665.

Cavallo, E., Daude, C., 2011. Public investment in developing countries: a blessing or a curse? J. Comp. Econ. 39 (1), 65–81. https://doi.org/10.1016/j.jce.2010.10.001.

Cerdeiro, D., Komaromi, A., 2017. Trade and Income in the Long Run: Are There Really Gains, and are They Widely Shared? IMF Working Papers, No. 17/231, International Monetary Fund, Washington, DC, https://doi.org/10.5089/9781484324851.001.

Cornia, G., 2011. Economic Integration, Inequality and Growth: Latin America vs. the European Economies in Transition. UN Department of Economic and Social Affairs (DESA), UN, New York, https://doi.org/10.18356/a6a4730a-en. Working Papers, No. 101.

Deichmann, U., Reuter, A., Vollmer, S., Zhang, F., 2019. The relationship between energy intensity and economic growth: new evidence from a multi-country multi-sectorial dataset. World Dev. 124, 104664. https://doi.org/10.1016/j.worlddev.2019.104664.

Devadas, S., Pennings, S., 2018. Assessing the Effect of Public Capital on Growth: An Extension of the World Bank Long-Term Growth Model. Policy Research Working Paper, No. 8604, World Bank, Washington, DC, https://doi.org/10.1596/1813-9450-8604.

Erden, L., Holcombe, R., 2006. The linkage between public and private investment: a co-integration analysis of a panel of developing countries. East. Econ. J. 32 (3), 479–492. Available from: https://www.jstor.org/stable/40326291.

Erenburg, S.J., Wohar, M.E., 1995. Public and private investment: are there causal linkages? J. Macroecon. 17 (1), 1–30. https://doi.org/10.1016/0164-0704(95)80001-8.

Faruqee, H., 2016. Regional Economic Outlook, April 2016, Western Hemisphere Department: Managing Transitions and Risks. International Monetary Fund, Washington, DC, https://doi.org/10.5089/9781498329996.086.
FGV Energia, 2016. A Comparative Analysis of Energy Transition in Latin America and Europe. pp. 1–72. http://www.fgv.br/fgvenergia/paper_kas-fgv_ingles/files/assets/common/downloads/Paper_KAS-FGV_Ingl_Web.pdf.
Fournier, J., 2016. The Positive Effect of Public Investment on Potential Growth. OECD Economics Department Working Papers, No. 1347, OECD Publishing, Paris, https://doi.org/10.1787/15e400d4-en.
Fuinhas, J.A., Marques, A.C., Koengkan, M., 2017. Are renewable energy policies upsetting carbon dioxide emissions? The case of Latin America countries. Environ. Sci. Pollut. Res. 24, 5044–15054. https://doi.org/10.1007/s11356-017-9109-z.
Gupta, S., Kangur, A., Papageorgiou, C., Wane, A., 2014. Efficiency-adjusted public capital and growth. World Dev. 57, 164–178. https://doi.org/10.1016/j.worlddev.2013.11.012.
Hacibedel, B., Mandon, P., Muthoora, P., Pouokam, N., 2019. Inequality in Good and Bad Times: A Cross-Country Approach. IMF Working Papers, No. 19/20, International Monetary Fund, Washington, DC, https://doi.org/10.5089/9781484392911.001.
Institute of the Americas (2016). Argentina's Energy Transition: The Macri Government's Vision.
IRENA, 2015. Renewable Energy Policy Brief: Uruguay. pp. 1–10. www.irena.org.
Jimenez, R., Mercado, J., 2014. Energy intensity: a decomposition and counterfactual exercise for Latin American countries. Energy Econ. 42, 161–171. https://doi.org/10.1016/j.eneco.2013.12.015.
Khan, Z., Sisi, Z., Siqun, Y., 2019. Environmental regulations an option: asymmetry effect of environmental regulations on carbon emissions using non-linear ARDL. Energy Sources Part A 41 (2), 137–155. https://doi.org/10.1080/15567036.2018.1504145.
Khan, Z., Ali, S., Umar, M., Kirikkaleli, D., Jiao, Z., 2020. Consumption-based carbon emissions and international trade in G7 countries: the role of environmental innovation and renewable energy. Sci. Total Environ. 730, 138945. https://doi.org/10.1016/j.scitotenv.2020.138945.
Koengkan, M., 2020. Capital stock development and their effects on investment expansion in renewable energy in Latin America and the Caribbean region. J. Sustain. Finance Invest., 1–20. https://doi.org/10.1080/20430795.2020.1796100.
Koengkan, M., Fuinhas, J.A., Poveda, Y.E.M., 2020. Globalisation as a motor of renewable energy development in Latin America countries. GeoJournal 85, 1591–1602. https://doi.org/10.1007/s10708-019-10042-0.
Koengkan, M., Fuinhas, J.A., Silva, N., 2021. Exploring the capacity of renewable energy consumption to reduce outdoor air pollution death rate in Latin America and the Caribbean region. Environ. Sci. Pollut. Res. 28, 1656–1674. https://doi.org/10.1007/s11356-020-10503-x.
Lopez, H., 2003. Macroeconomics and Inequality. World Bank, Washington, DC. Available from: http://documents.worldbank.org/curated/en/292721468319775386/Macroeconomics-and-inequality.
Martínez-Vázquez, J., Vulovic, V., Moreno-Dodson, B., 2012. The impact of tax and expenditure policies on income distribution: evidence from a large panel of countries. Hacienda Publica Esp. 200 (1), 95–130. https://doi.org/10.2139/ssrn.2188608.
Martorano, B., 2018. Taxation and inequality in developing countries: lessons from the recent experience of Latin America. J. Int. Dev. 30 (2), 256–273. https://doi.org/10.1002/jid.3350.
Presbitero, A.F., 2016. Too much and too fast? Public investment scaling-up and absorptive capacity. J. Dev. Econ. 120, 17–31. https://doi.org/10.1016/j.jdeveco.2015.12.005.
Pritchett, L., 2000. The tyranny of concepts: CUDIE (cumulated, depreciated, investment effort) is not capital. J. Econ. Growth 5 (4), 361–384. https://doi.org/10.1023/A:1026551519329.
Samargandi, N., 2019. Energy intensity and its determinants in OPEC countries. Energy 186, 115803. https://doi.org/10.1016/j.energy.2019.07.133.
Santiago, R., Fuinhas, J.A., Marques, A.C., 2020. The impact of globalization and economic freedom on economic growth: the case of the Latin America and Caribbean countries. Econ. Chang. Restruct. 53, 61–85. https://doi.org/10.1007/s10644-018-9239-4.
Solow, R.M., 1956. A contribution to the theory of economic growth. Q. J. Econ. 70 (1), 65–94. https://doi.org/10.2307/1884513.
Székely, M., Sámano-Robles, C., 2014. Trade and income distribution in Latin America: is there anything new to say? In: Cornia, A. (Ed.), Falling Inequality in Latin America. Policy Changes and Lessons. Oxford University Press, Oxford, https://doi.org/10.1093/acprof:oso/9780198701804.003.0011.
Tsounta, E., Osueke, A., 2014. What Is Behind Latin America's Declining Income Inequality? IMF Working Papers, No. 14/124, International Monetary Fund, Washington, DC, https://doi.org/10.5089/9781498378581.001.

Index

Note: Page numbers followed by *f* indicate figures, *t* indicate tables and *b* indicate boxes.

F

G

H

I

K

L

M

Printed in the United States
by Baker & Taylor Publisher Services